Current Topics in Physics

In Honor of
Sir Roger J. Elliott

CURRENT TOPICS IN PHYSICS

IN HONOR OF SIR ROGER J. ELLIOTT

Editors

R. A. Barrio
Institute of Physics
Universidad Nacional Autónoma de México
Mexico

K. K. Kaski
Laboratory of Computational Engineering
Helsinki University of Technology
Finland

ICP
Imperial College Press

Published by

Imperial College Press
57 Shelton Street
Covent Garden
London WC2H 9HE

Distributed by

World Scientific Publishing Co. Pte. Ltd.
5 Toh Tuck Link, Singapore 596224
USA office: 27 Warren Street, Suite 401-402, Hackensack, NJ 07601
UK office: 57 Shelton Street, Covent Garden, London WC2H 9HE

British Library Cataloguing-in-Publication Data
A catalogue record for this book is available from the British Library.

CURRENT TOPICS IN PHYSICS
In Honor of Sir Roger J Elliott

ISBN 1-86094-550-3

Typeset by Stallion Press
Email: enquiries@stallionpress.com

Printed by FuIsland Offset Printing (S) Pte Ltd, Singapore

CURRENT TOPICS IN PHYSICS
in Honor of Prof. Sir Roger James Elliott

CONTENTS

Participants of the Symposium "Current Topics in Physics," Mexico City, June 2003

PREFACE

In presenting this book, containing written versions of the invited talks delivered at the Symposium of *Current Topics in Physics*, held in Mexico City (June 2003) to celebrate the 75th birthday of Professor Sir Roger J. Elliott, we hope that the reader will benefit from the variety and high quality of each contribution.

Sir Roger is a remarkable human being in a number of ways. He has succeeded outstandingly in, at least, three significant endeavors. Firstly, as a scientist he has made a number of seminal contributions to the development of physics, by publishing at a relentless pace many highly cited scientific papers for several decades. Secondly, he has proved to be an excellent administrator, of both funds and people. Sir Roger was the head of the Theoretical Physics Department in Oxford for an extended period, and presided over various scientific organizations. Not only that, he has also directed other non-scientific and also commercial organizations with remarkable proficiency. Lastly and even more importantly, his ability to create life-long friendships with people is quite unparalleled. He is gentle and civilized, what you might think of a perfect English gentleman. To his friends, he offers a model of generosity, understanding, warmth and judicious advice.

Here, we do not take space to provide proof of all these statements, since a quite comprehensive scientific biography and a list of selected papers by him has been published before (*Disorder in Physics*, Oxford Univ. Press, 1989). Rather, we want to emphasize another aspect of his scientific achievements that is very seldom mentioned, namely his remarkable skill to 'produce' highly qualified scientists. A good number of his fifty or so former doctoral students are now scientists in various leading academic institutions round the world. One proof of this is this book, the chapters of which are mostly written by former students and research associates of his. All these outstanding qualities of his are not only due to his sharp mind and fruitful ideas, but also due to his eagerness to reach the bottom of the matter, together with his command of the English language which, apart from being a delight to hear, is usually very dense in concepts and clues.

It is also worth mentioning that his students and research associates are truly international and come from different parts of the world. He has always maintained close links with academic organizations in foreign countries to help with the healthy development of sciences there. For example and in particular, he has helped forming a lasting and fruitful relationship between the Royal Society and the Mexican Academy of Sciences, which simultaneously with his birthday celebration, decided to award him with the title of Miembro Correspondiente, in recognition of his contributions to Mexican science. In the inauguration ceremony, Sir Roger delivered a speech, which we reproduce here to allow the reader to share our delight in listening to Sir Roger's discourse.

Mexico City and Helsinki, September 2004

Rafael A. Barrio and Kimmo K. Kaski
Editors

ORDER AND DISORDER IN PHYSICS: INAUGURATION SPEECH BY PROFESSOR SIR ROGER ELLIOTT

1. Introduction

Mr. President, it gives me great pleasure to accept Membership of your Academy. I regard it as a great honor to belong to such a vigorous and growing institution which represents so well the broad sweep of Mexican science.

My own contacts with Mexican science began over 25 years ago when a group of research students began to come to our Department of Physics in the University of Oxford as part, I believe, of a policy by CONACYT to train a cohort of research scientists through outside experience in the US, in Continental Europe, and in particular in the United Kingdom. My first personal research student from this group was Julia Taguena, who, in her new role, is our host today in these splendid surroundings. She was followed by Rafael Barrio with whom I have maintained strong personal and scientific links through common research interests over the years. I am deeply indebted to him for arranging the scientific meeting which will follow in the next few days.

Twenty five years ago, the cohort of Mexican physicists was small but growing. Indeed I remember being told that at that stage there were no dead Mexican physicists — since earlier ones had been trained as engineers! I am afraid there are now, but there are also many who are fully alive and active.

In more recent years the flow of students to our department has been reduced as research in Mexico expanded and the facilities for training at home became better. Instead we have had a stream of post-docs and visitors to obtain broader experience after their initial research training. During this period, my group had the benefit of a significant grant from the European Union, arranged in collaboration with Professor Balkanski from Paris, which allowed us to build joint research programs.

My experience at the Royal Society of London and of other National Academies across the world has led me to appreciate the importance which such bodies carry in maintaining standards of scientific education and research in their countries. By creating a climate that rewards excellence, they can play an important role in ensuring that the best scientists and the best science is supported.

I therefore bring you fraternal greetings from the Royal Society, which seeks to maintain and improve old relationships for mutual help. It is true that the exchange program between our Academies has not been as active in recent years as it was in the past, but my colleagues believe that there may be new opportunities to expand these links when they meet you at the Inter-Academy Panel, which you are hosting here in December.

I also bring you greetings from ICSU, the International Council for Science, to which your Academy is now the National Member. Here again I have good Mexican contacts because one of my close colleagues is Ana Maria Cetto. My current position is Treasurer and am hence charged with looking after ICSU's finances.

I have also been involved with Mexico through two other organizations which are representative of British culture. The British Council, of which I was formerly a member of their Board, maintains an active office here, and in the past promoted a number of scientific exchanges from which we benefitted. The other is Oxford University Press, an arm of my University, which is as an international publisher has expanded its activities here in recent years. OUP takes responsibility for the definitive dictionary of the language and is deeply involved in disseminating the English language. The Oxford English Dictionary is celebrating this year the seventy-fifth anniversary of its first publication in 1928 when it was the result of over fifty years of work. In other European nations like France and Spain, this task of recording the language falls to the Academy. I do not know whether you have any similar projects here for preserving the specifically Mexican parts of the vocabulary, as OUP also does in other parts of the English speaking world.

So you see, Mr. President, my contacts with Mexico have been varied and wide-ranging but the strongest and deepest were in physics with my colleagues who are here today, and that is why I am particularly grateful for the honor which you have done me.

2. Research

I would therefore like to say a little about the physics which has interested me over my research career, which has now spanned more than fifty years. I chose my title "Order and Disorder in Physics" because almost everything that I have done can be related in some way to these concepts. It will also allow me the luxury of adding a few historical and philosophical comments.

These phenomena show up very clearly in the study of magnetism, a subject which has fascinated scientists since ancient times, and which still does today. Although this mysterious force encouraged many myths (such as its use to cure diseases or ascertain whether ones wife was faithful) it also attracted serious scientists in the Middle Ages such as Roger Bacon (arguably Oxford's first physicist) and Gilbert who wrote the first scientific treatise on the subject [1]. In the 19th century, the experiments of people like Oersted and Faraday had established the connection with electricity and shown that a current loop behaved like a small magnet. Since known magnetic materials like a lodestone and magnetized iron retained their properties when they were divided into smaller pieces, it became clear that the origin of such magnetism was in the atoms themselves.

At the beginning of the century, three great French physicists Langevin, Curie and Weiss had codified the different types of magnetic material and related them to induced currents and to magnetic dipoles. Weiss's brilliant concept of the molecular field not only allowed us to see a possible origin for macroscopically large ordered magnets, but also gave us a prototype for all order-disorder phase transitions [2].

But the fundamental origin of these effects did not become clear until the advent of quantum mechanics in the late 1920's. It is fascinating to look at the proceedings of the Solvay Congress of 1930 [3], which all the great names in theoretical physics attended. Even to them, the application of quantum mechanics to magnetic phenomena of a simplicity which we now teach our undergraduates, was clearly a struggle. But less than ten years later, in 1939, when the first modern style conference on magnetism was held in Strasbourg [4] all the essential features of modern day magnetism were in place. Van Vleck had explained from atomic theory the detailed value of magnetic moments found in solids. Stoner, Slater, and Mott were beginning to explain the more complex phenomena of metallic magnetism, while Neel, building on Weiss's theory, had predicted that there would be other forms of magnetic order to that of ferromagnetism where all the little magnets pointed in parallel.

When research on these topics resumed after the war (and my own research career began) there had been two dramatic developments in experimental techniques. One was microwave sources derived from the radar program, which allowed spectroscopy of be done within the energy levels of single magnetic ions using the technique of paramagnetic resonance [5].

This was widely exploited within the Physics Department at Oxford and my thesis provided a theoretical description of some of these effects when it was published almost exactly fifty years ago [6]. The other great advance came from the exploitation of slow neutrons from thermal reactors which because they have magnetic moments can, when scattered, determine both the nature of the magnetic order and the excitation spectrum of ordered magnetic materials [7]. My own interest turned particularly to the remarkable magnetic orderings which had been found within the rare earth metals, which, though complex, could be readily interpreted in terms of fairly simple concepts and field parameters [8].

The ordering which we see in these magnets occurs only at low temperatures and arises from the thermodynamics of the systems. The Weiss theory gives only a crude view of this effect neglecting as it does the gradual increase in local order as the phase transition is approached from higher temperatures. The fluctuations which increase in this region gave rise to a detailed and productive study of so-called critical phenomena in which my colleague Michael Fisher played an important role, but which I do not have time to expand on here [9].

As I have said, the existence of order and disorder in magnets is driven by thermodynamics with normally the ordered state appearing at lower temperatures. But many systems exist in nature where disorder becomes frozen in as the system is trapped into a local energy minimum and away from true equilibrium [10]. Generically there are two types, substitutional disorder such as exists in alloys, where a mixture of atoms is arranged on a more or less perfect lattice. The second is structural disorder, where the atoms lose the long-range correlated arrangement which is typical of crystals, although they usually maintain strong local order. Thus in materials which are fully comprised of atoms which have strongly directional covalent bonds, such as silica, we see the familiar glassy state.

The properties of such disordered systems needed to be investigated with different theoretical techniques to those which are available in the perfect symmetric case. The breaking of the symmetry means that the simplifying techniques associated with group theory no longer exist. One of my students, David Taylor, developed an approximate method for dealing with excitations in systems of the substitutional disordered types and this has been widely used by my other students to treat various excitations in such materials. It is called the coherent potential approximation and consists of finding an effective medium representing the average of the crystal in which scatterings from the deviation from that average gives zero total effects. It can be used for electrons in solids, for vibrations, for magnetic excitations, and was summarized by Krumhansl, Leath [11] and myself in another highly cited paper.

In materials such as glasses without long-range order different techniques are needed. These have usually consisted of good treatments of local clustering around any particular atom while approximating the rest of the material with something with a simple structure. Quite often this has been a Bethe lattice or Cayley tree which is continually branching and has no rings of atoms. One very satisfying application of this work with Rafael Barrio and Frank Galeener [12] showed that local rings of atoms treated correctly but within an approximate material could account for the sharp vibrational lines which appear in the Raman spectra of glasses.

Another type of disorder which leads to interesting effects, which I have studied, concerns those crystal lattices where atoms or couplings are removed at random. Eventually, as the concentration of missing atoms increases, the connected groups of remaining atoms become isolated. The lack of percolation which can take place in such circumstances has significant effects when translated into a magnetic problem since cooperative magnetism must vanish when the magnetic constituents become confined to isolated clusters [13].

A second strand to my research concerns the behavior of electrons in regular solid structures and derived from a post-doctoral position in Berkeley with Charles Kittel. At that stage no real energy band structures were known and attempts to calculate them with the computers then available were too imprecise. Much attention was focussed on the new semiconductors silicon and germanium following the discovery of the transistor and other potential uses of such materials. I recall that the best calculations always showed silicon to be a metal. But again resonance using microwaves was able to determine in detail the nature of the energy bands in which the effective electrons used in practical processes were contained. My own contribution [14] here was to point out that, in such detailed situations, it was important to include the spin orbit coupling which caused a splitting of otherwise degenerate bands. A second feature of this work was a closer investigation of the optical properties of such systems, where new structures, including sharp lines, were found to appear where the optical absorption first began. These were excitons, pairs of electrons and holes, which dominate these transitions, and which subsequently came to be crucial in developing semi-conductor lasers. The most remarkable material was cuprous oxide, which showed not one but two hydrogenic series of lines which I was able to explain in what still remains my most cited paper [15].

I will refer to one other strand of research before closing. When missing atoms or vacancies appear in a lattice it makes it possible for other atoms to move around by jumping into these vacant sites. This phenomenon of jump diffusion is important in irradiation damage and ignores a correlated motion of the atoms concerned. This is a topic much studied both analytically and numerically by my former student Kimmo Kaski [16], who has also maintained a good relationship with the physicists here in Mexico.

3. Reflections on Disorder

Let me end with a health warning to any social scientist who might be with us. I think it is clear from what I have said that order and disorder within physics have precise definitions. As with many terms in physics that use common words in a specific connotation, they can mislead if anyone attempts to use them in a different context. Attempts by social scientists and others to use concepts like uncertainty, chaos, relativity, or even spin and charm, which physicists have invented to deal with specific phenomena, can lead to rather peculiar conclusions.

Perhaps the most confusion amongst our non-scientific colleagues comes from attempts to associate entropy directly with disorder. Boltzmann himself was at least partially responsible because he sometimes compared entropy with aspects of disorder in the system. In fact, the principles of statistical mechanics state that in thermal equilibrium all situations with the same energy are equally likely and therefore the most probable state is what will be observed. One statement of the famous Second Law of Thermodynamics which relates thermal energy (heat) and work states that in a closed system entropy is a maximum in thermal equilibrium and increases as equilibrium is approached. But as we have seen in the magnetism examples, an ordered state can still have a lower energy than a disordered one and be preferred at lower temperatures. This idea that entropy will increase towards equilibrium has lead to some bizarre suggestions when it is implied out of context. For example, recently, Dennett in his book "Darwin's Dangerous Idea" seems to suggest that because evolution has favored the creation of more and more complex animals with more complicated chemical constitutions, it is some way in conflict with the Second Law. Long ago, Eddington gave the view of theoretical physicists on this point:

> "If someone points out to you that your pet theory of the universe is in disagreement with Maxwell's equations — then so much the

> worse for Maxwell's equations. If it is found to be contradicted by observation — well, these experimentalists do bungle things sometimes. But if your theory is found to be against the Second Law of Thermodynamics I can give you no hope; there is nothing for it but to collapse in deepest humiliation." Sir Arthur Eddington, *The Nature of the Physical World*

It seems, as it was in C.P. Snow's "two cultures," that understanding the Second Law of Thermodynamics is still a watershed between physicists and the rest of culture:

> "I believe the intellectual life of the whole of western society is increasingly split into two polar groups — literary intellectuals at one pole and at the other the physical scientists. Between the two is a gulf of mutual incomprehension. A good many times I have been present at gatherings of people who, by the standards of the traditional culture, are thought highly educated.... Once or twice I have been provoked and have asked the company how many of them could describe the Second Law of Thermodynamics. The response was cold: it was also negative." C.P. Snow, *Two Cultures in the Scientific Revolution*

I hope I have demonstrated, Mr. President, that ordered and disordered structures occur throughout physics. Sometimes that disorder is artificially created, frozen in a metastable state. Sometimes the disorder is thermally driven according to the laws of thermodynamics. In either case the properties of the system present fascinating challenges to the theoretical physicist — enough to keep this one at least entertained by their solution for a career of fifty years.

Mr. President, thank you again for this honor and thank you all for your attention.

References

[1] Gilbert, W., "*De Magnete*" (On the Magnet, Magnetic Bodies also, and on the Great Magnet the Earth, a New Physiology Demonstrated by Many Arguments and Experiments).

[2] Weiss, P., *J. de Phys.* **6** (1907) 666.

[3] "*Le Magnetisme*" — *Report of the 6th Solvay Conference* (Gauthier-Villars, Paris, 1932).

[4] "*Le Magnetisme*" — *3rd Collection Scientifique de CNRS* (Paris, 1940).

[5] Abragam, A. and Bleaney, B., *Electronic Paramagnetic Resonance of Transition Metals* (Oxford University Press, Oxford, 1970).

[6] Elliott, R.J. and Stevens, K.W.H., *Proc. Roy. Soc.* **A218** (1953) 553.
[7] Jensen, J. and Mackintosh, A.R., *Rare Earth Magnetism — Structures and Excitations* (Oxford University Press, Oxford, 1991).
[8] Elliott, R.J., *Phys. Rev* **124** (1961) 345.
[9] Fisher, M.E., *Rev. Mod. Phys.* **46** (1974) 597.
[10] Ziman, J.M., *Models of Disorder* (Cambridge University Press, Cambridge, 1979).
[11] Elliott, R.J., Krumhansl, J.A. and Leath, P., *Rev. Mod. Phys.* **46** (1974) 465.
[12] Elliott, R.J., Barrio, R.A. and Galeener, F.L., *Phys. Rev.* **B48** (1993) 15672.
[13] Stauffer, D., *Phys. Rep.* **54** (1979).
[14] Elliott, R.J., *Phys. Rev.* **96** (1954) 280.
[15] Elliott, R.J., *Phys. Rev.* **108** (1957) 1384.
[16] Balkanski, M. and Elliott, R.J. (eds.) *Atomic Diffusion in Disordered Materials* (World Scientific, Singapore, 1998).

PART 1

DISORDER AND DYNAMICAL SYSTEMS

CHAPTER 1

REFLECTIONS ON THE BEGUILING BUT WAYWARD SPHERICAL MODEL

Michael E. Fisher*

Institute for Physical Science and Technology, University of Maryland
College Park, Maryland 20742, USA

The talk fulfilled the promise of the title and led up to a sketch of recent work, with Jean-Noël Aqua, on criticality in multispecies and ionic spherical models. An extended summary touching on further aspects and applications of spherical models is recorded here.

1. Introduction

It is a pleasure to take part in an occasion honoring the achievements of a distinguished and well-loved scientist: and, in honoring Roger Elliott in the year in which he celebrates his 75th birthday, it may not be inappropriate to remark that while one cannot turn the clock back, one can and, perhaps, should look back! Indeed, that exercise can be fruitful provided that, as always in science, the enterprise then goes forward, building on past foundations and lessons. It is in that spirit that these purely personal reflections are offered.

First, the wording of the title should be justified. From the Oxford English Dictionary the meanings of "*beguiling*" to be understood are first, *charming* and *amusing*, but second, with overtones of *bewitching* and, as a verb, *to divert attention from*, and, finally, even *to delude* and *cheat*! Likewise for "*wayward*": first, *childishly self-willed*, but then, *perverse, capricious, unaccountable* and even *freakish*. So how does the spherical model warrant these descriptions?

---

*Presented on 17 June 2003 in Mexico City at a Symposium held at the Universidad Nacional Autónoma de Mexico in honor of Professor Sir Roger J. Elliott, with the Spanish title: "*Reflexiones sobre el Modelo Esférico, Hechicero pero Caprichoso.*"

2. The Ising Model: Ferromagnet and Lattice Gas

To answer, we must recall the origins of the spherical model. In 1944 [1] and in further masterly works [2], Lars Onsager calculated exactly many basic properties of the nearest-neighbor two-dimensional Ising model of a magnetic material in zero magnetic field $H = 0$. At each site, i, of an Ising model lattice in d dimensions sits a spin $s_i = \pm 1$ which interacts with a spin s_j separated by a lattice vector $\mathbf{R}_{ij}$ via the coupling term $-J(\mathbf{R}_{ij})s_i s_j$. In the interpretation as a lattice gas, an "up" spin, $s_j = +1$, corresponds to an empty site, a "down" spin to an occupied site; solving the model regarded as a magnet yields properties of the lattice fluid and *vice versa*. Thus, one learned that the critical exponents for the specific heat and spontaneous magnetization (or coexistence curve) for $d = 2$ were $\alpha = 0$ (log) and $\beta = 1/8$. (For critical phenomena notation, etc., see, e.g., [3] and [4].) Furthermore, Onsager's solution for the singular part of the reduced free energy could be written in the suggestive form

$$-F/k_{\mathrm{B}}T \propto \int d^d k \ln[u(T,H) + \Delta\hat{J}(\mathbf{k})], \tag{1}$$

where, in terms of Fourier transforms of $J(\mathbf{R})$, one has

$$\Delta\hat{J}(\mathbf{k}) = \hat{J}(0) - \hat{J}(\mathbf{k}) \propto \frac{1}{2}R_0^2 k^2 + \cdots, \tag{2}$$

while the smoothly varying function $u(T,H)$ vanishes at criticality. This form proves exact for $d = 1$ (where $T_{\mathrm{c}} = H_{\mathrm{c}} = 0$); but of course, for $d = 2$, is valid only with $H = 0$. The length R_0 represents the range of the interactions.

3. The Spherical Model

Before long, however, the profound difficulties of extending Onsager's exact results to $H \neq 0$ or to $d > 2$ were realized. Faced with this, Mark Kac, the mathematician, devised a "poor man's Ising model" by replacing the true Ising model constraint $s_i^2 = 1$ for $i = 1, \ldots, N$, by the much weaker "spherical constraint":

$$\sum_{i=1}^{N} s_i^2 = N \quad (-\infty < s_i < \infty), \tag{3}$$

the spin variables s_i being now taken as continuous, unbounded, real variables. Aided by Ted Berlin, the chemist (see [3], Sec. 8.4), this "*spherical model*" was then solved exactly [5] for general fields and the form (1) was recaptured! Indeed, this form is found to hold generally for *all* d and is not by any means restricted to nearest neighbor interactions.

However, while certainly beguiling, the spherical model harbored some surprises. Thus, while real three-dimensional fluids and (anisotropic) magnets are characterized by $\beta \simeq 0.33$, the spherical model yields only the classical, van der Waals, or mean-field value $\beta = 1/2$: and, again, this is true for all d.

At this point it must also be emphasized strongly that the so-called "mean spherical model," in which (3) is replaced by the "average constraint," $\langle s_i^2 \rangle = 1$ [enforced with the aid of a Lagrange multiplier, say, $\lambda(T, H)$], is *not* in the least "mean"! On the contrary, it is elegant, clean and much easier to analyze. Furthermore, in the thermodynamic limit, $N \to \infty$, which will be our main concern here, all the properties, both thermodynamic and correlational, are identical to those of the "canonical" Berlin–Kac version of the model.

Further investigation of the $d = 3$ spherical model reveals that the susceptibility (*alias* the compressibility of the related lattice gas) diverges (for $H = 0$) according to

$$\chi(T) \approx \frac{C^+}{t^\gamma} \quad \text{when} \quad t = \frac{T - T_\mathrm{c}}{T_\mathrm{c}} \to 0+, \tag{4}$$

with the exponent value $\gamma = 2$; but this seems quite "off base" given that for real three-dimensional systems one has $\gamma \simeq 1.24$ (while for the $d = 2$ Ising model $\gamma = 1\frac{3}{4}$) [3, 4]. For these reasons it was tempting initially to regard the spherical model as little more than a pretty mathematical plaything with rather little to teach us.

4. The n-Vector Models

But that changed dramatically in 1968 — sixteen years after Kac and Berlin had introduced the model — when Gene Stanley [6] investigated what were later often called the n-vector models, in which the scalar Ising spins, s_i, were generalized to n-component vectorial spins, $\vec{s}_i$, of fixed, bounded length $|\vec{s}_i|^2 = n$ (although other normalizations are permissible for $n < \infty$). He discovered and then proved [3, 6] that in the infinite-component limit, $n \to \infty$, the n-vector models reduce *exactly* to the corresponding spherical model! This brought the spherical model safely within (or, more precisely, to the borderline, but as a well-defined and sensible member, of) the important family of models including the XY model ($n = 2$), the (classical) Heisenberg model ($n = 3$), self-avoiding lattice walks ($n = 0$), etc., that in the early renormalization group studies, could well be regarded as forming, with the dimensionality, d, the basic "*Plane of Theory*," (n, d) [7]. It then became imperative to ask what lessons might be learned by studying the spherical model more thoroughly.

5. Some Lessons Learned

So what has the spherical model taught us? Quite a lot, indeed! First, is the unequivocal demonstration of the existence of *borderline dimensions*, specifically, the upper borderline (or critical) dimensionality, say, $d_>$, above which classical or mean field theory becomes valid, and the *lower borderline*, $d_<$, below which the phase transition and critical point disappear. For the short-range spherical models originally studied one has $d_> = 4$ and $d_< = 2$. Furthermore, the model can be extended rather naturally to *continuous dimensionalities* (as later capitalized upon [8]). Indeed, for the susceptibility exponent when $d \leq d_>$ one has

$$\gamma(d) = \frac{2}{d-2} = \frac{1}{1 - \frac{1}{2}\epsilon} = 1 + \frac{1}{2}\epsilon + \frac{1}{4}\epsilon^2 + \cdots , \tag{5}$$

where $\epsilon = 4 - d$ [8] and one sees that (for this special $n = \infty$ case) the ϵ expansion is actually convergent! But it must *not* be presumed that this remains true for finite values of n.

More engagingly still, the spherical model can be solved for long-range attractive power-law potentials [9, 10] of the form

$$J(\mathbf{R}) \approx \frac{J_\infty}{R^{d+\sigma}} \quad (0 < \sigma < 2). \tag{6}$$

One then finds $d_> = 2\sigma$ and $d_< = \sigma$ so that when σ is small enough criticality can be realized even in one-dimensional systems as, indeed, is true for Ising models with long-range forces. However, the susceptibility exponent for $\sigma < d \leq 2\sigma$ is now given by

$$\gamma(d; \sigma) = \frac{\sigma}{(d - \sigma)} = [1 - (\epsilon/\sigma)]^{-1}, \tag{7}$$

where a "generalized epsilon," namely, $\epsilon = d_> - d$, enters naturally [11].

When one examines the specific heat exponent $\alpha(d; \sigma)$ one finds that this, together with $\beta(= \frac{1}{2})$ and $\gamma(d, \sigma)$, satisfies the basic exponent relations [12]

$$\alpha + 2\beta + \gamma = 2, \quad \gamma + \beta = \beta\delta = \Delta, \tag{8}$$

where δ is defined *at* $T = T_c$ by the variation of the magnetization as $M \sim H^{1/\delta}$ (while Δ determines the scaling of H with t; see below). At the borderline $d = d_>$ logarithmic factors appear in the spherical model: e.g., one has $\chi(T) \sim \ln|t|/t$. This proves, in fact, to be much more generally true [7].

6. Thermodynamic Scaling

In more detail one may ask for a characterization of the *Equation of State.* If we use $t = (T - T_c)/T_c$, as above, and

$$m = \frac{M}{M_{\max}} \equiv \frac{\rho - \rho_c}{\frac{1}{2}\rho_{\max}} \quad \text{and} \quad h = \frac{H}{H_0} \sim \mu - \mu_0(T), \tag{9}$$

where ρ is the lattice gas density, H_0 is a reference field and μ is the chemical potential taking the value $\mu_0(T)$ at coexistence, one discovers the general form (for $d_<(\sigma) < d < d_>(\sigma)$)

$$c(h/m) \approx (B^2 t + m^2)^\gamma \quad \text{as} \quad t, m, h \to 0, \tag{10}$$

where c is a constant amplitude while B enters the spontaneous magnetization via $M_0(T) \approx B\,|t|^\beta$ with (here) $\beta = \frac{1}{2}$. Of course this simple form is special to the spherical model, i.e., to the $n \to \infty$ limit. Nevertheless, one easily sees (following Widom [13]) that it fully verifies the general scaling hypothesis

$$c'(h/\,|t|^\Delta) \approx Y_\pm(m/B\,|t|^\beta), \quad t \gtrless 0, \tag{11}$$

where $Y_\pm(y)$ is the universal scaling function. Normally $Y_\pm$ is hard to calculate and characterize; but the spherical model tells us that

$$Y_\pm(y) = y(y^2 \pm 1)^\gamma \quad \text{for} \quad n = \infty \tag{12}$$

which, incidentally, reduces to the universal classical or van der Waals result when $\gamma = 1$.

7. Some Signs of Waywardness

So far so good! The spherical model has charmed and, in fact, informed us! But, let us look at some matters more closely, especially from the original perspective of the poor-man's Ising model for a scalar ferromagnet or lattice gas. Consider, then, the magnetization $M(T, H)$ at fixed T below T_c or, equivalently, for a fluid, the density $\rho(T, \mu)$ as a function of μ (where μ may just as well be replaced, as is more traditional, by the pressure p). For small fields, $H \to 0+$ [or $\mu \to \mu_0(T)+$] one discovers from the equation of state (10) that the magnetization can be written

$$M(T, h) = M_0(T) + \chi(T)h^{\psi} + \cdots . \tag{13}$$

Since the (initial) susceptibility $\chi(T) = (\partial M/\partial H)_{T,H\to 0}$, or the compressibility $K(T)$ at coexistence, are well defined for an anisotropic ferromagnet, or for a fluid below T_c, one "obviously" expects $\psi = 1$ and then $X(T) \propto \chi(T)$.

However, that is *not* what the wayward spherical model has to say! Rather one finds $\psi = 1/\gamma = 1 - (\varepsilon/\sigma)$ (for $d < d_>$, $\sigma \leq 2$). Since this implies $\psi < 1$, the result means that $\chi = (\partial M/\partial H)_T$, or the compressibility $K(T, \mu)$, *diverges* on approach to the phase boundary (or to coexistence) below T_c . If nothing else, this means that as a model of a fluid the spherical model must be treated with caution below T_c [14].

But can one understand this seemingly perverse behavior? The answer is "Yes, provided one remembers that the spherical model is 'really' an n-vector model with $n \geq 2$ (indeed $n \gg 2$)." With this in mind, one merely has to consider the spontaneous magnetization as represented by a fixed-length vector, $\vec{M}_0$, which, in strictly zero field, may point anywhere over the corresponding sphere. An infinitesimal field H then initiates an unbounded response as $\vec{M}_0$ swings around to become parallel to H at essentially zero cost in free energy. What one "observes" is thus really the unbounded "transverse response" rather than the expected, finite "longitudinal response"! Put in other language, a (noncritical) finite-T calculation shows that spin-waves or "Goldstone bosons" must be present (for $n \geq 2$) and will, for $d < d_>$, always yield such a response. The value $\psi = 1 - \frac{1}{2}\epsilon$ (for $\sigma \geq 2$) is, indeed, the correct result, for all $n \geq 2$. Thus the waywardness is not so capricious as first appears!

8. The Yang–Lee Edge

To explore further let us consider the so-called *Yang–Lee edge singularities* first pointed out by Robert Griffiths [15, 16]. These arise *above* T_c when, at fixed T, one explores the *complex* plane of the field $h = h' + ih''$. On leaving the origin at $h = 0$ in a purely imaginary direction one soon encounters the Yang–Lee edge at $h'' = h''_{\mathrm{YL}}(T)$; on approaching the edge, the susceptibility diverges and one has

$$\chi \sim 1/(h''_{\mathrm{YL}} - h'')^{1-\sigma} \quad \text{and} \quad G(h'') \sim (h'' - h''_{\mathrm{YL}})^{\sigma}, \tag{14}$$

where $G(h'')$ denotes the density of Yang–Lee zeros that generate the cut in the complex h plane that terminates at $h''_{\mathrm{YL}}(T)$. These singularities are interesting because they represent the "most primitive" thermodynamic critical behavior since they are characterized by only a single relevant variable — the behavior is invariant, i.e., universal as T changes. Furthermore, they are described by an $i\varphi^3$ field theory [16] and as a function of d the exponent σ is known exactly for $d = 1$ [17] and 2, and rather accurately numerically for $3 \leq d < d_>$. (See, e.g., [18].) In fact one has $\sigma(d) = -\frac{1}{2}, -\frac{1}{6}, 0.087, 0.265$, for $d = 1, \ldots, 4$, and

$$\sigma(d) = \frac{1}{2} - \frac{1}{12}\varepsilon + \cdots, \quad \text{with} \quad \epsilon = 6 - d \geq 0, \tag{15}$$

for small ϵ which vanishes at $d_> = 6$. Evidently, $\sigma = \frac{1}{2}$ represents the mean field or classical result which may, indeed, be found directly from, e.g., the van der Waals equation.

So what insight may we hope to get from the spherical model? Sadly and perhaps surprisingly the answer is: none! For *all* spherical models only the classical answer $\sigma = \frac{1}{2}$ is delivered. (As a matter of fact this is also the case for the Dyson–Baker hierarchical models; see [19] and references therein.) Could this have been anticipated? Can it be rationalized? *Post facto* one can gain some understanding.

The essential feature of the spherical model *in zero field* is that the strong fluctuations induced by the many (indeed, infinitely many) "degenerate" components of the "true" or underlying order parameter dominate the critical behavior. However, the degeneracy is destroyed by the introduction of a nonzero field $\vec{H}$. Then one direction is favored and (for $\vec{H} \neq 0$) transverse fluctuations are suppressed and no longer play a dominant role. Indeed, even in an $n = 3$ Heisenberg *anti*ferromagnet, while the imposition of a field $\vec{H}$ does not destroy the transition, it does reduce the symmetry and the criticality becomes of XY ($n = 2$) character: see [20].

A finite complex field serves equally well to destroy the $n = \infty$ symmetry: but, alas, the spherical model's special resources are then exhausted and only the classical mechanism of "folding" free energy manifolds remains:[a] this means that nothing beyond Landau or van der Waals-type critical behavior can be generated.

---

[a] The Dutch mathematician D.J. Korteweg realized that the van der Waals theory of critical behavior and its extensions for mixtures could be regarded mathematically, both for critical points and higher-order classical singularities, as a consequence of the deformations of an analytical surface as a parameter such as the temperature varies smoothly. The oft-used words "plait" point to describe a critical point in fluid mixtures derives from his work, published in 1891, and has the meaning there of a smooth fold or pleat

9. Endpoints and Tricriticality

This conclusion is reinforced by a study of *critical endpoints* [22] in which a field-theoretic n-vector model with single spin potential

$$W(\vec{s}_i) = \frac{1}{2} D \frac{|\vec{s}_i|^2}{n} + \frac{1}{4} U \frac{|\vec{s}_i|^4}{n^2} + \frac{1}{6} V \frac{|\vec{s}_i|^6}{n^3}, \tag{16}$$

was investigated with $V > 0$ (for stability) but $U < 0$. As a function of $D \geq 0$ a lambda line, $T_\lambda(D)$, is generated with spherical model exponents. In many cases (see [22, 23]) this critical line then meets a first-order phase boundary, $D_\sigma(T)$, at a critical endpoint — which is of intrinsic interest in and of itself [24] and about which the spherical model has, once more, some valuable things to say [22]. However, the relevant point here is that in the spherical model limit, $n \to \infty$, the first-order line, $D_\sigma(T)$, does not respect the full n-vector symmetry. Thus for $n = \infty$ it ends in a critical point at some (D_c, T_c) but, once again, the behavior at this critical point is purely classical [22, 23].

On the other hand, for distinct choices of the interactions and of the parameters D, U, and V, families of *tricritical points* can be found. These prove to be highly nontrivial and can be investigated closely *at* the marginal (i.e., upper critical) dimensionality which is now only $d_> = 3$ [25–27]. Furthermore, in addition to the magnetic field H that couples linearly to the spins, $\vec{s}_i$, an independent and also relevant third-order field $\vec{H}_3$, that couples to $\vec{s}_i\,|\vec{s}_i|^2$ can (and should) be introduced and may be handled exactly when $n \to \infty$. Surprisingly, but *not* misleadingly, the thermodynamic scaling functions on the $d = 3$ borderline prove to be *nonuniversal*! However, one finds [25–27] that they can be parametrized by a *single-variable* (which actually vanishes in the Kac–van der Waals limit of infinite range interactions) [26]. Although $n = \infty$ may seem unphysical, this nonuniversal scaling description provides, in fact, a rather successful basis for analyzing real data

(without sharp edges or creases) on a singularity-free surface: see the interesting history set out by J. Levelt Sengers [21]. Nowadays one would regard this approach as part of "catastrophe theory" which, in turn, might well be viewed by physical scientists as merely a mathematization of Landau's renowned theory of phase transitions — to which, however, we also owe the crucial physical concept of the "order parameter" and its symmetry or tensorial character.

on antiferromagnets and helium 3-4 mixtures, corresponding to $n = 1$ and $n = 2$ systems [26].

However, the streak of waywardness shows up once more! The pair of "wing" critical lines (more generally surfaces) that branch off from the fully symmetric thermodynamic manifold are always described simply by classical exponents whereas in realistic $n < \infty$ systems, Ising-type behavior is realized. The explanation [25] again relates to the breaking of the $(n = \infty)$-fold symmetry by the fields H and H_3 — more explicitly in a manner that cannot be repaired merely by a shift in the spin fields.

10. Further Lessons from the Spherical Model

The spherical model, because it can be analyzed exactly, has had many applications beyond those mentioned so far. In the subsequent sections, some specific insights gained recently by studying spherical models for ionic systems will be described. But while this present exposition is most certainly *not* a systematic review of the spherical model and its uses, it seems not inappropriate to reflect on a few other, personally selected studies at this point.

Soon after the theory of finite-size scaling in the critical region was introduced [28, 29], the behavior of spherical models of finite thickness was studied analytically to check the theory [30]. A lattice of $d - 1$ infinite dimensions but finite with n layers in the dth direction was considered. Boundary conditions: (a) periodic, (b) antiperiodic, and (c) free-surface were imposed in the finite dimension.

The significant feature of the antiperiodic conditions [30, 31] is that they enable one to calculate exactly the *helicity modulus*, $\Upsilon(T)$, a characteristic of $n \geq 2$ systems which is the analog of the superfluid density, $\rho_s(T)$, in a fluid that undergoes Bose–Einstein condensation; see [32]. By the same token, the spherical model does *not* generate an *interface* between states of "opposite order" with a nonzero surface tension, $\Sigma(T)$, as would an *anisotropic* or Ising-type system. This fact serves, again, as a reminder that in using the spherical model to understand normal fluids, "anomalies" must be expected below T_c reflecting the existence of the $n - 1$ transverse degrees of freedom of the order-parameter.

Free-surface boundary conditions enable the wall or boundary free energies to be calculated explicitly and compared, successfully, with finite size scaling theory [28–30]. But, once more, caution must be exercised! The *shift* in the critical temperature is now asymptotically much larger than the basic finite-size scaling forms predict. This, in turn, may be traced to the long-range nature of the overall spherical constraint (3) which, since the local environments of spins near the boundary differ from those in the bulk, is sensitive to the number of layers, n, on the scale $1/n$.

Moreover, for this reason the $n \to \infty$ theorem of Stanley [3, 6] no longer applies directly: to relate the behavior of *nonuniform* n-vector models to a spherical model when $n \to \infty$, it is imperative to employ *independent* spherical constraints or, equivalently, spherical fields $\lambda_\tau(T, H)$, for each class, τ, of spins, s_i, with equivalent local environments. Hence an n-layer system requires n separate constraints (or $\sim\frac{1}{2}n$ if there is some symmetry) and the analysis becomes much harder: see, e.g., [33] and references therein.

Finite-size scaling also has significant statements to make about first-order transitions at fixed $T < T_c$; see [34] and references cited there. When the order parameter in zero field has an n-vector symmetry (with $n \geq 2$) the situation is complicated by the presence of spin waves — as already remarked in connection with (13) above. Nevertheless, the theory can be extended [35] and, yet again, the spherical model serves obediently to verify the detailed predictions [36].

Finally, with apologies for the many contributions employing spherical models not discussed here — a notable early contribution by Riedel and Wegner [37] on *crossover behavior* especially deserves mention — let us recall that *dynamic* or *kinetic spherical models* may be constructed and analyzed with instructive results. Some recent work is reported in [38–40]; particularly interesting are "two-time" correlation and response functions in which a system is quenched to a given initial state followed by a waiting time, t_w, and the subsequent responses to external stimuli are followed on time scales t_r.

11. What Makes the Spherical Model Tick?

Before moving on to discuss recent progress using spherical models to study criticality in charged systems with long-range Coulomb interactions, let us enquire more closely into the underlying mathematical mechanism that leads to the various results for critical exponents, etc., which we have reported so far — e.g., in Eqs. (5), (7) and (10). In fact, everything hinges on the behavior of the basic d-dimensional Fourier-space integral

$$\mathcal{J}_d^\sigma(u) = \int \frac{d^d k}{u + |\mathbf{k}|^\sigma + \cdots}, \tag{17}$$

where we suppose $0 < \sigma \le 2$, with $\sigma = 2$ describing short range interactions. This integral arises naturally from expression (1) for the free energy, combined with (2) and (6), when one takes a derivative with respect to u.

Now $\mathcal{J}_d^\sigma(u)$ becomes nonanalytic whenever $u \to 0+$: but the nature of the singularity depends dramatically on the dimensionality. Indeed, it is not hard to establish [22] the most crucial leading results, namely,

$$\mathcal{J}_d^\sigma(u) \to +\infty, \quad \text{for} \quad d \le d_< = \sigma, \tag{18}$$

as $u \to 0+$, and, writing $\gamma = \sigma/(d-\sigma)$ as in (7),

$$\mathcal{J}_d^\sigma(u) = I_d^0 - I_d' u^{1/\gamma} + \cdots, \quad \text{for} \quad \sigma < d < 2\sigma, \tag{19}$$

while at the upper borderline dimensionality one has

$$\mathcal{J}_{2\sigma}^\sigma(u) = I_>^0 - I_>' u\,|\ln u| + \cdots, \quad \text{for} \quad d = d_> = 2\sigma, \tag{20}$$

and, finally, for all larger dimensionalities,

$$\mathcal{J}_d^\sigma(u) = I_d^0 - I_d' u + \cdots, \quad \text{for} \quad d > d_>. \tag{21}$$

Evidently, it is the change in nature of this integral close to $u = 0$, from *divergent* for $d \le d_<$ (which leads to $T_c \equiv 0$), to *finite* with a *finite first derivative* at $u = 0$ for $d > d_>$ (leading to classical critical behavior), that generates the characteristic changes in the critical behavior of spherical models as d and σ vary. Notice, in particular, the appearance in (20) of the logarithmic factor on the borderline at $d = d_>$.

In anticipation of what *might* arise when charge–charge interactions play a role, let us notice that in d (>2) dimensions the Coulomb potential decays as $1/R^{d-2}$, which in the power-law expression (6) for $J(\mathbf{R})$ corresponds to $\sigma = -2$. This, in turn, for a repulsive potential ($J(\mathbf{R}) < 0$) leads to the integral form

$$\begin{aligned}\mathcal{J}_d^{-2}(u) &= \int \frac{d^d k}{u + k^{-2} + \cdots} \\ &= \int \frac{k^2 d^d k}{1 + uk^2 + \cdots}.\end{aligned} \tag{22}$$

In contrast to the cases $\sigma > 0$ examined above, this integral is *not* singular when $u \to 0$. (Strictly we have, here, assumed that the $\cdots$ terms contain no singular powers of k^2, etc., and have neglected a constant term that would normally add to u.) Indeed, if one inverts the integrand in (22) it yields an exponential decay in R space dominated by the factor $e^{-\kappa R}$ with $\kappa = 1/\sqrt{u}$, whereas Fourier inversion of the integrands in (17) when $d > d_<$ yields a power-law decay when $u = 0$.

12. Electrolytes, Plasmas and Ionic Fluids

The fact that many properties of electrolytes differ markedly from those of nonconducting fluid mixtures and solutions has been appreciated by physical chemists for well over a century. The differences, which appear already in very dilute solutions, are a direct consequence of the presence of "free" ions of, say, $S \geq 2$ species and the fact that ions carrying charges q_τ and q_υ interact via the long-range electrostatic potential $q_\tau q_\upsilon / DR^{d-2}$, where D is the dielectric constant of the "background medium": for an electrolyte the solvent may, in a first approach, be viewed as providing the medium in which the ions interact and move; for a plasma one may take $D = 1$ corresponding to vacuum.

The celebrated analysis by Debye and Hückel in 1923 [41] established that the presence of a neutral "gas" of free ions of (number) densities ρ_σ and overall ionic density

$$\rho = \sum_{\tau=1}^{S} \rho_\tau, \tag{23}$$

leads to an *exponential screening* of the "bare" interactions. This is most appropriately expressed in terms of the charge–charge correlation function which is predicted to behave as

$$G_{ZZ}(\mathbf{R}) \equiv \langle \rho_Z(\mathbf{0}) \rho_Z(\mathbf{R}) \rangle \sim e^{-R/\xi_{Z,\infty}} / R^{\frac{1}{2}(d-1)} \tag{24}$$

when $R \to \infty$, where the local charge density is simply

$$\rho_Z(\mathbf{R}) = \sum_{\tau=1}^{S} q_\tau \rho_\tau(\mathbf{R}). \tag{25}$$

The asymptotic screening length, $\xi_{Z,\infty}$, depends on the thermodynamic state and diverges when $\rho \to 0$ as

$$\xi_{Z,\infty}(T,\rho) \approx \xi_D(T,\rho) \propto \sqrt{T/\rho}, \tag{26}$$

where $\xi_D \equiv 1/\kappa_D$ is the Debye length [41, 42] (and $\kappa_D^2 = c_d \sum_\tau \rho_\tau q_\tau^2 / k_\mathrm{B} T$ with $c_3 = 4\pi$).

The constraints of electroneutrality and screening may be embodied most succinctly in the charge–charge structure factor

$$S_{ZZ}(\mathbf{k}) = 0 + \xi_{Z,1}^2 k^2 - \xi_{Z,2}^4 k^4 + \cdots, \tag{27}$$

which is essentially the Fourier transform of $G_{ZZ}(\mathbf{R})$. The leading zero in (27) reflects neutrality while the Stillinger–Lovett sum rule [43, 44] enforces $\xi_{Z,1}(T,\rho) \equiv \xi_D(T,\rho)$. (The behavior of the higher order moments, $\xi_{Z,2k}^{2k}(T,\rho)$, of S_{ZZ}, which need not, in general, be defined for integral k, are not so restricted.)

The short-range interactions and sizes of the various ionic species, which may be represented conveniently by the *interaction diameters*, say, $a_{\sigma,\tau}$, do not enter into the Debye length ξ_D. However, at higher densities the ionic diameters already play a role in Debye–Hückel theory [41, 42], and their importance at low temperatures, especially, it transpires, in the critical region, was already emphasized by Bjerrum in 1926 [45]. Notice that in the simplest continuum picture of a 1:1 electrolyte (or classical plasma), namely, the so-called *primitive model* of hard spheres carrying charges $q_\pm = \pm q_0$, *ionic symmetry* pertains when all the spheres have the same diameter. Conversely, ionic symmetry is *violated* whenever the hard-core diameters, a_{++} and a_{--} with $a_{+-} = \frac{1}{2}(a_{++} + a_{--})$, satisfy $a_{++} \neq a_{--}$ (as must be so in realistic systems).

13. Challenges of Ionic Criticality

Although one discovers [46, 47] that even the original Debye–Hückel theory [41] predicts a critical point in the restricted ($a_{++} = a_{--} = a$) primitive model (or RPM), albeit of classical nature and at a rather low temperature, namely $k_B T_c = \frac{1}{16}(q_0^2/Da)$, interest in critical behavior in the presence of strong ionic forces came to the fore only in 1990. At that time Pitzer [48] suggested that the character of liquid–liquid criticality in an electrolyte, previously expected to be of Ising nature, might instead become classical (or van der Waalsian). Indeed, Pitzer reported experiments indicating a coexistence-curve exponent β close to $\frac{1}{2}$ in place of the Ising value $\beta \simeq 0.32_6$; see [46–50]. Pitzer's proposal led to intense experimental and theoretical efforts [46–53]. The initial experiments tended to corroborate such a change in behavior or, at least, to support a rather rapid Ising-to-classical *crossover* [49, 51]; but the experimental situation was eventually resolved in favor of Ising-type criticality even when relatively strong Coulomb contributions drive the phase separation below T_c; see the 2001 review by Weingärtner and Schröer [51]. Nevertheless, it is not unfair to say that certain experimental puzzles do still remain.

Theoretical progress proved rather more elusive; some subsequent reviews are [52, 53]. As regards the character of criticality in the fully ion-symmetric hard-core RPM (restricted primitive model), the expected

Ising-type behavior was finally confirmed convincingly by extensive, high-precision simulations, combined with new finite-size scaling approaches; see [54–58]. At this point, however, the critical universality class in the presence of *ionic asymmetry*, either via $a_{++} \neq a_{--}$ in 1:1 models or in z:1 electrolyte models with $z > 1$, is *not* settled although Ising-type behavior may reasonably be anticipated and might soon be confirmed.

Nevertheless, as soon as one goes beyond the bulk thermodynamic properties to enquire about the behavior of the charge–charge, density–density and charge–density correlations near criticality, open questions abound!

Let us focus on some basic theoretical issues by recalling the behavior of the density–density correlation function

$$G_{NN}(\mathbf{R}) \equiv \langle \rho_N(\mathbf{0})\rho_N(\mathbf{R})\rangle - \langle \rho_N\rangle^2, \tag{28}$$

where $\langle \rho_N \rangle$ equals ρ as defined in (23), while, similarly, for an S-component system we simply have

$$\rho_N(\mathbf{R}) = \sum_{\tau}^{S} \rho_\tau(\mathbf{R}). \tag{29}$$

Away from criticality the density–density correlations will — when only short-range interactions are present — decay exponentially according to

$$G_{NN}(\mathbf{R}) \sim e^{-R/\xi_{N,\infty}}/R^{\frac{1}{2}(d-1)}. \tag{30}$$

This has the same form as the Debye screening displayed by the charge–charge correlations in (24); but the *density correlation length*, $\xi_{N,\infty}(T,\rho)$, is quite distinct from the charge screening length, $\xi_{Z,\infty}(T,\rho)$, which, of course, is significant only in ionic systems.

Typically, when the system is not close to criticality, $\xi_{N,\infty}$ is of magnitude corresponding to the range, say R_0, of the forces of interaction. Indeed, one may readily check this within the spherical model (for the case $S = 1$) where one finds that the definition of R_0 in (2) sets the relevant scale. However, on approach to gas–liquid or, more generally, fluid–fluid criticality, say along the critical isochore $\rho = \rho_c$, the density correlation length *diverges* as [3, 4]

$$\xi_{N,\infty}(T,\rho_c) \sim 1/t^\nu, \quad \text{when} \quad t \to 0+. \tag{31}$$

Precisely *at criticality* one then has a slow algebraic decay of the form [59]

$$G^{c}_{NN}(R) \sim 1/R^{d-2+\eta}, \tag{32}$$

where the normally nonnegative exponent η naturally depends on the dimensionality and the type of criticality. For the ($d = 2$) Ising model the famous work of Onsager and Kaufman [2, 60] yields $\eta = \frac{1}{4}$ [59, 61]. On the other hand, the classical Ornstein–Zernike or Landau theories give $\eta = 0$ [59].

The "trivial" value $\eta = 0$ always characterizes spherical models — unless long-range $1/R^{d+\sigma}$ interactions, as in (6), are present. In that case one has $\eta = 2 - \sigma > 0$ [9–11]. To this extent the $(d = 3)$-dimensional Ising model value, $\eta \simeq 0.035$ [62], can be mimicked in the long-range spherical model by a choice of σ. Quite generally, the spherical model yields the exponent relation

$$(2 - \eta)\nu = \gamma. \tag{33}$$

This was advanced originally for Ising-type systems on the basis of scaling arguments [59, 61] and is now known to be exact for the $(d = 2)$ Ising model (with $\nu = 1$ and $\gamma = 1\frac{3}{4}$).

The mechanism underlying the spherical model results can again be seen in the basic integral (17) where one finds that the integrand $1/[u + |\mathbf{k}|^\sigma + \cdots]$ is, in essence, simply the Fourier transform of $G_{NN}(\mathbf{R})$ and, hence, proportional to the *density–density structure factor* $S_{NN}(\mathbf{k}; T, \rho)$; see (28).

Now we can pose some major questions: In general terms, if fluid–fluid phase separation occurs in an ionic system — as experiment and simulation demonstrate is possible — how do the charge and density fluctuations interact near and at criticality? And, more specifically, if, as it must, the density correlation length, $\xi_{N,\infty}(T, \rho)$, diverges at criticality — in accord with (31) — what happens to the charge screening length $\xi_{Z,\infty}(T, \rho)$ on approaching the critical point? Does it diverge? If so, how? And then how will $G^{c}_{ZZ}(R)$ decay? Conversely, if $\xi_{Z,\infty}(T, \rho)$ remains finite at (T_c, ρ_c), does it display any sort of singularity when $t \to 0$?

More subtly, while charge neutrality must, surely, always be satisfied — in accord with the leading zero in (27) — will the Stillinger–Lovett relation, as expressed there by $\xi_{Z,1} = \xi_D$, remain valid at criticality? And are the answers to these questions sensitive to ionic symmetry — or to the presence of long-range ion interactions beyond the Coulomb coupling?

14. One-Component Spherical Model Plasma

To provide some guide to answering these questions it is natural to introduce Coulomb interactions into spherical models. The pioneering study was made by E.R. Smith [63] in 1988. He addressed what is, in fact, a *one-component plasma* (or OCP), i.e., an $S = 1$ model with only Coulomb interactions.

In the magnetic language, this means, in effect, that a positive spin value at a lattice site corresponds to the presence of positive charge while a negative value describes negative charge density. It is then rather clear that electroneutrality requires zero external field, i.e., $h = 0$ to ensure $m = 0$, in the notation of (9) and (10). As a consequence the overall ionic density is essentially fixed (by the spherical constraint) and only the temperature can be varied.

Nevertheless interesting features such as screening, the Stillinger–Lovett relation, and the effects of boundary conditions can be (and were) investigated. However, no critical point of gas–liquid character appeared. Rather, on lowering the temperature the systems "crystallized" on the lattice into an antiferromagnetic or spin-wave pattern with long-range $+ - + - + \cdots$ ordering.

More recently, in an endeavor to obtain gas–liquid criticality Smith [64] introduced short-range attractive, i.e., ferromagnetic couplings into his OCP spherical model. When all the charges are turned off this yields a standard spherical model critical point. However, as soon as any ionic charges are switched on, the critical point is destroyed! Roughly speaking this is simply because the mechanism sketched above in (22) comes into play: once a q^2/k^2 Coulomb-type term appears in $\hat{J}(\mathbf{k})$, it overwhelms the vanishing of the parameter u which, previously, located criticality and controlled the behavior. Although screening of charge is thereby ensured, there is no opportunity left for long-range, critical density fluctuations to build up.

15. Multispecies Spherical Model[b]

In contemplating the results of the one-component spherical model plasma and, in particular, its failure to exhibit gas–liquid criticality or a proper analog, one might first note that in any real fluid, or realistic model, the scale of the overall critical density, ρ_c, is primarily set by the range of the hard-core repulsions; it is the strength of the attractions (or effective attractions) that sets the critical temperature, T_c; see, e.g., [46, 49]. But in the one-component spherical model with $s(\mathbf{R}) \gtrless 0$ regarded as a charge density there is, in truth, no real "hard core" provided by the lattice.

This is in contrast to the usual Ising lattice where $s(\mathbf{R}) = +1$ describes an empty site at $\mathbf{R}$ while $s(\mathbf{R}) = -1$ specifies single occupancy *and* multiple

---

[b]The research reported hereon was performed in close collaboration with J.-N. Aqua; at this writing, the details have only been partly published [65, 66].

occupancy of any site is forbidden so that the lattice spacing, a, serves to measure the hard-core diameter. The problem in the OCP model is that positive charge and negative charge can mutually annihilate on the same lattice site leaving no trace! Said in other way s, Bjerrum pairs, reasonably modeled by significant + and − charge on adjacent sites — and known to be important in the critical region of, say, the restricted primitive model [45–47] — are unstable in the sphericalized OCP and will tend to collapse into a close-to-neutral, weakly interacting "ghost." Another aspect of the same issue is the inability in the OCP to change the density independently of the temperature.

Recognizing this feature, it is interesting to investigate multispecies spherical models [65, 66] in which one introduces a distinct sublattice for each of the $\tau = 1, 2, \ldots, S$ species of particle. The various sublattices may be interlaced in a spatially uniform manner as illustrated in Fig. 1 for a $d = 3$, $S = 2$ situation. With an application to a 1:1 electrolyte in mind, the particles occupying sublattice sites in Fig. 1 have been labeled by $\tau = +$ and $\tau = -$; but this does not imply that the two species need carry any charges. On each sublattice an Ising-type identification is made, i.e., $s(\mathbf{R}_i^\tau) = \pm 1$ means, respectively, either an empty site, $\mathbf{R}_i^\tau$, or occupation of the site by a single particle of species τ. Consequently, for two particles of distinct species, τ and υ, one always has $|\mathbf{R}^\tau - \mathbf{R}^\upsilon| \geq a_0$, where a_0 is the minimum intersublattice spacing [65, 66] which, thus, for $S \geq 2$ now acts as a true

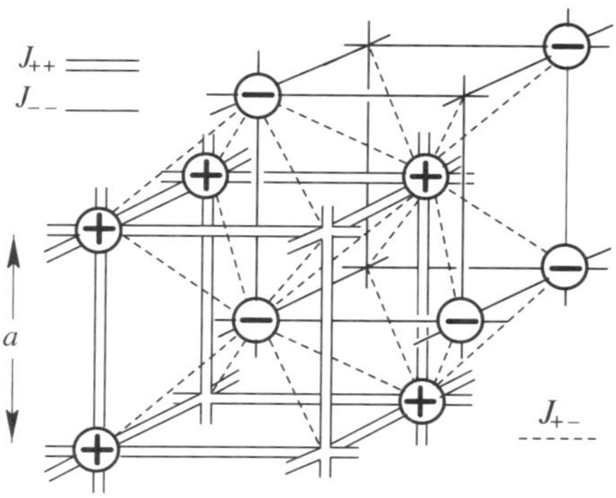

Fig. 1. Illustration of two interlaced simple cubic sublattices, one occupied by particles of species $\tau = +$, the other by particles of species $\tau = -$. The differently drawn bonds illustrate possible nearest-neighbor interparticle interactions of strengths J_{++}, J_{--} and J_{+-}. The closest approach of two particles of different species is $a_0 = \frac{1}{2}\sqrt{3}a$. (After J.-N. Aqua and M.E. Fisher [65].)

interparticle hard core (see Fig. 1). Similarly, the spherical, ordering field, and interaction terms entering the sphericalized Hamiltonian are then of the form

$$\lambda_\tau s_\tau(\mathbf{R}_i), \quad -h_\tau s_\tau(\mathbf{R}_i), \quad \text{and} \quad -J_{\tau\tau'}(\mathbf{R}_{ij})s_\tau(\mathbf{R}_i)s_{\tau'}(\mathbf{R}_j), \tag{34}$$

where, of course, interactions between particles of the same species on the same sublattice are allowed.

To solve the multispecies spherical model exactly it is appropriate to introduce the vector $\boldsymbol{\lambda} = (\lambda_\tau)_{\tau=1,\ldots,S}$ and the interaction matrix

$$\boldsymbol{\Lambda} = [\Lambda_{\tau\upsilon}] = [\lambda_\tau + \Delta\hat{J}_{\tau\tau}(\mathbf{k})]\delta_{\tau\upsilon} - \frac{1}{2}(1-\delta_{\tau\upsilon})\hat{J}_{\tau\upsilon}(\mathbf{k}), \tag{35}$$

in which $\hat{J}_{\tau\upsilon}(\mathbf{k})$ is the Fourier transform of $J_{\tau\upsilon}(\mathbf{R})$, while the $\Delta\,\hat{J}_{\tau\upsilon}$ are defined as in (2). Then the field-independent part of the free energy, which generalizes the original Ising-type form (1), is found to be [65]

$$-F_0/k_\mathrm{B}T \propto \int d^dk \ln\{\mathrm{Det}[\boldsymbol{\Lambda}(\mathbf{k};\boldsymbol{\lambda})]/(k_\mathrm{B}T)^S\}. \tag{36}$$

To complete the formal solution we may introduce the field vector $\mathbf{h} = (h_\tau)$, linearly related to the chemical potentials, μ_τ, and the reduced magnetization vector $\mathbf{m} = (m_\tau)$ where, generalizing (9), the densities of the various species are[c]

$$\rho_\tau = \frac{1}{2}\rho_{\max}(1+m_\tau). \tag{37}$$

Then the field dependence contributes

$$F_h/k_\mathrm{B}T \propto -\frac{1}{2}\langle\mathbf{h}|\boldsymbol{\Lambda}^{-1}(\mathbf{0};\boldsymbol{\lambda})|\mathbf{h}\rangle, \tag{38}$$

to the total free energy where, furthermore, one has

$$\mathbf{h} = \mathbf{2}\boldsymbol{\Lambda}(\mathbf{0};\boldsymbol{\lambda})\mathbf{m}. \tag{39}$$

Needless to say, when $S = 1$ these relations reduce to the standard spherical model expressions.

---

[c]For convenience, as done tacitly in (9), we have here reversed the occupancy convention so that a vacant site corresponds to $s_\tau(\mathbf{R}) = -1$ and an occupied site to $s_\tau(\mathbf{R}) = +1$.

16. Multispecies Equation of State

Mixtures, binary, ternary, etc., of fluids that undergo fluid–fluid phase separation are a commonplace and many then exhibit a critical point — sometimes also called a consolute point or *plait point* in this context [21]. To express the expectations of universality and scaling in such multicomponent systems the systematic formulation of Griffiths and Wheeler [67] is helpful. If, for simplicity, we consider only $S = 2$ species, say, 1 and 2, then, in the vicinity of a critical point one should anticipate the *same* single-component universal behavior — embodied, for the spherical model, in the equation of state (10) — provided one makes the replacements

$$h \Rightarrow \tilde{h} \approx b_1 h_1 + b_2 h_2, \quad m \Rightarrow \tilde{m} \approx c_1 m_1 + c_2 m_2, \tag{40}$$

in which b_1, b_2, c_1 and c_2 are suitable, nonuniversal *mixing coefficients.* The critical point itself will be drawn out into a critical line given by

$$T_\mathrm{c} \Rightarrow \tilde{T}_\mathrm{c}(m^\dagger) \quad \text{with, say,} \quad m^\dagger \approx (c_1' m_1 - c_2' m_2)_\mathrm{c}, \tag{41}$$

where this latter combination, $m^\dagger$, will, in the two-species spherical model, measure the *asymmetry* that reflects the degree to which species 1 and 2 differ in their composition and interactions at criticality.

Now, so it turns out, the spherical model again exhibits some of its waywardness! The mixing prescription (40) proves effective, as general scaling theory predicts, but in place of the original equation of state (10) one discovers [66] the *modified* equation

$$c(\tilde{h} - \mathcal{D}\tilde{m})/\tilde{m} \approx (B^2 t + \tilde{m}^2)^\gamma, \tag{42}$$

when t, $\tilde{m}$, and $\tilde{h} \to 0$. Evidently $\tilde{h}$ has been subjected to the *subtraction* of a term $\mathcal{D}\tilde{m}$, a totally unexpected result! One finds that $\mathcal{D} \sim m^{\dagger 2}$ so that this "anomalous" term does vanish in a symmetric situation, in particular, if $m_{1,\mathrm{c}} = m_{2,\mathrm{c}} = 0$.

But if $\mathcal{D}$ does not vanish at the critical point of interest, what is implied? And can one understand this behavior which, indeed, is most unwelcome for our interests since below T_c it describes a nonstandard first-order transition at which, in fluid language, the pressure (and mean chemical potential) isotherms increase monotonically with overall density in the two-phase region rather than remaining constant. Likewise, the compressibility on approaching T_c from above (with $m^\dagger \neq 0$) saturates at the value $1/\mathcal{D}$ rather than diverging.

As a first remark, one might notice that a ferromagnet endowed, as in the real world, with long-range, $1/r^d$, dipole–dipole interactions does, indeed, display a related behavior: because of the dipolar couplings, the magnetic field that is relevant for the critical behavior is *not* the *externally applied field*, $h_{\rm ext}$; rather it is the *internal* field given by [68, 69]

$$h_{\rm int} = h_{\rm ext} - \mathcal{D}m, \tag{43}$$

where m is the magnetization and $\mathcal{D}$ is the so-called *demagnetization factor* (which depends on the sample shape and is, in general, a tensor [68, 69]). Thus, if we were discussing a realistic ferromagnet, rather than a fluid mixture, we should, in (40), expect to take $h \Rightarrow h_{\rm int}$ as given by (43) with $h_{\rm ext} \equiv \tilde{h}$. But that is *not* actually the case!

Nevertheless, this magnetic analogy, although not apt, should again serve to remind us that the spherical model has a "vectorial character." How might that come into play here? Below $T_{\rm c}$ a mean-field picture is justified and helpful. Thus the density increments, m_1 and m_2, of the two species may be regarded as the z-components of two magnetization vectors, $\vec{m}_1$, and $\vec{m}_2$, of length, in leading approximation, $|\vec{m}_1| = |\vec{m}_2| = m_0$, subject to magnetic fields $\vec{h}_1$ and $\vec{h}_2$ parallel to the z-axis. The asymmetry $m^\dagger$ then translates into a fixed difference $h^\dagger = h_1 - h_2 \propto m^\dagger$. But one must also allow for the basic coupling, say j, of order $k_{\rm B}T_{\rm c}$, that tends to align $\vec{m}_1$ and $\vec{m}_2$.

In the symmetric situation one has $m^\dagger = h^\dagger = 0$ and the minimum energy is clearly achieved by taking $\vec{m}_1$ parallel to $\vec{m}_2$ with both parallel to $\vec{h}_1 = \vec{h}_2$ and so to the z-axis. Then when $\bar{h} = h_1 = h_2$ increases through zero the total magnetization, $m_1 + m_2$, switches abruptly, in the expected, standard first-order fashion, from $-2m_0$ to $+2m_0$. More generally, however, when $h^\dagger \neq 0$ one must allow for the possibility that $\vec{m}_1$ and $\vec{m}_2$ are *canted* strongly away from the z-axis, say by angles θ_1 and θ_2. If we suppose $\vec{m}_1$ and $\vec{m}_2$ remain coplanar, the mean-field energy then takes the form

$$E_{\rm eff} = -h_1 m_0 \cos\theta_1 - h_2 m_0 \cos\theta_2 - j\cos(\theta_1 - \theta_2). \tag{44}$$

To learn from this expression consider, first, the situation in zero field, i.e., $\bar{h} = \frac{1}{2}(h_1 + h_2) = 0$. On minimizing $E_{\rm eff}$ at fixed $h^\dagger$, which we may suppose relatively small, one immediately discovers that $\vec{m}_1$ and $\vec{m}_2$ swing over to a mean direction *perpendicular* to the z-axis so that $\theta_1 \simeq \theta_2 \simeq \frac{1}{2}\pi$, while one finds that $(\theta_1 - \theta_2)$ is of order $h^\dagger m_0/j$. Then, when $\bar{h}$ increases from a close-to-saturation negative value, $h_0 \approx -h^{\dagger 2} m_0/8j$, through zero and up to $+h_0$, the magnetization m, parallel to the z-axis, i.e., $\frac{1}{2}(\vec{m}_1 + \vec{m}_2)_z$, increases *linearly* from $-m_0$ to $+m_0$ just as implied by the equation of state (42). Furthermore, the slope of the $\tilde{m}$ versus $\bar{h}$ plot corresponds to $1/\mathcal{D}$ with, as expected $\mathcal{D} \propto h^{\dagger 2} \propto m^{\dagger 2}$.

Once again, therefore, the waywardness of the model can be seen to arise from the underlying, although generally hidden, vectorial nature. But, happily, as we will now explain, the waywardness is replaced by an exemplary performance when one considers *charged* species!

17. Ionic Spherical Models

The perverse behavior uncovered in the multispecies spherical models might make one apprehensive in approaching a nonsymmetric electrolyte by the same route: but, in fact, the requirements of electroneutrality will save the day! Thus, let us posit that, in addition to their short-range interactions, particles of species τ on the corresponding sublattice carry charges q_τ and that these interact with all other ions via the Coulomb potential $q_\tau q_\upsilon / R^{d-2}$ (for $d > 2$).[d] The simplest case to examine is a 1:1 electrolyte, i.e., $S = 2$ with species $+$ and $-$ (as in Fig. 1) and $q_\pm = \pm q_0$. It is appropriate to suppose that the short-range interactions, $J^0_{\tau\upsilon}(\mathbf{R})$, yield standard (spherical model) critical behavior even when the charges are switched off — as assumed in the extended OCP spherical model [64]. What then happens when the Coulomb interactions are turned on? To answer this we need the determinant of the interaction matrix $\mathbf{\Lambda} = [\Lambda_{\tau\upsilon}]$ as defined in (35). The matrix elements $\Lambda_{\pm\pm}$ follow from the couplings in Fourier space which evidently have the form

$$\hat{J}_{++} = \hat{J}^0_{++} - q_*^2/k^2, \qquad \hat{J}_{--} = \hat{J}^0_{--} - q_*^2/k^2, \tag{45}$$

and

$$\hat{J}_{+-} = \hat{J}_{-+} = \hat{J}^0_{+-} + q_*^2/k^2, \tag{46}$$

where for convenience we have defined a rescaled charge $q_* \propto q_0/a^{d/2}$. Note that the combinations $(\Lambda_{++} + \Lambda_{+-})$ and $(\Lambda_{--} + \Lambda_{+-})$ will be *independent* of the charges.

To simplify further consider **ion symmetric models**, like the RPM, in which $J_{++}(\mathbf{R}) = J_{--}(\mathbf{R})$. In (35) and (36) we may then take $\lambda_+ = \lambda_- = \bar{\lambda}$ whereupon the determinant reduces to $(\Lambda^2_{++} - \Lambda^2_{+-})$ which immediately

[d] Actually it is somewhat more appealing in a lattice system to replace $1/R^{d-2}$ by the solution of the corresponding discrete Laplace equation. This, however, has precisely the same long-distance behavior and makes only trivial changes at small k in Fourier space.

factorizes to yield

$$\ln\{\mathrm{Det}[\mathbf{\Lambda}]\} = \ln(\bar{\lambda}+\Delta\hat{J}^0_{++}-\hat{J}^0_{+-})+\ln(\bar{\lambda}+\Delta\hat{J}^0_{++}+\hat{J}^0_{+-}+2q_*^2/k^2). \quad (47)$$

The first term is charge-independent and clearly yields an Ising-type contribution like (1) to the total free energy. As a result, switching on the charges leaves the original spherical model critical behavior *unchanged* (provided q_0 is not too large [65]). Recall that this is *not* what happens in the extended OCP model [64].

By contrast, the second term in (47) embodies the ionic charges but does not directly drive criticality. However, it leads to exponential screening via the mechanism sketched in connection with (22) — as in the OCP model. Nevertheless, when the critical point is approached, the spherical field, $\bar{\lambda}(T, h)$, that enters both terms in (47), must pick up some singular dependence in light of the discussion leading from (17) to (19). Consequently, the screening length, while remaining finite at the critical point, varies on the critical isochore as

$$\xi_{Z,\infty}(T,\rho_{\mathrm{c}}) = \xi^{\mathrm{c}}_{Z,\infty}[1+c_Z t^{1-\alpha}+\cdots], \quad (48)$$

where we may recall that α is the specific heat exponent. This result confirms a structure expected on quite general grounds: see, e.g., [70, 71]. For the spherical model we find explicitly,

$$\alpha = \min\{0, -\varepsilon/(d-2)\} \quad \text{with} \quad \epsilon = 4 - d. \quad (49)$$

The singular behavior of $\xi_{Z,\infty}$ is reflected similarly in all the other moments, $\xi_{Z,k}$, that were introduced via (27) *except* for $\xi_{Z,1}$: this leading moment not only remains finite *at* criticality but, in addition, it *maintains* the appropriate Stillinger–Lovett value, namely, $\xi_D(T_{\mathrm{c}}, \rho_{\mathrm{c}})$. By comparison, the actual screening length in the simplest nearest-neighbor, symmetric ionic spherical model (with $J^0_{++} = J^0_{--} = 0$) is given by $\xi^{\mathrm{c}}_{Z,\infty} \simeq 2.4\xi^{\mathrm{c}}_D$ [65].

In reflecting upon these conclusions, one sees that the introduction of two distinct sublattices to keep apart the oppositely charged ions was a crucial step. Furthermore, in these ion-symmetric cases it is clear that the individual sublattice densities, ρ_+ and ρ_-, should always be equal. Thus the perverse "demagnetization" phenomena found in the general, nonsymmetric uncharged two-species spherical model — see (42) — does not materialize. But what about the more realistic situation in which ion symmetry does *not* pertain?

18. Criticality in Nonsymmetric Ionic Spherical Models

The general formulation of the multispecies spherical model is independent of any special symmetries. However, if in the 1:1 ionic model, one has $J^0_{++}(\mathbf{R}) \neq J^0_{--}(\mathbf{R})$, the simple factorization of Det$[\mathbf{\Lambda}]$ that led to (47) no longer applies. Nevertheless, it is trivial to diagonalize $\mathbf{\Lambda}$ and thereby to express the determinant as a product of the two eigenvalues $\Lambda_- \equiv \Lambda_N(\mathbf{k};\boldsymbol{\lambda})$ and $\Lambda_+ \equiv \Lambda_Z(\mathbf{k};\boldsymbol{\lambda})$. The reason for the subscript labels, denoting density and charge sectors, respectively, becomes apparent when one examines the small k behavior: indeed one finds [65]

$$\Lambda_N(\mathbf{k};\boldsymbol{\lambda}) = \lambda + j_0 R_N^2 k^2 + O(k^4), \tag{50}$$

$$\Lambda_Z(\mathbf{k};\boldsymbol{\lambda}) \propto \left(q_0^2/a^d\right)\left[1/k^2 + R_Z^2 + O(k^2)\right], \tag{51}$$

where λ simply measures the deviation of $(\lambda_1 + \lambda_2)$ from its critical value while $j_0 \propto k_{\mathrm{B}}T_0 \equiv \hat{J}_{+-}(0)$ sets the energy scale. Furthermore, provided q_0^2 is not too large, R_N is close to R_0, the range of the nonionic forces [see (2)], while one has $R_Z^2 \propto a^d j_0/q_0^2$. But, again, the crucial point, as in (47), is that two distinct terms contribute to the free energy, one from Λ_N being of the standard spherical model form, with no sign of the q_0^2/k^2 Coulomb divergence while the other, from Λ_Z, is dominated by the small-k divergence!

Before reporting the new results it is worthwhile to introduce a dimensionless measure of the *ionic asymmetry* via [65]

$$\delta_J = \max|\Delta\,\hat{J}_{++}(\mathbf{k}) - \Delta\,\hat{J}_{--}(\mathbf{k})|/k_{\mathrm{B}}T_0. \tag{52}$$

Then, even for $\delta_J \neq 0$, the first conclusion is that the nature of the criticality does *not* depend on the presence of Coulomb interactions (provided, again, that these are not too strong). Specifically, as already indicated, the "demagnetization" pathology of the nonsymmetric $S = 2$ spherical models does not arise because $\rho_+ = \rho_-$ must always be imposed to maintain electroneutrality, whether or not the + and – ions are related by symmetry.

So standard criticality survives in the presence of long-range Coulomb forces: but what happens to the charge correlations and screening near and at criticality? Naturally, exponential screening is preserved everywhere *away* from criticality; however, the screening length $\xi_{Z,\infty}$, is now "infected" by the divergent density fluctuations! Indeed, near criticality $\xi_{Z,\infty}$ tracks the density correlation length and thus diverges according to

$$\xi_{Z,\infty} \approx \xi_{N,\infty} \sim 1/t^{\nu} \tag{53}$$

when $T \to T_c+$ on the critical isochore $\rho = \rho_c$. Likewise, all the charge correlation moments $\xi_{Z,k}$ for $k \geq 2$ [see (27)] also diverge at criticality; specifically the analysis yields [65]

$$\xi_{Z,2}(T,\rho_c) \sim \sqrt{\delta_J}/t^{\frac{1}{2}\nu} \quad \text{as} \quad t \to 0+. \tag{54}$$

On the other hand $\xi_{Z,1}(T,\rho_c)$ remains finite even at the critical point; nevertheless, the Stillinger–Lovett sum rule now *fails at* criticality since one discovers [65] that

$$(\xi_{Z,1}/\xi_D)_c = 1 + c_D \delta_J^2 / q_0^2 + \cdots . \tag{55}$$

Hence the critical point of a nonsymmetric ionic spherical model yields an *anomalous conducting fluid.*

A further peculiarity is that, notwithstanding the divergence of $\xi_{Z,\infty}$ on approaching criticality, the charge–charge correlation function $G_{ZZ}(\mathbf{R})$ *at* the critical point still decays *exponentially* rather than algebraically — as does $G_{NN}^c(\mathbf{R})$; see (32). What happens is that the standard diverging, density-induced component that decays as $e^{-R/\xi_{N,\infty}}/R$, gains an amplitude which vanishes as $\delta_J^2 t^{4\nu}$ when $t \to 0$. Only a term varying as $-e^{-R/R_Z}/R$ then survives; see [65].

It should be mentioned that our exact spherical model results for ionic criticality are broadly consistent with a heuristic Ornstein–Zernike-based analysis advanced by Stell [50] for normal continuum electrolyte models (although there are some significant differences).

19. Scattering and Cross Correlations

The results for ionic criticality in our two-species spherical models follow directly from a remarkable decomposition of the various structure factors $S_{NN}(\mathbf{k})$, $S_{NZ}(\mathbf{k})$ and $S_{ZZ}(\mathbf{k})$. If X and Y stand for N or Z the analysis [65] reveals the formula

$$\frac{S_{XY}(\mathbf{k};\boldsymbol{\lambda})}{k_B T/4\rho a^d} = \frac{B_{XY}^N(\mathbf{k};\boldsymbol{\lambda})}{\Lambda_N(\mathbf{k};\boldsymbol{\lambda})} + \frac{B_{XY}^Z(\mathbf{k};\boldsymbol{\lambda})}{\Lambda_Z(\mathbf{k};\boldsymbol{\lambda})}, \tag{56}$$

in which Λ_N and Λ_Z are the density and charge eigenvalues of the interaction matrix — see (50), (51), and the text leading to (47). The B coefficients satisfy

$$B_{XX}^{N} + B_{XX}^{Z} = 1 \quad \text{and} \quad B_{NZ}^{N} + B_{NZ}^{Z} = 0, \tag{57}$$

while the mixing of charge and density fluctuations at small asymmetry δ_J is controlled by

$$B_{ZZ}^{N} = B_{NN}^{Z} \sim \delta_J^2 k^4 / q_0^4. \tag{58}$$

In addition to the principal features already mentioned, one discovers from (56)–(58) that the *cross-correlation* moments $\xi_{NZ,k}^{2k}$ diverge near criticality as $\xi_{N,\infty}^{2k}$, although, of course, they all vanish linearly with δ_J in the ion symmetric cases. More generally, $S_{NZ}(\mathbf{k} = \mathbf{0}; T, \rho)$ vanishes identically except *at* criticality where, however, it becomes proportional to δ_J^2.

Needless to say, the decomposition (56) has been derived only for the $S = 2$ spherical models. It is tempting, nonetheless, to speculate that some similar expression might be valid more generally. If true, it would be valuable to know.

20. Further "Spherical" Explorations

Many interesting extensions of the calculations described come to mind. In particular, one may ask, following the earlier results [9–11] already described above in (6)–(8), what effects *nonionic* $1/R^{d+\sigma}$ power-law couplings will have on the charge correlations? A case of practical interest is $d = \sigma = 3$ since $1/r^6$, van der Waals interactions characterize real ionic systems. As is well known, van der Waals attractions are a direct reflection of coupled quantum-mechanical charge fluctuations in polarizable atomic systems; but a not dissimilar $1/r^6$ tail arises even in the particle–particle correlations of a fully quantum-mechanical elementary-charge plasma [72–75]. Furthermore, in a quantal system the traditionally expected *exponential* Debye screening is modified at long distances leaving a slowly decaying algebraic tail [72–74]. One might hope to mimic these quantal effects in an ionic spherical model by including power-law potentials.

Such a study has, in fact, been carried out for the 1:1 models [66]. As observed — see (6)–(8), (17)–(21) and (32)–(33) — even in the absence of Coulomb interactions, the $1/R^{d+\sigma}$ couplings with $\sigma < 2$ change the spherical model critical exponents; thus one has $\eta = 2 - \sigma > 0$. Conversely, for $\sigma > 2$ the leading critical behavior with, in particular, the equation of state, remains unchanged; nevertheless, singular correction terms linked to σ appear and, in addition, the correlation function $G_{NN}(\mathbf{R})$ can never decay more rapidly than $1/R^{d+\sigma}$.

All these features survive the introduction of ionic charges. However, the screening is no longer exponential; rather, the charge–charge correlations decay in general as $1/R^{d+\sigma+4}$. Remarkably, the screening factor $1/R^4$ matches the true quantum-mechanical analyses which (with $d+\sigma = 3+3$) establishes $G_{ZZ}(\mathbf{R}) \sim 1/R^{10}$ [75]. In the *ion-symmetric* models the same behavior survives *at* criticality; but in the *nonsymmetric* models the coupling to the density fluctuations reduces the critical-point screening by a factor $R^{4-2\eta}$. Thus one finds

$$\begin{aligned} G^{c}_{ZZ}(R) &\sim 1/R^{d+4-\sigma} \quad \text{for} \quad \sigma < 2, \\ &\sim 1/R^{d+\sigma} \qquad \text{for} \quad \sigma > 2. \end{aligned} \tag{59}$$

Lastly, the Stillinger–Lovett condition is violated *at* criticality when $\eta = 0$ (or $\sigma > 2$) just as before; but it turns out to remain *valid* when, for $\sigma < 2$, one has $\eta > 0$. Since one expects $\eta > 0$ for $d = 3$ Ising-type systems, this suggests that the sum rule may well still hold at criticality in more realistic nonsymmetric systems.

Since, as seen above in (6) and (7), power-law interactions with $\sigma < 2$ or $\sigma < 1$, allow critical behavior in $d = 2$ or $d = 1$ dimensions, respectively, the effects of Coulomb-type couplings on criticality can likewise be examined in these low dimensions. Furthermore, one might, as is experimentally relevant when real charges are confined to layers, then examine $q_\tau q_\upsilon / r$ ionic interactions in planar, i.e., $(d = 2)$-dimensional systems.

Can the multisublattice spherical models be adapted to deal with 2:1 and 3:1 electrolytes while avoiding serious pathologies? It may be possible. Equally one might hope to treat models with $S = 3$ or more ionic species. Of especial interest would be an exactly soluble model with large ions carrying charges $Zq_0 \gg q_0$ in the presence of many small, counterions of charge $-q_0$ or, to represent a salt, additional small ions of charge $+q_0$. Regrettably, it is quite unclear how some spherical model might effectively represent the important geometrical features of such a "colloidal system." Nevertheless, as our look at the past and report on more recent developments suggests, it seems likely that spherical models, despite their waywardness, may yet have further lessons to teach us.

Acknowledgments

I am grateful to Professor Rafael Barrio for encouraging me to write-up my presentation and to him and Professor Kimmo K. Kaski for the invitation to speak at the Symposium in Mexico City in honor of Roger Elliott. It is a pleasure to thank Dr. Jean-Noël Aqua for his thoughtful and incisive comments on the draft manuscript and for his collaboration on our studies of the ionic spherical models which have been reported herein. The interest of Dr. Young C. Kim has been appreciated. The support of the National Science Foundation (through Grant No. CHE 03-01101) is gratefully acknowledged.

References

[1] Onsager, L., *Phys. Rev.* **65** (1944) 117.
[2] See Hemmer, P.C., Holden, H. and Kjelstrup Ratkje, S. (eds.), *The Collected Works of Lars Onsager (with commentary)* (World Scientific, Singapore, 1966) and, especially, the commentaries by Domb, C. and Yang, C.N., pp. 167–181.
[3] Stanley, H.E., *Introduction to Phase Transitions and Critical Phenomena* (Oxford University Press, New York, 1971).
[4] Fisher, M.E., *Rept. Prog. Phys.* **30** (1967) 615.
[5] Berlin, T.H. and Kac, M., *Phys. Rev.* **86** (1952) 821.
[6] Stanley, H.E., *Phys. Rev.* **176** (1968), 718; *J. Phys. Soc. Japan* **26S** (1969) 102.
[7] See the plots in Figs. 1–5 of the early review of renormalization group theory: Fisher, M.E., *Rev. Mod. Phys.* **46** (1974) 597.
[8] Wilson, K.G. and Fisher, M.E., *Phys. Rev. Lett.* **28** (1972) 240.
[9] Fisher, M.E., in *Proc. Conf.* "*Phenomena in the Neighborhood of Critical Points*," N.B.S., Washington D.C., 5–8 April 1965, eds. Green, M.S. and Sengers, J.V. (N.B.S. Misc. Publ. 273, Washington, 1966), pp. 21–25.
[10] Joyce, G.S., in *Phase Transitions and Critical Phenomena*, Vol. 2, eds. Domb, C. and Green, M.S. (Academic Press, New York, 1972), p. 375.
[11] Fisher, M.E., Ma, S.-K. and Nickel, B.G., *Phys. Rev. Lett.* **29** (1972) 917–920.
[12] Essam, J.W. and Fisher, M.E., *J. Chem. Phys.* **38** (1963) 802.
[13] Widom, B., *J. Chem. Phys.* **43** (1965), 3892.
[14] Fisher, M.E., Comment on: *J. Chem. Phys.* **93** (1990) 8405; *J. Chem. Phys.* **96** (1992) 3352–3354 [corrected version has an abstract: pp. 3352–3355].
[15] Kortman, P.J. and Griffiths, R.B., *Phys. Rev. Lett.* **27** (1971) 1439.
[16] Fisher, M.E., *Phys. Rev. Lett.* **40** (1978) 1610–1613.
[17] Fisher, M.E., *Prog. Theoret. Phys. Suppl.* **69** (1980) 14–29.
[18] Lai, S.-N. and Fisher, M.E., *J. Chem. Phys.* **103** (1995) 8144–8155.
[19] Baker, G.A. Jr., Fisher, M.E. and Moussa, P., *Phys. Rev. Lett.* **42** (1979) 615.
[20] Fisher, M.E. in *Magnetism and Magnetic Materials*, 1974, eds., Graham, C.D., Jr., *et al. AIP Conf. Proc.* **24** (1975) 273.
[21] Levelt Sengers, J., *How Fluids Unmix: Discoveries of the School of van der Waals and Kamerlingh Onnes* (Royal Netherlands Academy of Arts and Sciences, Amsterdam, 2003), Chaps. 4 and 5.
[22] Barbosa, M.C. and Fisher, M.E., *Phys. Rev. B* **43** (1991) 10635.
[23] Sarbach, S. and Schneider, T., *Phys. Rev. B* **16** (1977) 347.

[24] Fisher, M.E. and Barbosa, M.C., *Phys. Rev. B* **43** (1991) 11177.
[25] Sarbach, S. and Fisher, M.E., *J. Appl. Phys.* **49** (1978) 1350–1352.
[26] Fisher, M.E. and Sarbach, S., *Phys. Rev. Lett.* **41** (1978) 1127–1130.
[27] Sarbach, S. and Fisher, M.E., *Phys. Rev. B* **18** (1978) 2350–2363; *Phys. Rev. B* **20** (1979) 2797.
[28] Fisher, M.E. and Barber, M.N., *Phys. Rev. Lett.* **28** (1972) 1516.
[29] Fisher, M.E., in *Critical Phenomena: Proc. Enrico Fermi Int. Sch. Physics, Course No. 51,* ed. Green, M.S. (Academic Press, New York, 1971), pp. 1–99.
[30] Barber, M.N. and Fisher, M.E., *Ann. Phys.* (NY) **77** (1973) 1–78.
[31] Singh, S., Pathria, R.K. and Fisher, M.E., *Phys. Rev. B* **33** (1986), 6415–6422.
[32] Fisher, M.E., Barber, M.N. and Jasnow, D.M., *Phys. Rev. A* **8** (1973) 1111–1124.
[33] Singh, S., Jasnow, D.M. and Barber, M.N., *J. Phys. C: Sol. St. Phys.* **8** (1975) 3408.
[34] Privman, V. and Fisher, M.E., *J. Appl. Phys.* **57** (1985) 3327–3328.
[35] Fisher, M.E. and Privman, V., *Phys. Rev.* **32** (1985) 447–464.
[36] Fisher, M.E. and Privman, V., *Commun. Math. Phys.* **103** (1986) 527–548.
[37] Riedel, E.K. and Wegner, F., *Z. Physik* **225** (1969) 195.
[38] Picone, A. and Henkel, M., *J. Phys. A: Math. Gen.* **35** (2002) 5575.
[39] Picone, A., Henkel, M. and Richert, J., *J. Phys. A: Math. Gen.* **36** (2003) 1249.
[40] Henkel, M., Paessens, M. and Pleimling, M., *Europhys. Lett.* **62** (2003) 664–670; *Phys. Rev. E* (2004) [in press].
[41] Debye, P.W. and Hückel, E., *Physik. Z.* **24** (1923) 185.
[42] For a clear exposition of Debye–Hückel theory see, e.g., McQuarry, D.M., *Statistical Mechanics* (Harper and Row, New York, 1976), Chap. 15.
[43] Stillinger, F.H. and Lovett, R., *J. Chem. Phys.* **48** (1968) 3858.
[44] Hansen, J.-P. and McDonald, I.R., *Theory of Simple Liquids*, 2nd edn., (Academic Press, London, 1986), Chap. 10.
[45] Bjerrum, N., *Kgl. Danske Vidensk. Selsk. Mat.-Fys. Medd.* **7** (1926) 1.
[46] Fisher, M.E. and Levin, Y., *Phys. Rev. Lett.* **71** (1993) 3826–3829.
[47] Levin, Y. and Fisher, M.E., *Physica A* **225** (1996) 164–220.
[48] Pitzer, K.S., *Accts. Chem. Res.* **23** (1990) 333.
[49] Fisher, M.E., *J. Stat. Phys.* **75** (1994) 1.
[50] Stell, G., *J. Stat. Phys.* **78** (1995) 197.
[51] Weingärtner, H. and Schröer, W., *Adv. Chem. Phys.* **116** (2001) 1–66, and references therein.
[52] Fisher, M.E., *J. Phys. Cond. Matt.* **8** (1996) 9103–9109.
[53] Stell, G., in *New Approaches to Problems in Liquid-State Theory, Proc. NATO Adv. Study Inst.*, Sicily, Italy, July 1998, eds. Caccamo, C., Hansen, J.-P. and Stell, G. (Kluwer Academic Publishers, Dordrecht, 1999), pp. 71–90.
[54] Luijten, E., Panagiotopoulos, A.Z. and Fisher, M.E., *Phys. Rev. Lett.* **88** (2002) 185701.

[55] Kim, Y.C., Fisher, M.E. and Orkoulas, G., *Phys. Rev. E* **67** (2003) 051506: 1–21.
[56] Kim, Y.C. and Fisher, M.E., *Phys. Rev. E* **68** (2003) 041506:1–23.
[57] Kim, Y.C., Fisher, M.E. and Luijten, E., *Phys. Rev. Lett.* **91** (2003), 065701.
[58] Kim, Y.C. and Fisher, M.E., *Phys. Rev. Lett.* **93** (2004) 185703.
[59] Fisher, M.E., *J. Math. Phys.* **5** (1964) 944.
[60] Kaufman, B. and Onsager, L., *Phys. Rev.* **76** (1949) 1244.
[61] Fisher, M.E., *Physica* **25** (1959) 521.
[62] For recent, systematic and authoritative assessments of critical exponents for n-vector systems see: Guida, R. and Zinn-Justin, J., *J. Phys. A: Math. Gen.* **31** (1998) 8103–8121 and Pelissetto, A. and Vicari, E., *Phys. Repts.* **368** (2002) 549–727.
[63] Smith, E.R., *J. Stat. Phys.* **50** (1988) 813; *ibid.* **55** (1989) 127.
[64] Smith, E.R., *J. Stat. Phys.* (2004) [in press].
[65] Aqua, J.-N. and Fisher, M.E., *Phys. Rev. Lett.* **93** (2004) 135702.
[66] Aqua, J.-N. and Fisher, M.E., *J. Phys. A: Math. Gen.* **37** (2004) L241–248.
[67] Griffiths, R.B. and Wheeler, J.C., *Phys. Rev. A* **2** (1970) 1047.
[68] Landau, L.D. and Lifshitz, E.M., *Electrodynamics of Continuous Media* (Pergamon Press, Oxford, 1960), pp. 44–45, 169–170.
[69] Mattis, D.C., *The Theory of Magnetism* (Harper & Row, New York, 1965), pp. 123–124, 133.
[70] Fisher, M.E., *Phil. Mag.* **7** (1962) 1731.
[71] Fisher, M.E. and Langer, J.S., *Phys. Rev. Lett.* **20** (1968), 665.
[72] Brydges, D.C. and Seiler, E., *J. Stat. Phys.* **42** (1986) 405.
[73] Maggs, A.C. and Ashcroft, N.W., *Phys. Rev. Lett.* **59** (1987) 113.
[74] Alastuey, A. and Martin, Ph. A., *Phys. Rev. A* **40** (1989) 6485.
[75] Cornu, F., *Phys. Rev. E* **53** (1996) 4595–4631, and references therein.

CHAPTER 2

PHASE TRANSITIONS IN VECTOR SPIN GLASSES

A.P. Young
Department of Physics, University of California, Santa Cruz, California 95064, USA

We first give an experimental and theoretical introduction to spin glasses, and then discuss the nature of the phase transition in spin glasses with *vector* spins. Results of Monte Carlo simulations of the XY spin glass model in three dimensions are presented. A finite size scaling analysis of the correlation length of the spins and chiralities of both models shows that there is a single, finite-temperature transition at which both spins and chiralities order.

1. Introduction

It is a pleasure to present this paper on the occasion of Roger Elliott's 75th birthday and his induction as a "Miembro Correspondiente" of the "Academia Mexicana de Ciencias." Roger was both my undergraduate tutor and the supervisor for my D. Phil, so I had plenty of opportunity to learn from his great intuition for physics. In particular, I learned from his wonderfully clear lectures that the field of disordered systems is rich and interesting, and consequently disordered systems has been at the forefront of my research ever since. This talk will be about an area of disordered systems which has proved extremely challenging and where controversies continue: the *spin glass*.

A spin glass is a system with disorder and frustration. Figure 1 shows a toy example of frustration with a single square of Ising spins (which can only point up or down). The "+" or "−" on the bonds indicates a ferromagnetic or antiferromagnetic interaction, respectively. In this example, with one negative bond, it is impossible to minimize the energy of all the bonds so there is competition or "frustration."

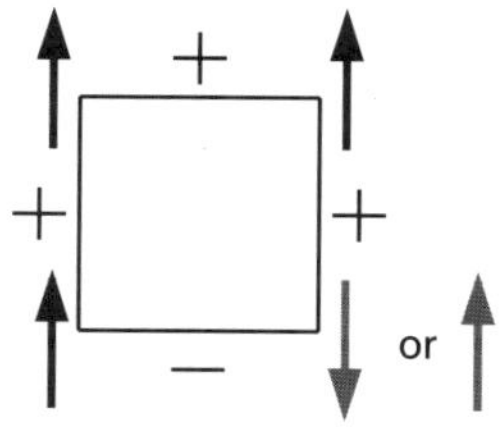

Fig. 1. A Toy model which shows frustration. If the interaction on the bond is a "+", the spins want to be parallel and if it is a "−" they want to be antiparallel. Clearly all these conditions cannot be met so there is competition or "frustration."

Most theoretical work uses the Edwards–Anderson (EA) model [1]

$$\mathcal{H} = -\sum_{\langle i,j \rangle} J_{ij} \mathbf{S}_i \cdot \mathbf{S}_j, \tag{1}$$

in which the spins $\mathbf{S}_i$ lie on the sites of a regular lattice, and the interactions J_{ij}, which we take to be between nearest neighbors only, are independent random variables with mean and standard deviation given by

$$[J_{ij}]_{\rm av} = 0; \quad \left[J_{ij}^2\right]_{\rm av}^{1/2} = J\,(=1). \tag{2}$$

A zero mean is chosen to avoid any bias towards ferromagnetism or anti-ferromagnetism, and we will follow common practice and take a Gaussian distribution for the J_{ij}. The $\mathbf{S}_i$ are of unit length and have m-components:

$$\begin{aligned} m &= 1 \quad \text{(Ising)} \\ m &= 2 \quad \text{(XY)} \\ m &= 3 \quad \text{(Heisenberg).} \end{aligned} \tag{3}$$

The Edwards–Anderson model is the simplest one which includes the necessary ingredients of randomness and frustration.

Different types of experimental systems have these ingredients:

- Metals:
  Diluted magnetic atoms, e.g., Mn, in a non-magnetic metal such as Cu, interact with the RKKY interaction,

  $$J_{ij} \sim \frac{\cos(2k_{\rm F} R_{ij})}{R_{ij}^3}, \tag{4}$$

  where $k_{\rm F}$ is the Fermi wavevector. We see that J_{ij} is random in magnitude and *sign*, so there is frustration. Note that Mn is an S-state ion and so has little anisotropy. It should therefore correspond to a *Heisenberg* spin glass.

- Insulators:
  An example is $Fe_{0.5}Mn_{0.5}TiO_3$, which comprises hexagonal layers. The spins align perpendicular to layers (hence it is *Ising*-like). Within a layer the spins in pure $FeTiO_3$ are ferromagnetically coupled while spins in pure $MnTiO_3$ are antiferromagnetically coupled. Hence, the mixture gives an *Ising spin glass with short range interactions.*
- Other systems where spin glass ideas have proved useful are:
  — Protein folding
  — Optimization problems in computer science
  — Polymer glasses, foams, ...

An important feature of spin glasses is that they undergo a sharp thermodynamic phase transition at temperature $T = T_{\rm SG}$, such that for $T < T_{\rm SG}$ the spin freeze in some random-looking orientation. As $T \to T_{\rm SG}^+$, the spin glass correlation length $\xi_{\rm SG}$, which we will discuss in detail below, diverges. Here we just note that the defining feature of the correlation length is that the correlation function $\langle S_i S_j \rangle$ becomes significant for $R_{ij} < \xi_{\rm SG}$, though the *sign* is random. A quantity which diverges, therefore, is the *spin glass susceptibility*:

$$\chi_{\rm SG} = \frac{1}{N} \sum_{\langle i,j \rangle} \left[\langle \mathbf{S}_i \cdot \mathbf{S}_j \rangle^2 \right]_{\rm av} \tag{5}$$

(notice the square), which is accessible in simulations. It is also essentially the same as the *non-linear susceptibility*, $\chi_{\rm nl}$, which can be measured experimentally and is defined by the coefficient of h^3 in the expansion of the magnetization m:

$$m = \chi h - \chi_{\rm nl} h^3 + \cdots, \tag{6}$$

where h is the magnetic field. We expect that $\chi_{\rm nl}$ diverges at $T_{\rm SG}$ as

$$\chi_{\rm nl} \sim (T - T_{\rm SG})^{-\gamma}, \tag{7}$$

where γ is a critical exponent.

This divergent behavior has been seen in many experiments. Figure 2 shows the results of Omari *et al.* [2] on 1% Mn in Cu. They define $m = a_1 h - a_3 c_3 h^3 + a_5 c_5 h^5$ and choose units (and constants $c_3 = 1/15, c_5 = 2/305$) such that $a_i = 1$ for independent Mn spins. It follows that a_3 is $\chi_{\rm nl}$ in dimensionless units. We see that $\chi_{\rm nl}$ becomes very large, ($>10^3$), and presumably diverges. A fit gives $\gamma = 3.25$.

An important feature of spin glasses at low temperature is that the dynamics becomes very slow, and below $T_{\rm SG}$ the system is never fully in

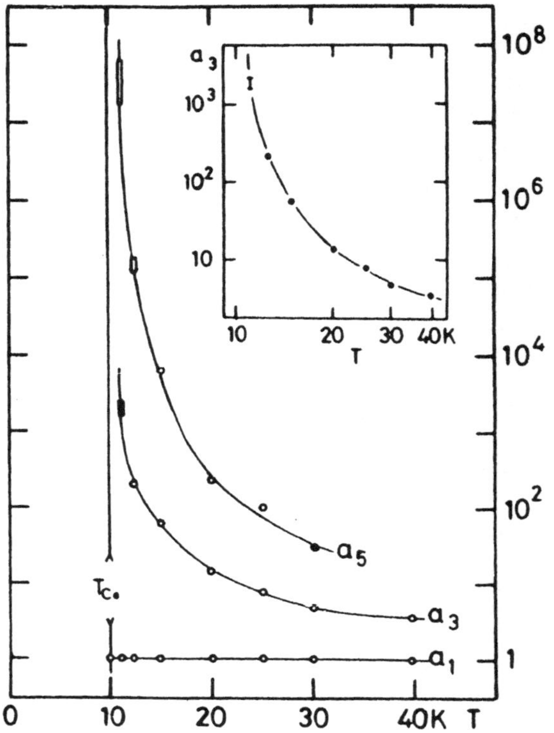

Fig. 2. Results for the non-linear susceptibility of 1% Mn in Cu from Omari *et al.* [2]. The quantity a_3 is the non-linear susceptibility in dimensionless units.

equilibrium. This is because the "energy landscape" becomes very complicated with many "valleys" separated by "barriers." The (free) energies of the valleys can be very similar and yet the spin configurations rather different. Hence, there are large-scale, low-energy excitations in spin glasses.

This non-equilibrium behavior has been extensively studied in recent years. Of particular note has been the study of "aging" in spin glasses, pioneered by the Uppsala group [3]. One cools the system below $T_{\rm SG}$ and waits for a "waiting time" $t_{\rm w}$. The system is then perturbed in some way, e.g., by applying a magnetic field, and the subsequent response is measured. It is found that the nature of the response depends on $t_{\rm w}$, providing clear evidence that the system was not in equilibrium.

More complicated temperature protocols are possible, which have led to surprising results. For example, one can cool smoothly below $T_{\rm SG}$ and wait at a temperature T_1, say, before cooling further, and then warming back up through T_{SG} this time without waiting at T_1. While waiting at T_1 during the cooling process, the data shows a drift with time, and on warming, one finds a similar feature at T_1 even though the system did not wait there. This "memory" effect [4] is still not well understood, and neither is "rejuvenation," the fact that aging at one temperature does not help equilibration at a lower temperature [4].

On the theoretical side, there is a mean field solution due to Parisi [5, 6] which following Sherrington and Kirkpatrick [7], is the exact solution of an EA-like model with infinite range interactions. One finds a finite spin glass transition temperature T_{SG}.

Most of what we know about short range short-range (EA) models in three dimensions has come from simulations on Ising systems, which also indicate a finite T_{SG}, as we will see below. However, less is known about vector spin glass models and these will be the main focus of the rest of the talk.

While the existence of a phase transition in three-dimensions is not in serious dispute, the nature of the equilibrium state below T_{SG} has been much more controversial. While an experimental system is not in equilibrium below T_{SG}, to develop a theory for the non-equilibrium behavior we presumably need to know the equilibrium state towards which it is trying to get to but never reaches. There are two main scenarios:

- "Replica Symmetry Breaking" (RSB), which is like the Parisi [5, 6] mean field solution, and
- The "droplet picture" (DP) of Fisher and Huse [8, 9].

These differ in the nature of the large-scale, low-energy excitations, whose energy ΔE scales as

$$\Delta E \propto \ell^{\theta}, \tag{8}$$

where ℓ is the linear size of the excitation and θ is a "stiffness" exponent. RSB and DP have different predictions for θ:

- RSB, $\theta = 0$ for some excitations;
- DP, $\theta > 0$ (but small, around 0.2 for 3d Ising).

Hence, a lot of cancellation occurs in the calculation of the energy to flip a cluster of spins. A characteristic feature of spin glasses, then, is the presence of excitations which involve a large number of spins but which cost very little energy.

There are two main sets of issues in spin glasses:

- the nature of the phase transition;
- the nature of the spin glass phase below T_{SG}.

For both problems, most theory has been on Ising systems though the vector nature of the spins may be relevant. In the rest of this talk I will discuss the nature of the phase transition in *vector* spin glass models.

2. Vector Spin Glasses

Most theory has been done for the Ising ($S_i = \pm 1$) spin glass, where there is clear evidence for a finite T_{SG}. The best evidence is from finite size scaling (FSS) of correlation length by Ballesteros *et al.* [10]. This technique is discussed further below. However, many experimental systems, such as CuMn described above, are closer to an isotropic vector spin glass ($\mathbf{S}_i$ is a vector), where the theoretical situation is less clear.

Old Monte Carlo simulations [11] found that T_{SG}, if it occurs at all, must be very low, and this was interpreted as being evidence for $T_{\mathrm{SG}} = 0$. Motivated by this, Kawamura [12–15] argued that $T_{\mathrm{SG}} = 0$ but there can be a glass-like transition at $T = T_{\mathrm{CG}}$ in the "chiralities" (i.e. vortices). This implies *spin–chirality decoupling*. However, the possibility of finite T_{SG} has been raised by various authors, e.g. Maucourt and Grempel [16], Akino and Kosterlitz [17], Granato [18], Matsubara *et al.* [19, 20], and Nakamura *et al.* [21].

The situation seemed confusing and so we decided to try to clarify it by an FSS analysis of the correlation lengths of *both* the spins and chiralities for the XY and Heisenberg spin glasses. We expected this to be useful because:

- it was the most successful approach for the Ising spin glass [10];
- it probes directly *divergent* quantities;
- if spin–chirality decoupling occurs then eventually the spin glass correlation length must exceed the chiral glass correlation length. Can we see this?

Next we discuss how to define chirality in spin glasses. In unfrustrated systems the ground state is collinear and so chirality needs to be thermally excited. Such *thermally activated* chiralities (vortices) are responsible for the Kosterlitz–Thouless–Berezinskii transition in the 2d XY ferromagnet. However, in spin glasses, an important difference is that chiralities are *quenched in* at low-T because the ground state is non-collinear as a result

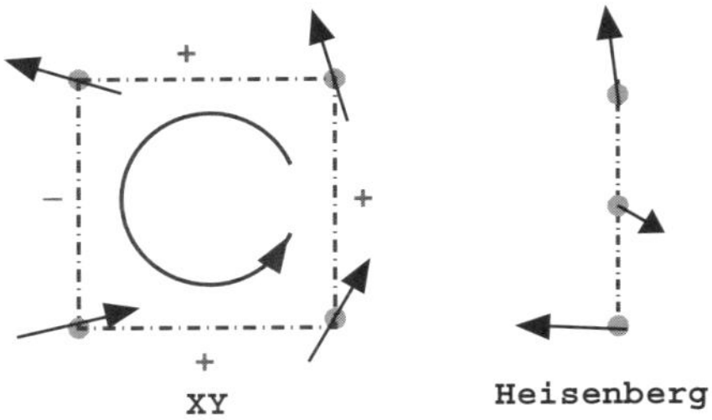

Fig. 3. An illustration of chirality for XY and Heisenberg spin glasses.

of the disorder and frustration. Following Kawamura [13, 14] we define chirality by:

$$\kappa_i^\mu = \begin{cases} \dfrac{1}{2\sqrt{2}} \sum\limits_{\langle l,m \rangle}' \mathrm{sgn}(J_{lm}) \sin(\theta_l - \theta_m), & \text{XY } (\mu \perp \text{square}), \\ \mathbf{S}_{i+\hat{\mu}} \cdot \mathbf{S}_i \times \mathbf{S}_{i-\hat{\mu}}, & \text{Heisenberg} \end{cases} \tag{9}$$

(see Fig. 3).

Next we discuss the various quantities that will be calculated in the simulations. To determine the correlation lengths of the spins and chiralities, we need to Fourier transform the appropriate correlation functions:

$$\begin{aligned} \chi_{\mathrm{SG}}(\mathbf{k}) &= \frac{1}{N} \sum_{i,j} \left[\langle \mathbf{S}_i \cdot \mathbf{S}_j \rangle^2 \right]_{\mathrm{av}} e^{i\mathbf{k}\cdot(\mathbf{R}_i - \mathbf{R}_j)} \quad \text{(spins)}, \\ \chi_{\mathrm{CG}}^\mu(\mathbf{k}) &= \frac{1}{N} \sum_{i,j} \left[\langle \kappa_i^\mu \kappa_j^\mu \rangle^2 \right]_{\mathrm{av}} e^{i\mathbf{k}\cdot(\mathbf{R}_i - \mathbf{R}_j)} \quad \text{(chiralities)}. \end{aligned} \tag{10}$$

Note that $\chi_{\mathrm{nl}} \sim \chi_{\mathrm{SG}}\ (\mathbf{k} = 0)$, which is essentially the "correlation volume" of the spins.

We determine the spin glass correlation length of the finite-size system, ξ_L, from the Ornstein Zernicke equation:

$$\chi_{\mathrm{SG}}(\mathbf{k}) = \frac{\chi_{\mathrm{SG}}(\mathbf{0})}{1 + \xi_L^2 \mathbf{k}^2 + \cdots}, \tag{11}$$

by fitting to $\mathbf{k} = 0$ and $\mathbf{k} = \mathbf{k}_{\min} = \frac{2\pi}{L}(1, 0, 0)$. The precise formula is

$$\xi_L = \frac{1}{2\sin(k_{\min}/2)} \left(\frac{\chi_{\mathrm{SG}}(0)}{\chi_{\mathrm{SG}}(\mathbf{k}_{\min})} - 1 \right)^{1/2} \tag{12}$$

The chiral glass correlation length of the system, $\xi_{c,L}^\mu$, is determined in an analogous way.

The results for the correlation lengths will be analyzed according to finite-size scaling (FSS). The basic assumption of FSS is that the size dependence comes from the ratio $L/\xi_{\rm bulk}$, where

$$\xi_{\rm bulk} \sim (T - T_{\rm SG})^{-\nu} \tag{13}$$

is the *bulk* correlation length. In particular, the *finite-size* correlation length is expected to vary as

$$\frac{\xi_L}{L} = X(L^{1/\nu}(T - T_{\rm SG})), \tag{14}$$

since ξ_L/L is dimensionless (and so has no power of L multiplying the scaling function X). Hence data for ξ_L/L for different sizes should intersect at $T_{\rm SG}$ and splay out below $T_{\rm SG}$. Similarly, data for $\xi_{c,L}$ should intersect at $T_{\rm CG}$.

3. Results

Let's first see how FSS scaling of the correlation length works for the Ising SG. The data in Fig. 4 shows clear intersections, and hence evidence for

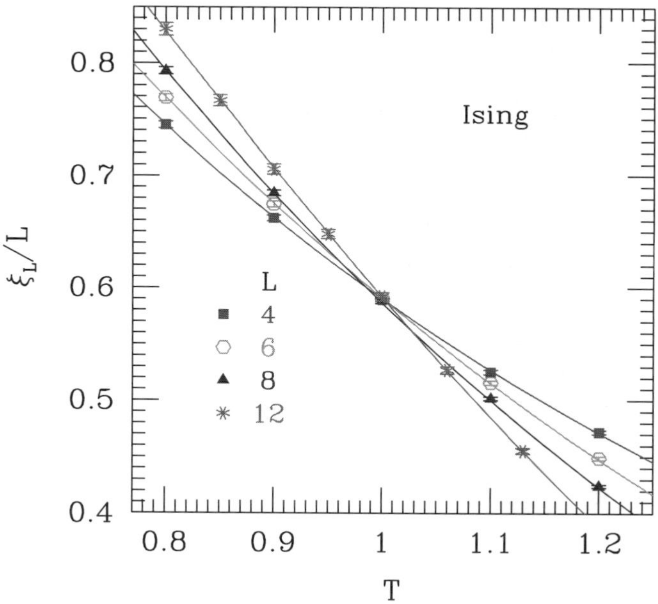

Fig. 4. Data for the correlation length of the Ising spin glass showing clear evidence for a transition at $T_{\rm SG} \simeq 1.00$.

a transition, at $T_{\rm SG} \simeq 1.00$, and the data splay out again on the low-T side demonstrating that there is spin glass order below $T_{\rm SG}$. This is data for the Gaussian distribution. The technique of determining $T_{\rm SG}$ by FSS of ξ_L was first used by Ballesteros *et al.* [10] who took the "$\pm J$" distribution in which $J_{ij} = \pm 1$ with equal probability. This has a somewhat higher transition temperature, $T_{\rm SG} \simeq 1.14$.

Prior to the work of Ballesteros *et al.*, determination of $T_{\rm SG}$ generally used the "Binder ratio," a dimensionless ratio of the moments of the order parameter distribution which has a finite size scaling of the same form as in Eq. (14). However, this gives a much less convincing demonstration of a transition, see Fig. 5 which shows data from Marinari *et al.* [22] for the Gaussian distribution.

We have seen that the best method for studying the transition in the Ising spin glass is FSS of the correlation length. We now apply this to the spin glass with vector spins. Similar results were obtained [23] for both

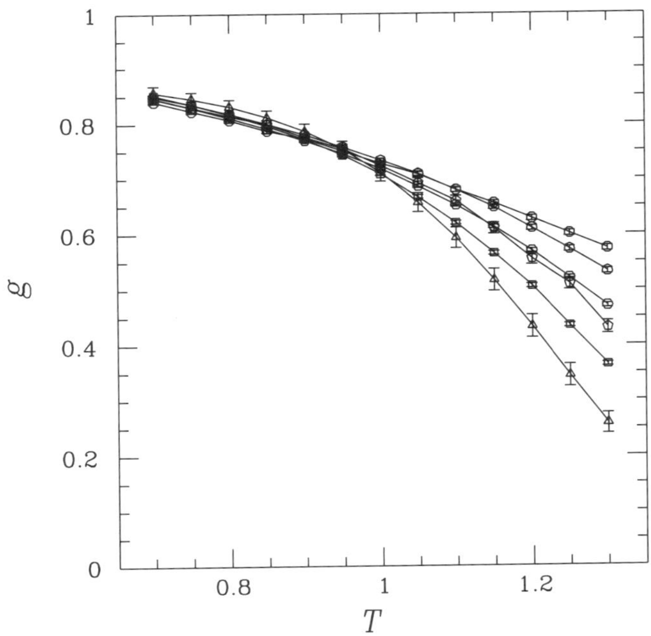

Fig. 5. Data for the Binder ratio length of the Ising spin glass with Gaussian interactions, from Marinari *et al.* [22]. The data merge but do not clearly splay out on the low-T side, unlike the results for the correlation length shown in Fig. 4.

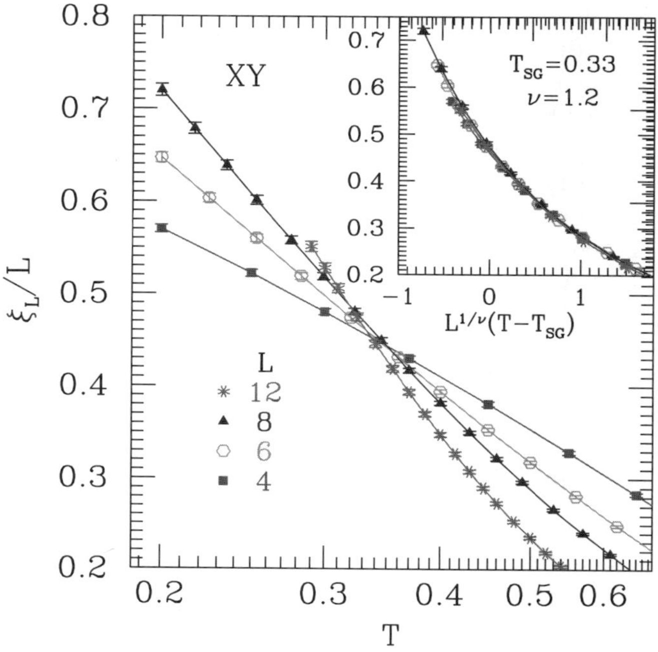

Fig. 6. Data for the spin glass correlation length of the XY spin glass (from [23]).

the XY and Heisenberg models. Here, for conciseness, we just present results for the XY case.

Figure 6 shows data for ξ_L/L. While the intersections are not quite as clean as those for the Ising model, the data does intersect and splay out again at low temperatures indicating a finite-temperature spin glass transition. The inset shows that the data can be collapsed reasonably according the FSS form in Eq. (14) with $T_{\mathrm{SG}} \simeq 0.33$, $\nu \simeq 1.2$.

Figure 7 shows data for the chiral correlation length. There are actually two such lengths depending upon whether the wavevector $\mathbf{k}_{\mathrm{min}}$ in Eq. (12) is parallel or perpendicular to the normal to the plaquettes. The main figure shows the perpendicular correlation length and the inset the parallel correlation length. Apart from the smallest size, the data intersect pretty well. Furthermore, the transition temperature T_{CG} seems to be equal to T_{SG}, namely about 0.33.

We conclude that a *direct* study of the correlation lengths indicates that there is a single phase transition at which both spins and chiralities order in vector spin glasses.

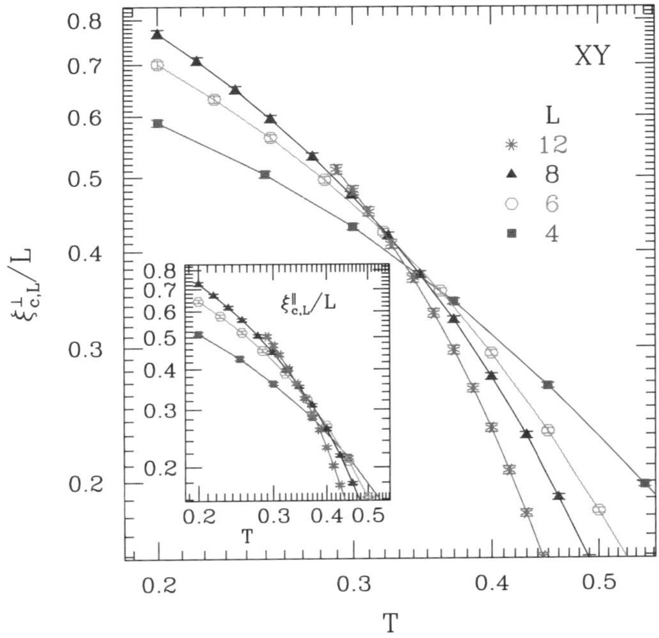

Fig. 7. Data for the chiral glass correlation length of the XY spin glass (from [23]).

4. Conclusions

It is interesting to see how the spin glass transition temperature varies with the number of spin components m. To compare different values of m it is necessary to note that there is an m dependence for T_{SG} even in mean field theory: $T_{\text{SG}}^{\text{MF}} = \sqrt{z}/m$ where z (=6 here) is the number of neighbors. Hence, we show below values for $T_{\text{SG}}/T_{\text{SG}}^{\text{MF}}$ determined from the numerics:

m	Model	$T_{\text{SG}}^{\text{MF}}$	T_{SG}	$T_{\text{SG}}/T_{\text{SG}}^{\text{MF}}$
1	Ising	2.45	1.00	0.41
2	XY	1.22	0.34	0.28
3	Heisenberg	0.82	0.16	0.20

We see that $T_{\text{SG}}/T_{\text{SG}}^{\text{MF}}$ is small and decreases further with increasing m. Physically, this means that fluctuation effects are large and get larger with increasing m. The data suggest that perhaps $T_{SG} = 0$ for $m = \infty$. This is currently under investigation.

To conclude, spin glasses continue to present serious challenges. In this talk, I have presented results which, in my view, resolve one of the controversies, whether there is a finite temperature phase transition in a vector spin glass without anisotropy. The answer appears to be "yes." However, the nature of the putative equilibrium state below T_{SG}, towards which the system evolves but never reaches, as well as non-equilibrium phenomena such as aging and rejuvenation, remain to be fully understood.

Acknowledgments

This work was done in collaboration with Lik Wee Lee and I would like to thank him for many valuable discussions. I acknowledge support from the National Science Foundation under Grant No. DMR 0337049.

References

[1] Edwards, S.F. and Anderson, P.W., *J. Phys. F* **5** (1975) 965.
[2] Omari, R., Prejean, J.J. and Souletie, J., *J. de Physique* **44** (1983) 1069.
[3] Nordblad, P. and Svendlidh, P., in *Spin Glasses and Random Fields*, ed. Young, A.P. (World Scientific, Singapore, 1998).
[4] Jonason, K., Vincent, E., Hammann, J., Bouchaud, J. and Nordblad, P., *Phys. Rev. Lett.* **81** (1998) 3243, cond-mat/9806134.
[5] Parisi, G., *J. Phys. A* **13** (1980) 1101.
[6] Parisi, G., *Phys. Rev. Lett.* **50** (1983) 1946.
[7] Sherrington, D. and Kirkpatrick, S., *Phys. Rev. Lett.* **35** (1975) 1972.
[8] Fisher, D.S. and Huse, D.A., *Phys. Rev. Lett.* **56** (1986) 1601.
[9] Fisher, D.S. and Huse, D.A., *Phys. Rev. B* **38** (1988) 386.
[10] Ballesteros, H.G., Cruz, A., Fernandez, L.A., Martin-Mayor, V., Pech, J., Ruiz-Lorenzo, J.J., Tarancon, A., Tellez, P., Ullod, C.L. and Ungil, C., *Phys. Rev. B* **62** (2000) 14237, cond-mat/0006211.
[11] Jain, S. and Young, A.P., *J. Phys. C* **19** (1986) 3913.
[12] Kawamura, H. and Tanemura, M., *Phys. Rev. B* **36** (1987) 7177.
[13] Kawamura, H. and Li, M.S., *Phys. Rev. Lett.* **87** (2001) 187204, cond-mat/0106551.
[14] Kawamura, H., *Phys. Rev. Lett.* **80** (1998) 5421.
[15] Hukushima, K. and Kawamura, H., *Phys. Rev. E* **61** (2000) R1008.
[16] Maucourt, J. and Grempel, D.R., *Phys. Rev. Lett.* **80** (1998) 770.
[17] Akino, N. and Kosterlitz, J.M., *Phys. Rev. B* **66** (2002) 054536, cond-mat/0203299.
[18] Granato, E., *J. Magn. Magn. Matter.* **226** (2000) 366, cond-mat/0107308.
[19] Matsubara, F., Shirakura, T. and Endoh, S., *Phys. Rev. B* **64** (2001) 092412.

[20] Endoh, S., Matsubara, F. and Shirakura, T., *J. Phys. Soc. Jpn.* **70** (2001) 1543.

[21] Nakamura, T. and Endoh, S., *J. Phys. Soc. Jpn.* **71** (2002) 2113, cond-mat/0110017.

[22] Marinari, E., Parisi, G. and Ruiz-Lorenzo, J.J., *Phys. Rev. B* **58** (1998) 14852.

[23] Lee, L.W. and Young, A.P., *Phys. Rev. Lett.* **90** (2003) 227203, cond-mat/0302371.

CHAPTER 3

TRANSITIONS, DYNAMICS AND DISORDER: FROM EQUILIBRIUM TO NONEQUILIBRIUM COLLECTIVE SYSTEMS

Robin Stinchcombe
Theoretical Physics, Physics Department, Oxford University,
1 Keble Road, Oxford, OX1 3NP, UK
E-mail: stinch@thphys.ox.ac.uk

The paper describes the development of current theories of nonequilibrium collective phenomena from earlier seminal work on equilibrium systems, particularly that of Roger Elliott and collaborators. A brief overview is first given of ordering in equilibrium systems, and their representation using lattice gases and spin or quantum spin models obtained by using pseudospin representations. Techniques such as mean field approximations and exact mappings to free fermion systems are referred to, which were first introduced to treat equilibrium collective behavior, especially classical and quantum phase transitions. Effects of disorder are also mentioned. After an introduction to nonequilibrium collective phenomena, the basic nonequilibrium collective models (particle exclusion models) are obtained by adding prescribed nonequilibrium kinetic processes to the equilibrium lattice models. The resulting models are mapped to quantum spin systems using pseudospin representations. This maps nonequilibrium steady state transitions into quantum phase transitions, which is already a conceptual advance. It also provides various exact results by exploiting Goldstone symmetries or free fermion equivalences or the Bethe ansatz. Direct approaches to nonequilibrium transitions and dynamics are also illustrated by a mean field treatment of a collective nonequilibrium flow/traffic model. The paper concludes with a discussion of recent advances for nonequilibrium systems, concerning exact solutions, including an exact operator algebra approach, disorder effects, and glassy dynamics.

1. Introduction

Earlier pioneering work concerning the study of equilibrium cooperative phenomena in physical systems has turned out to provide many fundamental concepts which carry over to collective behavior in general, including nonphysical systems, and nonequilibrium situations.

The consequent developments in the theory of nonequilibrium systems are the subject of this paper: we attempt here to show how current advances in the theory of nonequilibrium phenomena have emerged from the earlier work on equilibrium collective systems.

The problem of the description of nonequilibrium was already posed by Gibbs, yet until the last decade or so the progress has been largely limited to noncollective or to near-equilibrium behavior.

Most interesting real nonequilibrium systems are collective, and in addition, stochastic rather than deterministic (see Sec. 4). For such systems the basic equation (the master equation) has long been in existence, and with detailed balance relations between transition rates (Sec. 4) it takes the system towards normal Boltzmann–Gibbs equilibrium. The study of the approach to equilibrium has produced many important advances over a long period.

The recent developments in our understanding of collective out-of-equilibrium phenomena have largely come from the taking-over of ideas/concepts and techniques concerning collective behavior from equilibrium systems studies.

Many of these ideas were pioneered or developed in the work of Roger Elliott and collaborators, and some of his fundamental contributions to our understanding of equilibrium collective systems necessarily figure prominently in early sections of this paper.

The carry-over from equilibrium to nonequilibrium studies was not fully realized until specific generalizations of the equilibrium models were studied and certain mappings between equilibrium and nonequilibrium models were established.

Among the fundamental concepts which apply at least qualitatively in the new context are those of equilibrium phase transitions (order, order parameters, correlation lengths, criticality, and universality, etc.). Also widely applicable and generalizable have been particular models (e.g., lattice gases and spin systems) and techniques (including mean field methods, and pseudospin and fermion mappings).

Because the master equation can be recast in the form of a Schrödinger equation, nonequilibrium problems correspond to quantum ones. A very powerful and specific example of this, arising from the use also of a pseudospin description, is the exact equivalence of nonequilibrium lattice gases to quantum spin models.

Among the first mappings to quantum pseudospin systems and studies of quantum collective behavior and transitions were those by Elliott and coworkers. These arose from pseudospin representations of structural phase transitions, particularly in certain cooperative Jahn–Teller systems and in tunnelling systems such as potassium dihydrogen phosphate (KDP). The most famous of these quantum spin models is the transverse Ising model (Ising model in a field transverse to the ordering direction).

The quantum spin models have provided many new features and concepts, which have been important in other contexts, particularly for collective nonequilibrium particle models. An important example is the "quantum phase transition," i.e., the phase transition occurring at absolute zero as one changes a parameter such as the tunnelling or transverse field (e.g., by application of pressure in KDP).

These quantum transitions are associated with some change in the character of the ground state, typically the development of long-range order. Since, in the mapping between quantum spin and nonequilibrium systems, the quantum spin ground state becomes the nonequilibrium steady state, the quantum transitions imply the existence of steady state nonequilibrium transitions, and we learn about their character from the understanding of the quantum ones.

Some of this understanding comes from exact relationships of some quantum spin models to solved or solvable ones, as in the case of the transverse Ising chain [1, 2], which is equivalent to a higher-dimensional Ising model and to a free fermion system. Such relationships have been important for nonequilibrium models and are possible because of the mappings to quantum spin models. Because of those mappings we may also learn from equilibrium studies much about a huge variety of other nonequilibrium phenomena. Among these are disorder effects: a great deal is known about them in equilibrium, from work of Elliott and others, and generalizations are now being made for the nonequilibrium analogs.

The following sections of this paper first present the ideas we need from the studies of equilibrium collective phenomena, particularly those of Elliott and co-workers, and the subsequent sections proceed to develop the nonequilibrium generalizations.

We begin with spin ordering, lattice models and pseudospin mappings in Sec. 2, followed by quantum spin models and transitions, and then disorder effects in Sec. 3. Then, after an introduction to basic aspects of nonequilibrium collective behavior (Sec. 4), we introduce (Sec. 5) the generalization of the equilibrium lattice models and mappings which provide the representative models of nonequilibrium systems, and their mappings. Section 6 treats a particular fundamental nonequilibrium flow model which has both intuitive connections with everyday experience (such as traffic flow and jamming), and connections with quantum transitions and with generalized spin order in equilibrium systems. It provides the simplest example of a nonequilibrium (steady state) phase transition, and of the application of mean field techniques. The paper concludes with a discussion of exact results and a survey of further developments, particularly regarding recent work on disorder and glassy behavior (Sec. 7).

2. Spin Ordering, Lattice Models and Pseudospin

This section presents fundamental ideas concerning equilibrium collective systems which have wider implications, particularly regarding nonequilibrium phenomena.

The first concerns the equilibrium transition to an ordered state and in particular its characterization by an order parameter (non-zero in the ordered state). The original ideas of parallel (and antiparallel) spin alignment in the ground states of ferromagnets (and antiferromagnets) were generalized by Elliott to twisted order (helical and fan, commensurate and incommensurate, etc.) in his theory of rare earth magnetism [3–5]. Figure 1 gives a schematic representation of such types of order. It will be seen that such generalizations are important for nonequilibrium systems: only the simplest of these (such as symmetric hard core diffusion and dimer deposition/evaporation processes) have steady states which correspond to the untwisted structures. Cases with twisted order are of particular interest, and examples are given in Secs. 3 and 6).

Just as in equilibrium theories appropriate forms of mean field approximation interpolate through the transition between such forms of ground state order and the disordered state, so mean field theories of nonequilibrium transitions interpolate between (or do the equivalent of) different ground state orders in quantum phase transitions (see Secs. 3 and 6).

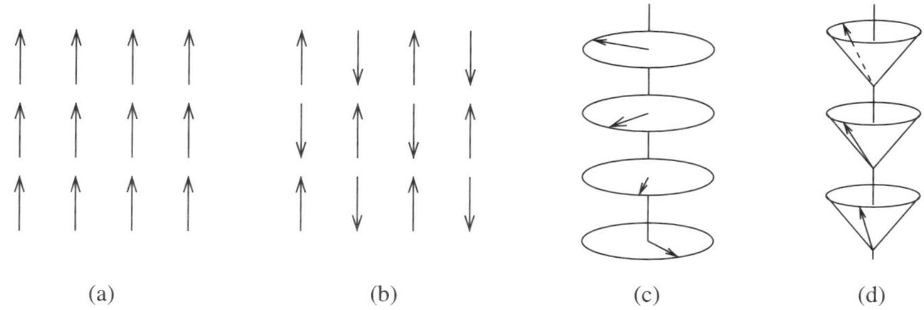

Fig. 1. Ground state order in spin systems (and in steady states of nonequilibrium particle models in their pseudospin or quantum spin representations): (a) ferromagnet [symmetric hard core diffusion, etc.]; (b) antiferromagnet [dimer evaporation and deposition, etc.]; (c), (d) helical ordering in e.g., rare earths [asymmetric hard core diffusion, etc.].

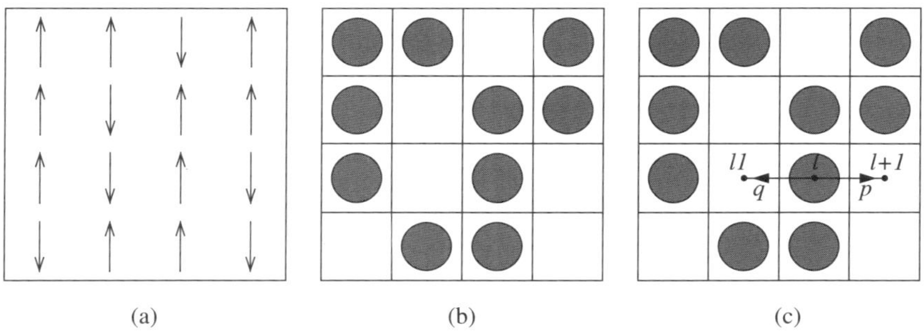

Fig. 2. Configurations of (a) a spin-1/2 Ising system ($\sigma_l^z = \pm 1$), (b) a lattice gas ($n_l = 1, 0$), and (c) a kinetic lattice gas, with stochastic asymmetric hopping.

Pseudospin mappings have been implicit from early on in exploitations of the equivalence of Ising systems to lattice gases and binary alloys (see, e.g., [6]). Figures 2(a) and (b) illustrates this. The equivalence corresponds to having a pseudospin variable σ_l^z at each site l, related to the occupation number n_l (=1 or 0 for particle or vacancy at site l) through the equation

$$\sigma_l^z = 2n_l - 1. \tag{1}$$

The pseudospin description was generalized by Elliott and others, in the context of structural phase transitions, to systems with generalized order and possibly quantum spins [7–11].

One such system is the spin 1/2 Ising model in a transverse field [7–10, 12–14] whose Hamiltonian can be written as follows in terms of components of Pauli operators:

$$H_{\mathrm{TI}} = -\sum_{ij} J_{ij}\sigma_i^z\sigma_j^z - \gamma\sum_i \sigma_i^x . \tag{2}$$

An example where such a form results is in cooperative Jahn–Teller systems, such as dysprosium and thulium vanadate [8, 15–17], where $\sigma_i^z = \pm 1$ represents two low-lying electronic states at site i, which are split by tunnelling (rate proportional to γ) or applied field, and J is an effective coupling between sites resulting from phonon exchange.

A generalized explanation of this sort applies to the hydrogen-bonded ferroelectrics such as KDP, where the tunnelling is between the two double well states on the hydrogen bond [9, 10, 12].

The pseudospin methods, and the phase transitions in these "quantum" systems are of great interest in their original context, since the signatures of quantum phase transitions have been observed, for example, in KDP under pressure (which changes the tunnelling rate) [18]. But they are also of central importance in the development of nonequilibrium theories.

We develop further the quantum phase transition aspects in the next section and apply much of Secs. 2 and 3 thereafter.

3. Quantum Spin Models and Transitions, Disorder

As well as quantum spin models and transitions, this section briefly discusses pioneering work treating disorder in collective equilibrium phenomena, since extensions of those ideas have been of importance in very recent work on the typically severe effects of disorder in nonequilibrium systems.

The transverse Ising model is the archetype for quantum spin transitions. The mean field discussion of the zero (as well as finite) temperature transition in that model is relatively simple [8, 9, 13, 14, 19]. It involves a rotation of the direction of the average of the total spin vector, its z-component becoming non-zero in the ground state below some critical value of $h = \gamma/J$.

Mean field theory predicts phase transitions in any dimension. This is incorrect in one-dimensional finite range systems for finite temperature phase transitions but not for typical zero temperature quantum transitions. Similarly, steady state phase transitions occur in $d = 1$ nonequilibrium systems with finite range processes. An example is given in Sec. 6.

The d-dimensional transverse Ising model is in fact related to the $d+1$-dimensional Ising model so there is already a transition for $d = 1$ [1]. Full details of the quantum transition are provided by the exact solution of the transverse Ising chain [2]. The solution was achieved using the Jordan–Wigner procedure in which σ^z and the conjugate pair of raising and lowering operators, σ^+, σ^-, are mapped to fermion operators [20].

This transformation takes the one-dimensional, nearest neighbor version of the Hamiltonian (2), after a rotation between x-and z-axes, into free fermion form, hence yielding the exact solution. Corresponding procedures apply for particular classes of one-dimensional nonequilibrium model (Sec. 7).

The phase diagram of a generalized transverse Ising model [21] is shown in Fig. 3. The generalization is the addition of an effective field ζ, which drives an energy current J. The case $\zeta = 0$ is that solved by Pfeuty [2], and the point $h = 1$ on the h-axis in Fig. 3 is the quantum transition between disordered and ordered ground states. Power law correlations occur at $h = 1$ (and in the current-carrying phase with non-zero J (Sec. 6)). Many other quantum spin models are now known, including many which have quantum phase transitions. From now on we will restrict our attention to ones which occur in mappings from non-equilibrium particle systems (see particularly Secs. 5–7). These include the Heisenberg model and generalizations with uniaxial anisotropy or Dzyaloshinsky–Moriya interactions, etc., as well as many previously unknown ones.

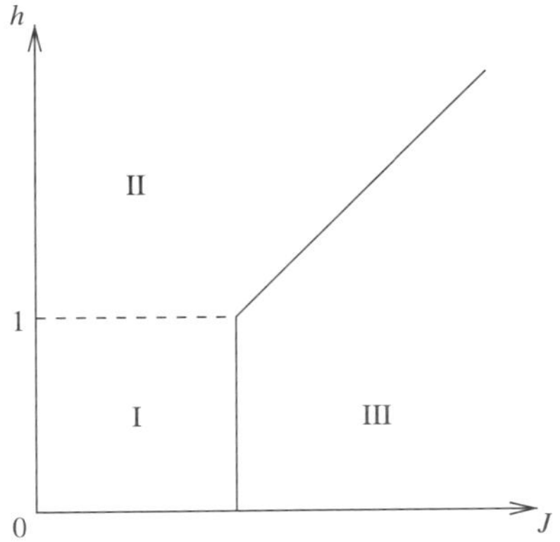

Fig. 3. Ground state phase diagram of Ising model in transverse field h and field ζ driving an energy current J. The phases are characterized as follows. Phase I: $\langle\sigma^z\rangle = 0$, $J = 0$; Phase II: $\langle\sigma^z\rangle \neq 0$, $J = 0$; Phase III: $J \neq 0$. (After [21].)

Finally, in this review of equilibrium collective considerations having important bearing on nonequilibrium ones, we turn to disorder effects.

In equilibrium situations it is well known that in many contexts homogeneous disorder can be represented in terms of an effective medium generalization of the original pure system. That can be quantified using the virtual crystal or, better, coherent potential approximation. The latter can allow for scattering and decay of excitations in disordered media, and has been very successfully used for a wide range of systems from semiconductors to magnets [22].

The earlier studies show that particular sorts of disorder can have their own effects. A striking example is the percolation transition in diluted lattices, where connectivity is lost at a finite bond (or site) concentration. This removes magnetic and other cooperative order below the percolation threshold [23–27] in diluted finite-range magnets, etc. And near the percolation transition, new critical effects occur.

Even weak disorder of any type can modify equilibrium critical behavior, according to a criterion of Harris [28], and it can give rise to localization of excitations.

All these effects have analogs or repercussions in nonequilibrium collective systems. Their study has been crucially dependent on the foregoing work, and on much else we have no space to mention.

4. Nonequilibrium Collective Phenomena

The commonest and most noticeable everyday and natural occurrences are nonequilibrium phenomena, and they are typically also collective and stochastic.

Examples are weather conditions and events, such as winds, precipitation etc., river flow, sea movements, erosion, traffic flow and jamming, crowd behavior, social segregation, supply gluts and shortages, financial market moves, insect population changes, biological growth, etc.

Some of these (weather, etc.) are determined by strictly physical laws while many others have some element of agents (e.g., drivers or traders) and their ability to choose.

Laboratory and production processes give many examples of strictly physical nonequilibrium systems: examples are chemical reactions, epitaxial growth, fluid and granular flow, compaction, energy production and conversion, etc.

All those mentioned above are both stochastic and collective. That is, the microscopic processes are probabilistic or noisy, and the systems involve many constituents (e.g., molecules or agents) which affect or interact with each other.

The collective aspect is crucial for all the most striking effects. Without direct (or effective) interactions between many constituents one would have the independent behavior of small groups, which does not produce phase transitions, jams, market crashes, population extinction, etc. It is the proper inclusion of this collective feature (by generalizing equilibrium many body descriptions) that has given the major advances of the last decade or so.

Though stochasticity is typically present, there are situations where its effects are subdominant. A flow model discussed in Sec. 6 is one such case: here the phase transition occurring with full allowance for effects of the stochasticity is quite well described in the noiseless approximation resulting from a mean field approximation neglecting fluctuations. This type of approximation gives reductions to deterministic descriptions, including some which are closely related to those introduced long ago by Turing in connection with pattern formation.

To describe the rich and challenging nonequilibrium phenomena in everyday systems, the collective models usually need generalization to allow for the effects of agents' choice. Few-player game theory approaches have been available but they clearly are not general enough to combine with collective approaches. However, the recent introduction of the minority game [29, 30] and other such many-player descriptions dramatically widens the possibilities. Significant progress could come from combining the minority game with collective nonequilibrium particle models such as those for simple financial markets, e.g., the related models recently introduced for limit order markets [31–33].

However, except for illustrative purposes, this account is confined to simple physical processes.

A basic question is "What are the features which keep a system away from equilibrium?"

One, occurring for example in weather and traffic systems, is continual feeding, e.g., of energy into the earth's atmosphere from the sun's heating, or of cars onto motorways (see Sec. 6). Another is a driving field, e.g., an electric or diffusion field on ions in a cell, or the effects of gravity on flow through a hopper.

Systems are also kept out of equilibrium if equilibration times become excessively long. An example occurs in window glass. Here the relaxation times grow very rapidly as temperature decreases, so cooling at constant rate reaches a temperature (the "glass temperature") at which the internal rates become slower than the cooling rate, and the system "freezes" into a nonequilibrium state (see Sec. 7).

The foregoing examples suggest some of the interesting and complex nonequilibrium phenomena: phase transitions (as in weather or in traffic jamming); pattern formation (as in chemical reactions and growth); self organization (as in avalanching, and formation of river networks); freezing and ageing (as in glasses).

For any understanding of such phenomena, one has as in equilibrium phenomena first to identify the essential constituents, their possible states, and their interactions. In addition we now have to characterize the basic dynamical processes. Typically a simple model is built [34–36]. For collective behavior the simplest models are lattice gases with typically stochastic kinetic processes. An example is biased hopping, at rates p, q (i.e., the particle-conserving process depicted in Fig. 2(c)). Or the model might involve stochastic pair annihilation of neighboring particles. This (for example) is described in terms of an annihilation probability for any neighboring pair, $\varepsilon'\delta t$ say, in any time interval δt. The rates, such as ε' (or the biased hopping rates p, q, etc., then occur in transition probabilities $W_{C\to C'}$ in the master equation for the evolution of probability (P_C) of configurations (C) for the whole system:

$$dP_C/dt = \sum_{C'}(W_{C'\to C}P_{C'} - W_{C\to C'}P_C). \tag{3}$$

If these transition probabilities satisfy a detailed balance condition, the system will approach normal Boltzmann–Gibbs equilibrium, and will then be described by a thermodynamic formulation involving some free energy functional.

That is not the case with nonequilibrium systems. They are defined by dynamic rules which do not satisfy detailed balance, even though the master equation still applies, and no thermodynamic formulation exists in general for their steady state. To find that, and also more generally to discuss the evolution, the master equation has to be solved, which is typically highly nontrivial. It is an additional level of complication beyond finding a free energy for a collective equilibrium system.

However, there are some simplifications from mappings. The most important of these will be discussed in the next section. But we already note here that the form of the master equation (3) makes it possible to consider it as a Schrödinger equation for the evolution of a state vector formed from the configuration probability P_C under the action of a Hamiltonian with matrix elements

$$H_{CC'} = W_{C' \to C} - \delta_{CC'} \sum_{C''} W_{C \to C''}. \tag{4}$$

So the problem maps quite generally to a quantum (many body) problem.

5. Nonequilibrium Models and Mappings

A simple class of nonequilibrium models are the kinetic lattice gases, or particle exclusion models [35, 36]. In these, particle configurations are of the type depicted in Figs. 2(b) and (c), where each cell, or lattice site, l, has a vacancy ($n_l = 0$) or a particle ($n_l = 1$). The implicit hard core interaction equivalent to the exclusion constraint ($n_l \not> 1$) is the source of the collective behavior.

In addition the models are defined in terms of the prescribed stochastic kinetic processes. Two examples have already been given: asymmetric (hard core) hopping [37–40], as in Fig. 2(c), and pair annihilation (or equivalently dimer evaporation [41]). These, together with their rates, can be denoted respectively by the entries (b)(iii), (b)(iii)′, and (b)(i) in the list below. Here ϕ represents vacancy and A a particle. Entry (b)(ii) involves pair creation/dimer deposition. These are all two-site processes. A number of other representative processes (and rates, where we shall later need them) are given in the list, including the one-site processes (a)(i)/(a)(i)′: single-particle deposition/evaporation (creation/annihilation), (a)(ii): conversion of a single particle between two species; (b)(iv): two-species interchange; and the three-site processes (c)(i): dimer diffusion [42], (c)(ii): trimer evaporation [41].

The list of representative processes is as follows:

(a)(i) $\phi \to A$ at rate α,
(a)(i)′ $A \to \phi$ at rate β,
(a)(ii) $B \to A$, $A \to B$,
(b)(i) $AA \to \phi\phi$ at rate ε',
(b)(ii) $\phi\phi \to AA$ at rate ε,
(b)(iii) $A\phi \to \phi A$ at rate p,
(b)(iii)′ $A\phi \leftarrow \phi A$ at rate q,
(b)(iv) $AB \to BA$, $BA \to AB$,
(c)(i) $AA\phi \to \phi AA$, $\phi AA \to AA\phi$,
(c)(ii) $AAA \to \phi\phi\phi$.

These, and many other such processes, can be combined in various ways to produce a great variety of models providing minimal descriptions of processes ranging from physical processes and chemical reactions in bulk or on surfaces to traffic, etc.

Because of mappings between particles and spins (or pseudospins) these processes are still more general than may appear. For example the single-spin-flip Glauber dynamics of the Ising chain is equivalent to diffusion, pair creation and pair annihilation of domain walls [43], so is exactly equivalent to a particle process combining (b)(i), (b)(ii), (b)(iii) and (b)(iii)′, actually with the rates related by

$$p + q = \varepsilon + \varepsilon' \tag{5}$$

and $p = q$. This is actually the condition for detailed balance, and arises because Ising Glauber dynamics defines an approach to equilibrium.

The biased case (with $p \neq q$) corresponds to a generalized class of nonequilibrium models. There (5) is sufficient to make a solution possible, by a fermion mapping [44, 45] (Sec. 7).

A further example which emphasizes the extended generality of the particle exclusion processes is their mapping to surface growth processes, of which an example is given in Sec. 6 (Fig. 5).

Further justification of the use of such minimal models is provided by universality: these models exhibit a wide range of critical effects. And as far as those are concerned the models are exactly representative of universality classes each of which contains diverse examples of real systems which differ only in details corresponding to irrelevant variables which have no influence on the critical properties.

The pseudospin picture provides the very powerful mapping of the single-species particle exclusion models to quantum spin systems (multi-species models need a generalization we shall not discuss here). The equivalence of particle configurations to spin ones, quantified by (1), allows us to consider the processes of moving a particle to or from a site l as spin flip operations, produced by the action of raising and lowering operators σ_l^+ or σ_l^-.

In that way we can build up a stochastic evolution operator e^{-Ht} for any kinetic single-species particle exclusion process [35, 37, 41, 46]. The Hamiltonian so constructed corresponds to a quantum spin system, since it involves noncommuting spin operators. It is in general nonhermitian. The minus sign in the exponent is chosen for convenience because the processes are dissipative. Then the eigenvalues of H have a nonnegative real part. As well as giving the time evolution of the nonequilibrium system the Hamiltonian gives its steady state as the zero "energy" (ground) state $|0\rangle$ of H, that is $H|0\rangle = 0$. Probability conservation implies a simple form for the zero energy left eigenstate $\langle S|$ but the right eigenstate $|0\rangle$ is typically nontrivial.

It turns out that the Hamiltonian for each individual process involves as well as the spin flip operators the following projection operators at each lattice site l:

$$P_l^{\pm} \equiv \frac{1}{2}(1 \pm \sigma_l^z). \tag{6}$$

This is because the Hamiltonian consists of two parts, necessary for probability conservation, corresponding to the two parts of (4) (from the "in" and "out" processes). One can see it alternatively by considering the evolution over a small time interval δt, say, caused by a specific stochastic process, e.g., the single-particle creation process (a)(i) where $\phi \to A$ at rate α at, say, site l. There, in $e^{-H\delta t} = 1 - H\delta t + \cdots$, a term $-\sigma_l^+ \alpha\delta t$ in $H\delta t$ corresponds to particle production at empty site l with probability $\alpha\delta t$, and a term $-P_l^- \alpha\delta t$ in $H\delta t$ corresponds to having the empty site, but with probability $1 - \alpha\delta t$ not producing the particle there in the time interval δt. The resulting Hamiltonian can be written:

$$H_l^+(\alpha) \equiv -\alpha\left[\sigma_l^+ - P_l^-\right]. \tag{7}$$

This gets summed over l if the process can occur at any site. Similarly, for the pair annihilation/dimer evaporation process (b)(i) where $AA \to \phi\phi$ on, say, bond $l, l+1$, at rate ε', the Hamiltonian is $H_{l,l+1}^{--}(\varepsilon') \equiv -\varepsilon'\left[\sigma_l^- \sigma_{l+1}^- - P_l^+ P_{l+1}^+\right]$. Over the bulk (all bonds) of a chain it becomes $H^{--}(\varepsilon') \equiv \sum_l H_{l,l+1}^{--}(\varepsilon')$ (and similarly for other lattices).

With an obvious generalization of the notation introduced here, the Hamiltonian constructed in a similar way for trimer evaporation [(c)(ii)] on three adjacent sites of a chain is $H_{l,l+1,l+2}^{---}$, and that for dimer hopping [(c)(i)] to the right on a chain involves $H_{l,l+1,l+2}^{-+-}(p) \equiv -p\left[\sigma_l^- P_{l+1}^+ \sigma_{l+2}^+ - P_l^+ P_{l+1}^+ P_{l+2}^-\right]$.

For the well-known fully and partially asymmetric hard core hopping models (the ASEP and PASEP), i.e., processes (b)(iii), (b)(iii)$'$ in the list, in the one-dimensional nearest neighbor case, the Hamiltonians are

$$H_{\mathrm{ASEP}} = H^{-+}\{p_l\} \equiv \sum_l H_{l,l+1}^{-+}(p_l) \equiv -\sum_l p_l\left[\sigma_l^- \sigma_{l+1}^+ - P_l^+ P_{l+1}^-\right], \tag{8}$$

$$H_{\mathrm{PASEP}} = H^{-+}\{p_l\} + H^{+-}\{q_l\}. \tag{9}$$

Here we have allowed for the rates to be bond-dependent, since disordered models, particularly (8), will be discussed in (Sec. 7).

Most of the resulting Hamiltonians describe previously unknown quantum spin systems. But some are familiar, one example being the hard core hopping Hamiltonian (9) in the symmetric case $p = q$. This is the (isotropic) Heisenberg model.

A more complicated example of a model and its quantum spin equivalent [35] is provided by the exclusion process shown for the pure one-dimensional nearest neighbor case in Fig. 4. This combines single particle creation and annihilation at "open" boundaries with bulk processes of asymmetric hopping and pair creation and annihilation. The Hamiltonian is

$$H = H_{\mathrm{PASEP}} + H^{++}(\varepsilon) + H^{--}(\varepsilon') - \alpha\left(\sigma_0^+ - P_0^-\right) - \beta\left(\sigma_L^- - P_L^+\right). \quad (10)$$

This contains as special cases: the (isotropic) Heisenberg model ($p = q$, $\varepsilon = \varepsilon' = \alpha = \beta = 0$), the uniaxially anisotropic Heisenberg model ($p = q$, $\varepsilon = \varepsilon'$, $\alpha = \beta = 0$), the Ising model ($p = q = \varepsilon = \varepsilon'$, $\alpha = \beta = 0$), and the Heisenberg model with added complex staggered Dzyaloshinsky-Moriya ($i(-1)^l \mathbf{z} \cdot (\sigma_l \times \sigma_{(l+1)})$) and boundary field terms ($p \neq q$, $\varepsilon = \varepsilon' = 0$, α and β non-zero).

This last case is the asymmetric exclusion process with open boundaries: the bulk term in the Hamiltonian is (8) and the extra boundary field terms correspond to boundary injection (c.f. (7)) and ejection. This is the most fundamental driven system, and it has a nonequilibrium steady state transition [38, 39]. It is sometimes taken as a basic agentless traffic model [40]. Section 6 explores all these aspects.

The Hamiltonian description of the nonequilibrium system is a real advance. First of all, it enables symmetries to be seen. An example is the Goldstone symmetry of the Heisenberg Hamiltonian, which consequently plays a role in the symmetric hard core diffusion process (c.f. Fig. 1(c), with $p = q$). This symmetry persists with bond disorder, including dilution. A far less obvious Goldstone symmetry is that in the Hamiltonian for the dimer or trimer (etc.) evaporation deposition process on lattices with

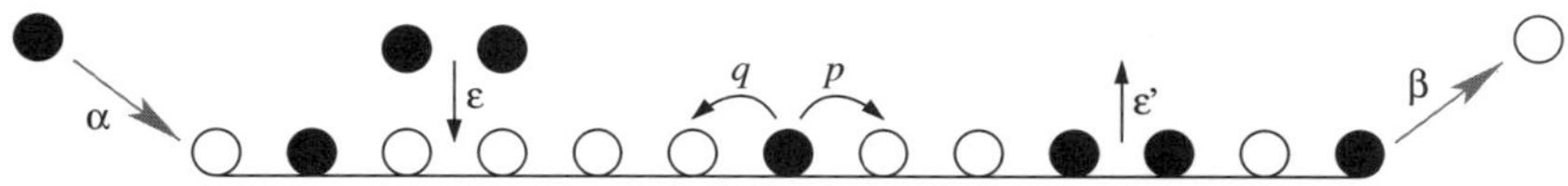

Fig. 4. Basic particle exclusion processes on a chain: boundary injection, asymmetric hopping, pair creation/annihilation (or dimer deposition/evaporation), boundary ejection. Filled circles denote particles, open circles are vacancies.

appropriate sublattice decompositions [41]. In the simplest (dimer) case, the Neel "antiferromagnetic" steady state (Fig. 1(b), for the case of the square lattice) is part of a degenerate multiplet. The consequent gapless spectrum gives rise to power law, rather than exponential, time decay.

Symmetries are related to conservation laws, and their identification is an important step in the understanding of any system. An extreme example of this is the n-mer ($n \geq 3$) evaporation/deposition process on a chain of L sites. This has $O(\mu^L)$ macroscopic conserved quantities (where $\mu > 1$ is a generalized Fibonacci number) [41], and this leads to infinitely many dynamic critical exponents in the infinite system and to other interesting phenomena collectively known as "dynamic diversity" [42].

A still more important advantage of the quantum spin Hamiltonian description is the possibility it gives to use techniques previously developed for quantum spins, or to use on quantum spin systems techniques, such as the operator algebras discovered for nonequilibrium exclusion models. Section 7 discusses such approaches, which include exact solutions obtained by Jordan–Wigner mappings to fermions, and by Bethe ansatz and operator algebra methods.

6. Flow Model: Transitions and Dynamics

This section illustrates some of the ideas introduced in preceding sections by considering the totally asymmetric exclusion process (ASEP), i.e., fully biased hard core diffusion (process (b)(iii) of the list) with open boundaries in one-dimension.

As mentioned earlier, this system has a steady state phase transition [39] and many other characteristics of nonlinear flow and highway traffic systems, which makes it suited to intuitive interpretations.

The phase transition is captured even in the mean field approach [38]. So after some further remarks we shall turn to that description.

The case showing the steady state transition and having highly correlated non-Gibbsian steady states is that with open boundaries at which particles are injected/ejected. So we consider the sub-case of the process in Fig. 4 in which $q = \varepsilon = \varepsilon' = 0$, and p, α, and β are non-zero.

The steady state transition is between maximal and low current phases and the density profiles in these two phases are, respectively, monatonically decreasing or increasing in the direction of the bias. In the mean field approximation their dependences on lattice site position are, respectively, of tan or tanh form (see below). The latter is a fixed soliton and the position of its centre is related to the average total number of particles in the system. The dynamics includes moving solitons and is actually asymptotically subdiffusive (dynamic exponent $z < 2$).

The fixed solitons can be qualitatively understood in terms of the corresponding reduced form of the quantum spin Hamiltonian (10) which applies to it. The combination of complex staggered Dzyaloshinsky–Moriya and complex field terms introduces a twist into the quantum spin ground state, similar to that in the rare earths (Sec. 2), and the projection of this onto the spin z-direction (which gives the density profile ρ_l) results in the soliton steady states.

It was mentioned earlier that certain particle exclusion processes map to surface growth models. The ASEP provides an example of this, and the mapping is illustrated in Fig. 5. Here an occupied/vacant site of the ASEP is mapped into a downward/upward bond of the growth model. The ASEP hop becomes a corner flip (up) and stochastic particle flow becomes stochastic surface growth [47].

We now turn to the mean field approximation. This is most easily effected using the lowest member of the hierarchy of equations for correlation functions of the occupation variable n_l. Its average over histories, ρ_l, satisfies the lowest (continuity) equation:

$$\partial \rho_l / \partial t = J_{l-1,l} - J_{l,l+1}. \tag{11}$$

The right-hand side is the difference of particle currents into and out of the site l (i.e., on the neighboring bonds), whose exact and approximate (mean field) values are

$$J_{l,l+1} = \langle p n_l (1 - n_{l+1}) \rangle \sim p \rho_l (1 - \rho_{l+1}). \tag{12}$$

The mean field reduction is the usual neglect of fluctuations/correlations and this uncouples the hierarchy of equations.

In the steady state the left-hand side of (11) vanishes and $J_{l,l+1}$ becomes independent of l (and t), J say. That converts (12) into a relationship

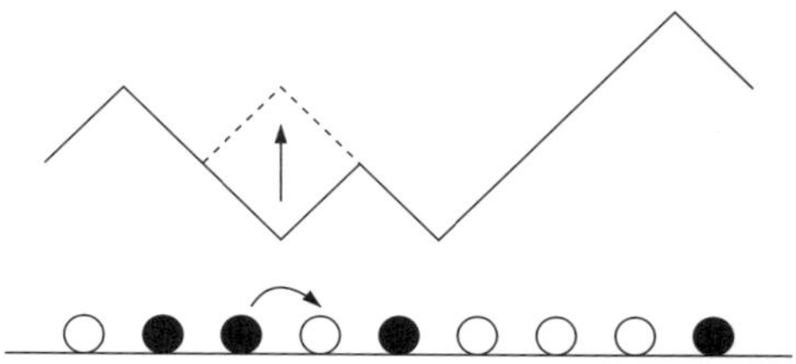

Fig. 5. Mapping of biased hard core hopping (asymmetric exclusion process) to a discrete interface growth model.

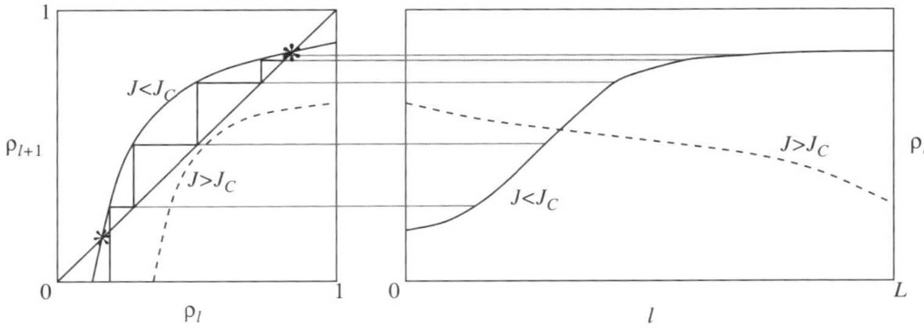

Fig. 6. Steadystate profile map, and density profile ρ_l resulting from its iteration, for $J < J_c$ (full lines) and for $J > J_c$ (dashed lines).

between ρ_{l+1} and ρ_l, which maps ρ_l into ρ_{l+1}. This profile map [38] is shown in Fig. 6, and its iteration is indicated by the zigzag trail.

For $J_c \equiv p/4 > J(< J)$ the map (solid (dashed) line in the figure) has two (no) fixed points and the iteration provides the increasing kink (decreasing) shape shown. The profiles are easily obtained analytically in the form

$$\rho_l = \frac{1}{2} + k \tanh k(l - l_0) \tag{13}$$

for $J < J_c$, and similarly (with tanh replaced by tan) for $J > J_c$. Here J_c is the critical current dividing the high and low current phases, and k is the characteristic inverse length ($k \propto |J - J_c|^{\frac{1}{2}}$). The profile centre l_0 is related to the total number of particles in the system. In a system of large length L, J can only exceed J_c by $O(1/L)$.

With boundary injection (Fig. 4) the rates α, β determine both J (hence k) and l_0 and lead to a phase diagram in α, β space with the high current phase (where α, β are both $>1/2$) divided by second order boundaries from the low current phase. That itself is divided by a first order line ($\alpha = \beta$) into high and low density regions. The saturation is due to the non-linearity of the model coming from exclusion, and can easily be understood by reference to the homogeneous case ($\rho_l = \rho$ independent of l) where $J = p\rho(1 - \rho)$ is a parabola with a maximum at $\rho = 1/2, J = p/4$.

These results, and their dynamic generalization, can be obtained from a continuum approach, in which the continuous variable x replaces the discrete l, $\rho_l(t) \to \rho(x, t)$, and finite differences become derivatives. Then (11) with the mean field reduction (12) results in the mean field continuum equation

$$\partial\rho/\partial t = \partial/\partial x(D\partial\rho/\partial x - \lambda\rho(1 - \rho)), \tag{14}$$

where D, λ are continuum forms of $p + q$, $p - q$, respectively. (14) is the noiseless Burgers equation [48] (the mean field approximation has removed the stochasticity). The transformation from ρ to a height variable h related by $\rho - \frac{1}{2} = \partial h/\partial x$ (the continuum version of the transformation in Fig. 5) takes (14) into the noiseless version of the equation for the KPZ nonlinear growth model [49]. This itself is linearizable by the Cole–Hopf transformation [50] $h = (D/\lambda)\ln u$ which takes the noiseless KPZ equation to the diffusion equation. This means the full mean field continuum dynamic solution can be obtained. The solutions contain propagating solitons, related by the Galilean invariance of the continuum model to the steady state form (13).

The moving kinks are indicative of a phenomenon often noticed by drivers, namely of passing between a high density low current neighborhood to a low density high current one without any sign of a bottleneck or accident that could have caused the effect. The analysis above shows that no such mechanism is needed: the phenomenon is usually an intrinsic property of the "pure" nonlinear system.

Since in collective equilibrium systems mean field approximations typically predict a transition in low dimensions where fluctuations would have prevented it, the above discussion is not by itself convincing. Happily, the steady state ASEP can be treated exactly by an operator algebra method [39] (see Sec. 7), and that confirms the existence of the transition just described, and shows that the mean field phase boundaries are exact. Some details, e.g., of pinning of profiles and of kink width, are incorrect in mean field theory, and we also know from a Bethe ansatz solution [37] that the dynamic exponent is $z = 3/2$ rather than the trivial mean field value 2.

The twisted spin ground state of the quantum version of the ASEP was mentioned at the beginning of this section. That can be seen in a mean field approximation directly on the quantum spin model, and it can also be inferred from the exact solution.

Similar features can be seen in a hybrid model, referred to in Sec. 3, in which a term is added to the transverse Ising Hamiltonian containing a field ζ which drives an energy current J [21]. The Hamiltonian does not have the proper probability conserving form of the quantum spin representation of a nonequilibrium process, but it is exactly soluble by a generalization of the methods introduced by Pfeuty [2]. In the resulting phase diagram, shown in Fig. 3, the $J \neq 0$ phase is characterized by a correlation function of the form

$$\langle \sigma_l^x \sigma_{l+n}^x \rangle = a n^{-\frac{1}{2}} \cos kn, \tag{15}$$

which corresponds to a twisted spin state.

7. Further Developments

This account of the evolution from collective equilibrium treatments to nonequilibrium theories continues here to exact solutions and then concludes with a brief discussion of some current developments (disorder effects and glasses).

Exact treatments of nonequilibrium models can be found both from the quantum spin mapping and directly from the master equation. We illustrate these two alternatives in turn.

Apart from cases where the quantum spin equivalent has a known solution (see, e.g., [44]), the two principal techniques which proceed from the quantum spin starting point are mappings to fermions via the Jordan–Wigner transformation [20], and Bethe ansatz procedures [51]. Both are limited to one-dimensional models.

The fermion mapping sometimes results in free (noninteracting) fermion systems. An example is when the rate relation (5) applies in the class of models obtained by combining the processes (b)(ii), (b)(iii) (see Sec. 5); i.e., for the combination of asymmetric hopping and pair creation and annihilation shown in Fig. 4, but with periodic rather than open boundary conditions. As remarked in Sec. 5, this contains a special case equivalent to the Glauber–Ising model. The resulting free fermion system is not particle-conserving, so the reduction involves a Bogoliubov transformation after a momentum space diagonalization. In this way dynamic and steady state properties have been exactly obtained [44, 45, 52], which also yield properties for a range of related systems [53]. The excitations are gapless if one of $\varepsilon, \varepsilon'$ is zero, and otherwise gapped. This distinguishes power law from exponential decay in time. The method is a generalization of that developed by Pfeuty for the one-dimensional transverse Ising quantum spin system, and the one used by Lieb, Schultz and Mattis for the two-dimensional Ising model (these models are themselves related by a Hamiltonian mapping (see, e.g., [14]).

The Bethe ansatz provides the spectrum and excitations of certain quantum spin systems by an explicit construction of wave functions, allowing for phase shifts, or by representation-independent generalizations. The method has its best known application in nonequilibrium systems on the quantum spin equivalent of the one-dimensional ASEP, with periodic boundary conditions [37]. This is the work which yielded the dynamic exponent $z = 3/2$.

A very powerful approach starting from the master equation is the operator algebra technique [34–36, 39, 54]. This uses a string representation of system configurations, and so is restricted to $d = 1$. With D, E standing respectively for particle, vacancy, $\cdots DEDDE \cdots$ then represents a particle configuration ($\cdots$ particle–vacancy–particle–particle–vacancy $\cdots$). The ansatz is made that if D and E are operators with an appropriate algebra, an appropriate scalar $\langle \cdots DEDDE \cdots \rangle$ formed from the operator string can give the probability of the represented configuration. For cyclic boundary conditions $\langle \cdots \rangle$ is the trace.

This is the method by which the exact steady state solution for the ASEP with open boundary conditions was obtained [34, 39] (giving the exact treatment of the transition referred to in Sec. 6). In that case the algebra is $\Lambda = C$, where

$$\Lambda \equiv DE, \tag{16}$$

$$C \equiv D + E, \tag{17}$$

and the scalar is a particular matrix element of the operator string.

Subsequently, it was shown [54] how to generalize the method to represent the full dynamics of the ASEP. The algebra then becomes

$$dD/dt = \Lambda C^{-1} - C^{-1}\Lambda, \tag{18}$$

$$\Lambda C^{-1} D = D C^{-1} \Lambda. \tag{19}$$

Λ, C can be interpreted respectively as current and shift operators. Then the first equation can be interpreted as an operator generalization of the continuity equation (11), and the second equation is a constraint equation which "quantizes" the operators. It is easy to see that the reduced algebra $\Lambda = C$ satisfies (18), (19) in the steady state.

Algebras of this type can be written down for any stochastic particle exclusion model in one dimension [35]. The reduction of the full dynamic algebras is very difficult. An example where this has been fully accomplished is symmetric hard core hopping with injection and ejection [54]. This is a special case (not free fermion) of the process in Fig. 4, equivalent to the Heisenberg chain with "twisting" boundary fields. The operator algebra solution in this case involves (from the constraint equation) an operator relation providing phase shifts. This has subsequently been obtained by generalizing the Bethe ansatz.

Many important results are currently being obtained for nonequilibrium systems. Examples are the exact derivations of a type of local free energy functional and of probability distributions of macroscopic variables, both for a particular model, the ASEP. Further examples are developments concerning fluctuation dissipation relations, symmetry breaking, and scaling, and field theoretic renormalization group treatments. Significant advances are also being made in the areas of disorder effects and glasses, and we now briefly turn to those.

Earlier work on disorder effects in equilibrium collective phenomena has strongly influenced perspectives on disordered nonequilibrium systems. Some things carry over directly from the quantum spin mappings. These include bond disorder in symmetric hard core diffusion because of its mapping to the Heisenberg model. The spin wave stiffness in this model becomes the hard core diffusion constant, which is related by an Einstein relation to the conductance. This was first treated using CPA for spin waves by Elliott and Pepper [55] and there are also earlier discussions on special lattices, such as the Cayley tree [6], and by scaling methods [56]. Ideas developed in such earlier work (see, e.g., [27]) permeate more recent discussions on nonequilibrium systems, and these include percolation and finite cluster effects in processes on diluted networks (see, e.g., [57]).

Intrinsic nonlinearities of such nonequilibrium systems as the ASEP (Secs. 5 and 6) make their disordered generalizations richer [58–60]. One way [60, 61] of treating the bond disordered ASEP in $d = 1$ is by a generalization of the steady state profile map (Fig. 6), which varies with each iteration because of its dependence on the hopping rate of each bond. This generates profiles similar to those seen in simulation results. A typical example is shown in Fig. 7: the nonmonatonic profile is due the

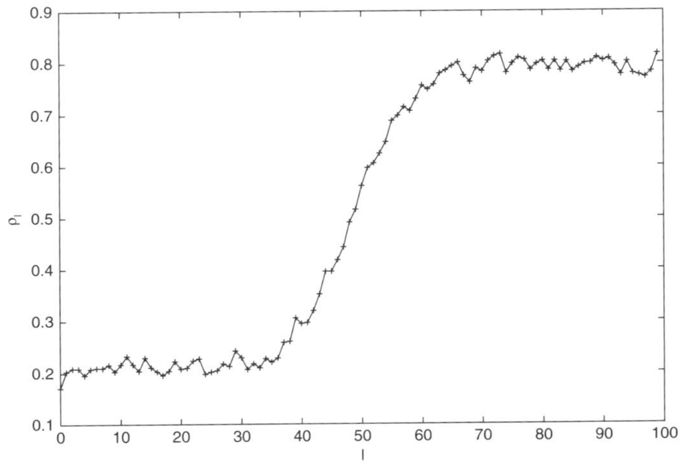

Fig. 7. Density profile in the steady state of a bond-disordered biased hard core hopping model.

analog of Griffiths singularities first encountered in the equilibrium states of disordered magnets [62]. Another approach uses a generalization of the continuum approach in which $J(x)$ in (12) involves an x-dependent rate p. A generalized Cole–Hopf linearization is still possible which, combined with a scaling technique first developed for excitations in a Mattis spin glass ([63]), yields both steady state and dynamic descriptions [61]. Localization plays a role in both of these, and in the steady state it corresponds to a profile shift already present in the disordered profile map. A generalized Harris criterion has also been shown to apply at the steady state phase transition.

For the nonequilibrium systems soluble by free fermion techniques their disordered generalizations are still noninteracting provided rate relations like (5) apply everywhere. But the lack of a momentum space diagonalization prevents the direct use of a Bogoliubov transformation. A generalization of such techniques is however possible which exhibits the combined effects of disorder on generalized phases (giving localization) and on Bogoliubov angles [64]. This allows an interpretation of results of computer simulation of the substitutionally disordered nonequilibrium systems.

Concerning the final topic, glasses, the brief discussion given here addresses only idealized glassy models whose Hamiltonians contain no disorder, but which, e.g., under slow cooling freeze into an inhomogeneous state as the (activated) internal rates become slower than the cooling rate. Simple models such as the binary fluid can have such nonequilibrium dynamic transitions.

Similar characteristics are seen over a range of real glassy systems, from structural glasses to foams and granular assemblies [65]. There is no evidence in such real systems for a diverging static length, as is encountered at continuous transitions in equilibrium systems (and in the steady state transition in simple nonequilibrium models like the ASEP, where the diverging length is $1/k$, where k is given in (13)).

So universal glass characteristics and diverging dynamic lengths are currently being discussed mainly in highly idealized models not unlike those introduced in Sec. 5. Some of these, e.g., the kinetically constrained models [66–68] show severe slowing with often internal rates having a Vogel–Tamm–Fulcher dependence on parameters like temperature, as well as other glass characteristics such as ageing, etc. Fluctuation dissipation theorems [69] have also been discussed for certain of these idealized glass models.

The granular systems are an interesting subcase showing in controlled compaction experiments [65] all the typical glass characteristics.

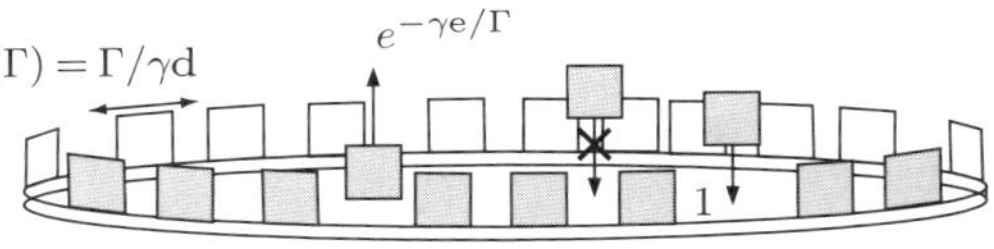

Fig. 8. An excluding block model for granular compaction, involving diffusion, evaporation, and deposition.

A particularly simple model for these experiments involves diffusion and deposition, not of particles on a lattice (Sec. 5) but of blocks on a continuum (Fig. 8). This exhibits the V-T-F law (which arises from a marginality in dynamic scaling, which is possibly a universal cause [70]). It also shows simple ageing and other glass characteristics, and also has a fluctuation dissipation relation as well as a consistent thermodynamic description of its steady state properties in terms of an entropy (proportional to the logarithm of the number of blocked states) proposed by Edwards and co-workers [71]. Though highly idealized, such models are helping to determine the necessary ingredients for a theory of glasses.

References

[1] Elliott, R.J., Pfeuty, P. and Wood, C., *Phys. Rev. Lett.* **25** (1970) 443.
[2] Pfeuty, P., *Ann. Phys.* **57** (1970) 79.
[3] Elliott, R.J., *Phys. Rev.* **124** (1961) 346.
[4] Elliott, R.J., in *Magnetism*, eds. Rado, G.T. and Suhl, H., Vol. IIA (1963), p. 385.
[5] Elliott, R.J. (ed.), *Magnetic Properties of Rare Earth Metals* (Plenum, London, 1972).
[6] Ziman, J.M., *Models of Disorder* (CUP, Cambridge, 1979), especially Secs. 1, 5, 9 and 12.
[7] Elliott, R.J., in *Structural Phase Transitions and Soft Modes*, eds. Samuelsen, E.J., Andersen, E. and Feder, J. (Universitetsforlaget, Oslo), (1971).
[8] Elliott, R.J., Harley, R.T., Hayes, W. and Smith, S.R.P., *Proc. Roy. Soc. A* **328** (1972) 217.
[9] Elliott, R.J. and Young, A.P., *Ferroelectrics* **7** (1974) 23.
[10] deGennes, P.G., *Solid State Comm.* **1** (1963) 132.
[11] Stinchcombe, R.B., in *Electron–Phonon Interactions and Phase Transitions*, ed. Riste, T. (Plenum, New York, 1977).
[12] Blinc, R., *J. Phys. Chem. Solids* **13** (1960) 204.
[13] Stinchcombe, R.B., *J. Phys. C* **6** (1973) 2459.
[14] See for example Chakrabarti, B.K., Dutta, A. and Sen, P., *Quantum Ising Phases and Transitions in Transverse Ising Models* (Springer, 1996).

[15] Elliott, R.J., *Proc. Int. Conf. Light Scattering in Solids*, ed. Balkanski, M. (Flammarion Press, Paris, 1971), p. 354.
[16] Elliott, R.J., Gehring, G.A., Malozemoff, A.P., Smith, S.R.P., Staude, W. and Tyte, R.N., *J. Phys. C* **4** L179 (1971).
[17] Elliott, R.J., Hayes, W., Palmer, S.B., Sandercock, J.R., Smith, S.R.P. and Young, A.P., *J. Phys. C* **4** (1971) L317.
[18] Samara, G.A., *Phys. Rev. Lett.* **27** (1971) 103.
[19] Brout, R., Muller, K.A. and Thomas, H., *Solid State Comm.* **4** (1966) 507.
[20] Jordan, P. and Wigner, E., *Z. Phys.* **47** (1928) 631.
[21] Antal, Racz, Z. and Sasvari, *Phys. Rev. Lett.* **78** (1997) 167.
[22] Elliott, R.J., Leath, P. and Krumhansl, J.A., *Rev. Mod. Phys.* **46** (1974) 465.
[23] Brout, R., *Phys. Rev.* **115** (1959) 824.
[24] Elliott, R.J., Heap, B.R., Morgan, D.J. and Rushbrooke, G.S., *Phys. Rev. Lett.* **5** (1960) 366.
[25] Elliott, R.J., *J. Phys. Chem. Solids* **16** (1960) 165.
[26] Elliott, R.J. and Heap, B.R., *Proc. Roy. Soc. A* **265** (1962) 264.
[27] For dilution in magnetic systems, see Stinchcombe, R.B., in *Phase Transitions and Critical Phenomena*, eds. Domb, C. and Lebowitz, J., Vol. 7 (1983).
[28] Harris, A.B., *J. Phys. C* **7** (1974) 1671.
[29] Challet, D. and Zhang, Y.-C., *Physica A* **246** (1997) 407.
[30] Challet, D., http://www.unifr.ch/econophysics/minority.
[31] Challet, D. and Stinchcombe, R., *Physica A* **300** (2001) 285, cond-mat/0106114.
[32] Daniel, M.G., Farmer, J.D., Iori, J. and Smith, E., cond-mat/0112422 (2001).
[33] Bouchaud, J.-P. and Potters, M., *Theory of Financial Risks* (CUP, Cambridge, 2001).
[34] Derrida, B. and Evans, M.R., in *Non-Equilibrium Statistical Mechanics in One Dimension*, ed. Privman, V. (CUP, Cambridge, 1997), and references therein.
[35] See, e.g., Stinchcombe, R.B., *Advances in Physics* **50** (2001) 431, and references therein.
[36] See, e.g., Schutz, G.M., in *Phase Transitions and Critical Phenomena*, eds. Domb, C. and Lebowitz, J., Vol. 19 (2001), and references therein.
[37] Gwa, L.-H. and Spohn, H., *Phys. Rev. Lett.* **68** (1992) 725; *Phys. Rev. A* **46** (1992) 844.
[38] Derrida, B., Domany, E. and Mukamel, D., *J. Stat. Phys.* **69** (1992) 667.
[39] Derrida, B., Evans, M.R., Hakim, V. and Pasquier, V., *J. Phys. A* **26** (1993) 1493.
[40] Schmittmann, B. and Zia, R.K.P., in *Phase Transitions and Critical Phenomena*, eds. Domb, C. and Lebowitz, J., Vol. 17 (1995).
[41] Barma, M., Grynberg, M.D. and Stinchcombe, R.B., *Phys. Rev. Lett.* **70** (1993) 1033; Stinchcombe, R.B., Grynberg, M.D. and Barma, M., *Phys. Rev. E* **47** (1993) 4018.
[42] Barma, M. and Dhar, D., *Phys. Rev. Lett.* **73** (1994) 2135.

[43] Family, F. and Amar, J.G., *J. Stat. Phys.* **65** (1991) 1235.
[44] Grynberg, M.D., Newman, T.J. and Stinchcombe, R.B., *Phys. Rev. E* **50** (1994) 957.
[45] Grynberg, M.D. and Stinchcombe, R.B., *Phys. Rev. Lett.* **74** (1995) 1242; *Phys. Rev. Lett.* **76** (1996) 851.
[46] Alcaraz, F.C., Droz, M., Henkel, M. and Rittenberg, V., *Ann. Phys.* (New York) **230** (1994) 250.
[47] Krug, J. and Spohn, H., in *Solids far from Equilibrium*, ed. Godreche, C. (CUP, Cambridge, 1991).
[48] Burgers, J.M., *The Non-Linear Diffusion Equation* (Riedel, Boston, 1974).
[49] Kardar, M., Parisi, G. and Zhang, Y.-C., *Phys. Rev. Lett.* **56** (1986) 889.
[50] Hopf, E., *Commun. Pure. Appl. Math* **3** (1950) 201; Cole, J.D., *Quart. Appl. Math.* **9** (1951) 225.
[51] Bethe, H., *Z. Phys.* **71** (1931) 205.
[52] Santos, J., Schutz, G.M. and Stinchcombe, R.B., *J. Chem. Phys.* **105** (1996) 2399; Santos, J., *J. Phys. A* **30** (1997) 3249.
[53] Henkel, M., Orlandini, E. and Schutz, G.M., *J. Phys. A* **28** (1995) 6335.
[54] Stinchcombe, R.B. and Schutz, G.M., *Europhys. Lett.* **29** (1995) 663; Stinchcombe, R.B. and Schutz, G.M., *Phys. Rev. Lett.* **75** (1995) 140.
[55] Elliott, R.J. and Pepper, D.E., *Phys. Rev. B* **8** (1973) 2374.
[56] See, e.g., Stinchcombe, R.B., in *Scaling Phenomena in Disordered Systems*, eds. Pynn, R. and Skjeltorp, A. (Plenum, New York, 1985), pp. 1,465.
[57] Grynberg, M.D. and Stinchcombe, R.B., *Phys. Rev. E* **61** (2000) 324.
[58] Tripathy, G. and Barma, M., *Phys. Rev. E* **58** (1998) 1911.
[59] Krug, J., *Braz. J. Phys.* **30** (2000) 97; Kolwankar, K.M. and Punnoose, A., *Phys. Rev. E* **61** (2000) 2453.
[60] Stinchcombe, R., *J. Phys. Condens. Matter* **14** (2002) 1.
[61] Harris, R.J. and Stinchcombe, R.B., to be published.
[62] Griffiths, R.B., *Phys. Rev. Lett.* **23** (1969) 17.
[63] Pimentel, I.R. and Stinchcombe, R.B., *Europhys. Lett.* **6** (1988) 719; Stinchcombe, R.B. and Pimentel, I.R., *Phys. Rev. B* **38** (1988) 4980.
[64] Harris, R.J. and Stinchcombe, R.B., to be published.
[65] Liu, A. and Nagel, S.R. (eds.), *Jamming and Rheology* (Taylor and Francis, London and New York, 2001).
[66] Jackle, J., *Rep. Prog. Phys.* **49** (1986) 171.
[67] Jackle, J. and Eisinger, S., *Z. Phys. B* **84** (1991) 115.
[68] Sollich, P. and Evans, M.R., *Phys. Rev. Lett.* **83** (1999) 3238.
[69] Cugliandolo, L. and Kurchan, J., *J. Phys. A* **27** (1994) 5749.
[70] Stinchcombe, R. and Depken, M., *Phys. Rev. Lett.* **88** (2002) 125701; Depken, M. and Stinchcombe, R., to be published.
[71] Edwards, S.F. and Oakeshott, R.B.S., in *Jamming and Rheology*, eds. Liu, A. and Nagel, S.R. (Taylor and Francis, London and New York, 2001).

CHAPTER 4

TWO-DIMENSIONAL GROWTH IN A THREE COMPONENT MIXTURE WITH COMPETING INTERACTIONS

C. Varea

Instituto de Física, Universidad Nacional Autónoma de México, Apartado Postal 20-364, 01000, D.F., Mexico

We study numerically the dynamics, in two dimensions, of phase separation in ternary mixtures with competing interactions which lead to the formation of modulated phases. Inside the crystal structures (of hexagonal or lamellar symmetry) an additional phase separation occurs "coloring" the texture. The lamellar phase does not evolve toward parallel lamellae, and the phase separation inside the channels proceeds until they reach a grain boundary. The hexagonal bubble phase is never formed due to the phase separation that forms an interface of bubbles which blocks the contact between the two phases. In its place we find an unsuspected lamellar phase.

The domain coarsening which occurs when a system is subjected to a rapid quench is a problem of interest and importance [1]. There are two interesting cases when the order parameter is scalar: simple fluids and fluids with competing interactions. In simple fluids, an initially homogeneous binary mixture that is quenched into a two phase-region, phase-separates starting to form droplets of the minority phase, which grow in size and number until the volume fraction occupied by the droplets attains its equilibrium value. In the late stages of the separation, the larger droplets grow at the expense of the small ones. This Ostwald ripening follows a universal growth law driven by surface tension. Lifshitz and Slyozov [2] have developed the phenomenology of this growth in which the average domain size R scales with time t as $R = At^{1/3}$, this has been confirmed experimentally in two-dimensional systems [3, 4]. Even when the system is more complex and contains three components, with the possibility of three phase equilibria, droplets of two different phases grow, when quenched from a uniform phase, which at late times follow the Lifshitz–Slyozov growth law [5].

When short-ranged attractive interactions and long-ranged repulsive interactions are present (as is the case of Langmuir monolayers of polar molecules), undulating phases become stabilized [6]. McConnell [7] has developed an effective interface free-energy for molecular films in the water–air interface where the long-ranged repulsive interactions are dipolar interactions among the single amphiphilic molecules. This free-energy contains surface and electric terms which in the case of an isolated circular domain of radius R has the form

$$F = 2\pi R \upsilon^2 \left(\ln \frac{e^2 \delta}{4R} + \lambda \right), \tag{1}$$

where υ is the dipole density in the monolayer, λ is the line tension, and δ is a short distance cutoff length. Equation (1) has a minimum when $R_{\text{eq}} = \left(e^3\delta/4\right)\left[e^{\lambda/\vartheta^2}\right]$ so that, in this case, domain coarsening is suppressed by the dipolar repulsion. From a different point of view Andelman [6] and Sagui and Desai [8] have analyzed a free-energy density with short ranged attractive interactions and long ranged non-local repulsive interactions. Solving the Euler–Lagrange equations in restricted symmetries (lamellar and hexagonal) they obtain a phase diagram with first-order phase transitions from a uniform gas phase to a droplet hexagonal phase, from the droplet phase to a lamellar phase, from the lamellar phase to a bubble hexagonal phase, and from the bubble phase to a dense liquid uniform phase as the concentration of polar molecules is increased at constant temperature. Sagui and Desai study the time evolution of these systems through Langevin simulations when the system is quenched and find that after an initial shape transition into the hexagonal phase the system forms modulated patterns broken up by topological defects which anneal away as the system orders. In near critical quenches and in a closely related system Boyer and Viñals [9] show that transient lamellar configurations do not achieve long ranged orientational order but rather evolve into glassy configurations with very slow dynamics.

Here we present the results for the time evolution of a model that predicts several equilibria between two different modulated phases. The free-energy of the model contains two coupled order-parameters with long ranged interaction terms that involve only one order-parameter. The result is a competition between modulation and phase separation due to excess free-energy at interfaces with line tension that is not compensated by long ranged repulsive interactions, and the formation of new glassy metastable states that prevent the system from achieving equilibrium.

Our model is a three component lattice model with both short-ranged and long-ranged interactions. Let u_i^α be the occupation number of species $\alpha = 1, 2, 3$ at site i, the grand potential Ω for our model is

$$\Omega = \sum_{i,\alpha} kT u_i^\alpha \ln u_i^\alpha + \frac{1}{2} \sum_{i,j;\alpha,\beta} V^{\alpha,\beta}(r_{i,j}) u_i^\alpha u_j^\beta - \sum_{i,\alpha} \mu_\alpha u_i^\alpha, \quad (2)$$

where T is the temperature, $V^{\alpha,\beta}(r_{i,j})$ is the interaction potential between molecules of class α and β located at a distance $r_{i,j}$, and μ_α is the chemical potential for species α. We consider a fully occupied lattice so that $\sum_{\alpha=1}^3 u_i^\alpha = 1$. The interaction potential is attractive for nearest-neighbor sites, zero for second nearest neighbor sites, and repulsive from third nearest neighbors on, with $V^{\alpha,\beta}(r_{i,j}) \sim \nu^\alpha \nu^\beta / r_{i,j}^3$ which corresponds to dipolar interactions with dipolar moment ν^α for species α. Since the lattice is fully occupied there are three independent interaction parameters for the short ranged interactions and we use the notation in [10] for them. For a square lattice this interaction term is

$$\frac{1}{4} \sum_{i,j} a u_i^2 u_j^3 + b u_i^3 u_j^1 + c u_i^1 u_j^2, \quad (3)$$

where the sum over j is over the four first neighbors of i. Also we assume $a + b + c = 1$, which sets the temperature and dipolar moments scale. In addition we use $b = a$, $\nu^1 = \nu^2$ and $\nu^3 = 0$ so that the mixture is symmetric. This leaves three independent parameters: the dimensionless temperature kT, the dimensionless interaction parameter c and the dimensionless dipolar moment ν. In this work we choose $c = 0.285$, $kT = 0.08$, and $\nu = 0.2$. When $\nu = 0$, this mixture, at this temperature, shows a triple point for a wide range of concentrations in the composition triangle [5]. With $\nu \neq 0$ this model is approximate to a Langmuir monolayer with two different segregating polar molecules in the air–water interface. Notice that Eq. (2) may be rewritten in terms of two new order parameters; the concentration of dipolar molecules $\rho_i = u_i^1 + u_i^2$ and $\chi_i = u_i^1 - u_i^2$, which we call color. In terms of these the repulsive part of the interaction depends only on the concentration ρ_i.

The Euler–Lagrange equations, $\delta\Omega / \delta u_i^\alpha = 0$, for the two independent occupation numbers u_i^1 and u_i^2 may be cast in the mean field form!

$$u_i^\alpha = \frac{e^{-(v_i^\alpha - \mu^\alpha)/kT}}{1 + \sum_{\beta=1}^2 e^{-(v_i^\beta - \mu^\beta)/kT}}, \quad (4)$$

where $v_i^\alpha = \sum_{i,j} V^{\alpha,\beta}(r_{i,j})\, u_j^\beta$ is the functional derivative of the interaction term in Ω that is calculated by means of Fourier transforms. The Euler–Lagrange equations were solved by successive iterations with a global error $\leq 10^{-10}$ and $\mu_1 = \mu_2$ starting from a random configuration in a lattice of size 100^2.

In Fig. 1 we show the resulting phase diagram. We see that for small global concentration ρ of polar molecules there is a first order transition from a uniform phase to a hexagonal phase where the droplets are of a non-segregated liquid $(\chi_i = 0)$; at larger values of ρ we find a second-order phase transition followed by two phase equilibria between two hexagonal phases with droplets rich in the 1 and 2 components in region I of the figure. At larger ρ there is a first order phase transition into a region of two phase equilibria of two lamellar phases (in region II of Fig. 1) where stripes rich in component 1 (2) alternate with stripes rich in component 3. In region III of Fig. 1, there is equilibrium between two bubble hexagonal phases. In one of them the background liquid is rich in component 1 while in the other it is rich in component 2. Finally, in region IV we find two phase equilibria between two uniform phases.

The time evolution of the system, after a quench from a uniform phase is described by the equations:

$$\frac{du_i^\alpha}{dt} = \nabla^2 \frac{\delta\Omega}{\delta u_i^\alpha} \tag{5}$$

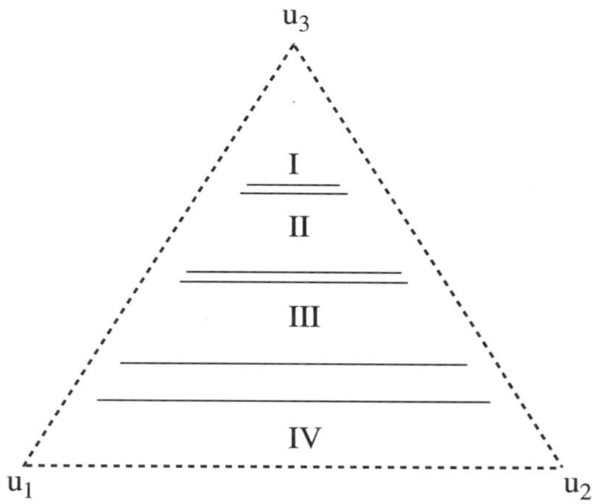

Fig. 1. Phase diagram for the three component mixture.

for a system with conserved order parameter. We used several initial conditions with $u_i^\alpha = u_0 + \Delta u_i^\alpha$, where Δu_i^α is a fluctuation with zero mean and the same average composition u_0 for components 1 and 2. The evolution equations were solved by a simple Euler method with a time step $\Delta t = 0.006$.

Figure 2 shows typical configurations after 250,000 iterations for $u_0 = 0.15$, 0.27, 0.33, and 0.38. Inside region I, for $u_0 = 0.15$, linear analysis (see [5] for the properties of the spinodal regions of this model) shows that the unstable fluctuations are concentration fluctuations with $\Delta u_i^1 = \Delta u_i^2$; following the quench the system forms a complex pattern of interconnected domains. After around 30,000 iterations, the system already shows a short-ranged liquid-like hexagonal structure of dense droplets with $\chi_i = 0$ and $\rho = 0.9$. At this density fluctuations with $\Delta u_i^1 = -\Delta u_i^2$ become unstable and the droplets start to decompose into droplets rich in components 1 or 2. Since the line tension between the dense droplets and the gas is larger than that of the decomposed droplets, these grow in size. The growth law in this regime is $R \sim t^{1/2}$, where R is the first zero of the pair correlation function $g_{\alpha,\alpha}(r,\theta)$ averaged over angles, θ, and over components $\alpha = 1$ and $\alpha = 2$ of the mixture (see Fig. 3).

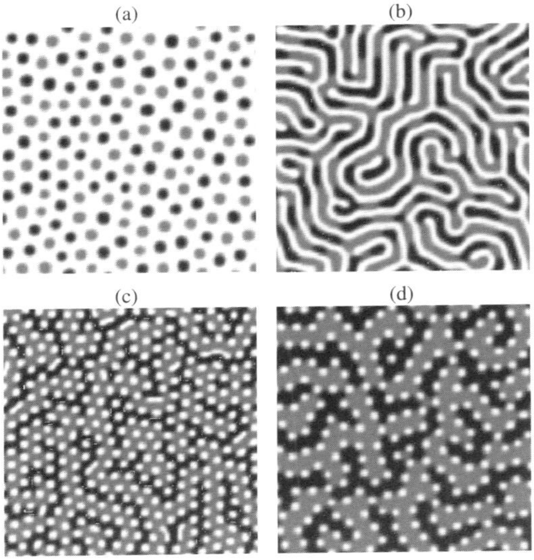

Fig. 2. Pictures obtained from simulation of the model after 250,000 iterations for: (a) $u_0 = 0.15$, (b) 0.27, (c) 0.33, and (d) 0.38. The tones of gray represent the values of the parameter $1 - u_1 - 0.5u_2$.

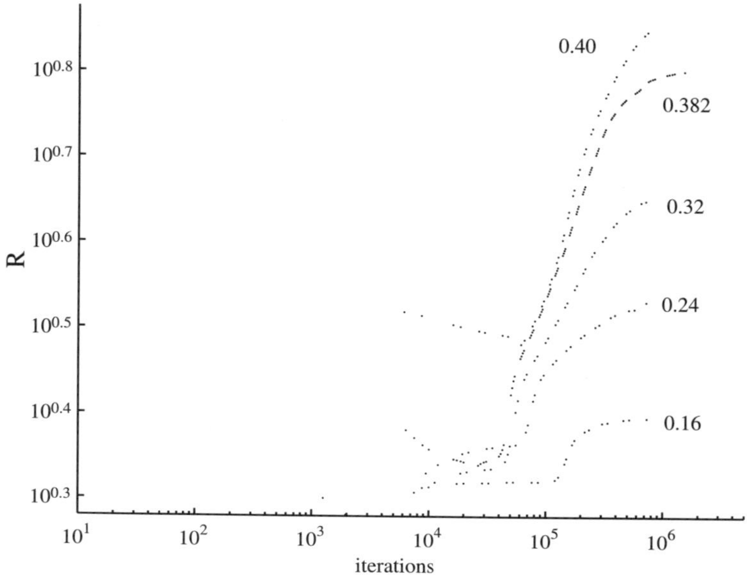

Fig. 3. Plot of the time evolution of the first zero of the correlation function for several values of the concentration u_0. R is in units of the lattice constant.

We have analyzed the evolution of the underlaying hexagonal structure using Voronoi and triangular representations and followed the evolution of the number of sites with coordination z. We find that, very soon, there are only sites with $z = 5, 6$ and 7. The defects with $z = 5$ and 7 pair and evolve very much in the same way that Sagui and Desai [8] describe; through T_1 and T_2 processes forming boundaries between the different hexagonal grains in this polcrystalline structure. There is, however, a difference: the average number n of defects first decreases but as the droplets decompose n increases since the smaller droplets with coordination $z = 5$ evaporate to feed the growing segregated droplets. When the droplets are mature, n starts to decrease again as the liquid system orders both positionally and orientationally. We have measured the time evolution of the orientational order parameter $f_6 = \left|\sum_k \sum_\theta e^{i6\theta} S_{3,3}(k,\theta)\right|$, where $S_{3,3}(k,\theta)$ is the normalized structure factor of the third component of the mixture. Its evolution is also non-monotonic revealing an initial ordering followed by the disordering effects of the segregation and then by ordering again when the droplets are mature. In Fig. 2(a) $n = 0.35$ and $f_6 = 0.17$ and the structure is still liquid-like. There is a purely kinetic effect in the structure that induces an additional correlation; droplets rich in component 1 are surrounded, on average, by four droplets rich in component 2 and two droplets rich in component 1 because growing droplets feed from its neighbors and a growing fluctuation with $\chi > 0$ induces fluctuations with $\chi < 0$ in its neighbors. Since there is no interfacial free-energy associated to the mixing of the two equilibrium phases of droplets there is no reason for them to unmix.

For $u_0 = 0.27$ inside region II of Fig. 1 (where there is equilibrium between two lamellar phases) linear analysis shows again that the unstable fluctuations are of the form $\Delta u_i^1 = \Delta u_i^2$, very soon alternating lamellae of a dense unmixed fluid and a fluid rich in component 3 are formed with a structure that is full of disclinations and dislocations. The subsequent evolution is a complex mixture of annihilation of defects and the unmixing of components 1 and 2 in the dense lamellae (see Fig. 2(b)). At this stage $R \sim t^{1/2}$ (see Fig. 3). The unmixing inside the lamellae creates interfaces with a linear tension and no dipolar forces to contrarrest an Oswald ripening inside them. To study this we have run simulations with initial conditions with random noise larger along the direction $(1, 1)$ than in the perpendicular one $(1, -1)$. The structure of lamellae, formed along the $(1, -1)$ direction, has very few defects. We have measured the second moment in the direction $(1, -1)$, of the u^1, u^1 structure factor and found that the length L of the droplets inside the lamellae grow like $L \sim t^{1/3}$ as in the Lifshitz–Slyozov universal law.

The final result is a lamellar structure with liquids rich in the 1, 3 and 2 components alternating with no interfaces of liquids rich in 1 and 2 present. In an undirected deep quench the defects in the stripes pattern are pinned [9] and the phase separation continues until the length of the stripes is the same as the channels that the dense fluid makes, in this disordered structure. This kinetics is very slow and in Fig. 4(a) we show the appearance of the pattern after 1.5×10^6 iterations. We have convinced ourselves that the disordered structure is a solution of the Euler–Lagrange equations within an error $<10^{-10}$, and thus it is metastable. The appearance of this structure is very similar to that which Pomeau [11] classifies as labyrinths, but its structure factor lacks the long tail towards long wave-lengths that characterizes the labyrinthine structures.

For $u_0 = 0.33$ inside region III of Fig. 1 there are two hexagonal bubble phases in equilibrium. The backgrounds of these two bubble phases are liquids, L_1 and L_2, rich in components 1 and 2, respectively. The early time behavior shows growing fluctuations with $\Delta u_i^1 = \Delta u_i^2$ so that an hexagonal bubble phase with $u^1 = u^2 = 0.0003$ forms in a sea of a non-segregated

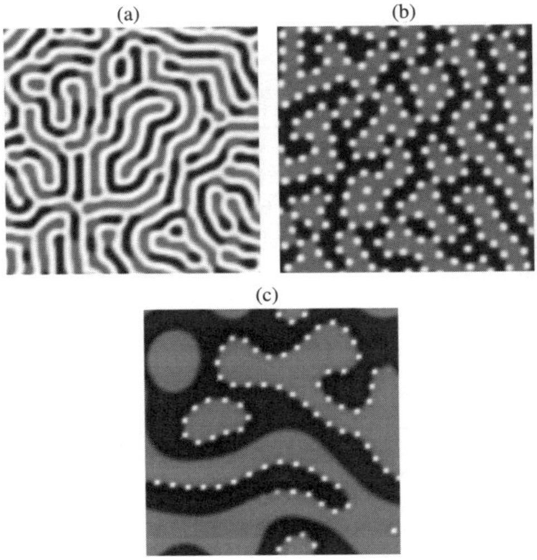

Fig. 4. Pictures obtained from simulation of the model after 1.5×10^6 iterations. In (a) the concentration is $u_0 = 0.27$, in (b) $u_0 = 0.382$ and in (c) $u_0 = 0.45$.

phase with $u^1 = u^2 = 0.4967$. This non-equilibrium structure evolves eliminating defects, but after about 50,000 iterations the unstable dense fluid segregates forming the complex structure shown in Fig. 2(c). The two segregated liquids form interfaces with excess line tension and the bubbles are dragged into that interface to shield the two liquids from each other. As a result, the number of defects n increases with time ($n = 0.3$ for the configuration shown in Fig. 2(c)). As we increase the value of u_0 the density of bubbles, in the early time regime, is smaller leaving more space for the two liquids that again segregate leading to the formation of a new lamellar structure (see Fig. 2(d)). There, the lamellae are of liquids L_1 and L_2 are separated by the bubbles in a structure that resembles a two-dimensional microemulsion with the bubbles taking the role of a surfactant. In Fig. 4(b) we show the late time configuration (after 1.5×10^6 iterations) for $u_0 = 0.382$ close to the boundary of region III. In Fig. 3 we monitor the growth of the regions of liquids L_1 and L_2 and we find that $R \sim t^{1/3}$ for early times with a crossover into an asymptotic regime where this structure freezes and is a metastable solution of the Euler–Lagrange equations.

For $u_0 = 0.45$ in the four phase region, and close to the L_1–L_2 coexistence region, linear analysis shows that the unstable fluctuations are of the form $\Delta u_i^1 = -\Delta u_i^2$ and the two liquids immediately separate. At early times the interface is very long. Since part of component 3 is absorbed in the interface, the bulk phases that condense have less component 3 than those at the four phase equilibrium. As the structure ripens and the interface is reduced, bubbles start to form in the interface. The average size of the domains that was growing at a rate in accord with the Lifshitz–Slyozov law is also reduced. In Fig. 4(c) we show the pattern that is forming after 1.5×10^6 iterations.

In summary, we have presented results for the evolution of a novel model characterized by short-ranged attractive interactions and long-ranged repulsions with two order parameters. We have concentrated on a very symmetric model where two modulated symmetric phases coexist and found that the system never attains equilibrium in the sense that the equilibrium phases are never reached. Moreover, the system often ends up in new unsuspected metastable modulated states. We have generalized the model to include two-dimensional Coulomb interactions instead of dipolar interactions, and found the same behavior. We believe that its properties are those of a wide range of models including a generalization of the Swift–Hohemberg model with two active order parameters. We expect that this findings will encourage experiments in mixtures of polar molecules in the water–air interface and related problems.

Acknowledgments

The author acknowledges support from the CONACyT Grant No. 27643-E, and helpful discussions with Denis Boyer.

References

[1] For reviews, see: Gunton, J.D., San Miguel, M. and Sahni, P.S., in *Phase Transitions and Critical Phenomena*, Vol. 3, eds. Domb, C. and Lebowitz, J.L. (Academic Press, New York, 1983) Komura, S. *Phase Transitions* **12** (1998) 3.

[2] Lifshitz, I.M. and Slyozov, V.V., *J. Phys. Chem. Solids* **19** (1961) 35.

[3] Krichevsky, O. and Stavans, J., *Phys. Rev. Lett.* **70** (1993) 1473.

[4] Seul, M., Morgan, N.Y. and Sire, C., *Phys. Rev. Lett.* **73** (1994) 2284.

[5] Varea, C., to be published in *Phys. Rev. E* (2003).

[6] Andelman, D., Brochard, F. and Joanny, J.-F., *J. Chem. Phys.* **86** (1987) 3673.

[7] McConnell, H.M., *Ann. Rev. Phys. Chem.* **42** (1991) 171.

[8] Sagui, C. and Desai, R.C., *Phys. Rev. Lett.* **71** (1993) 3995; Sagui, C. and Desai, R.C., *Phys. Rev. E* **49** (1994) 2225.

[9] Boyer, D. and Viñals, J., *Phys. Rev. E* **64** (2001) 0501001; *ibid.* **65** (2002) 0406119.
[10] Furman, D., Dattagupta, S. and Griffiths, R.B., *Phys. Rev. B* **15** (1977) 441.
[11] Le Berre, M., Ressayre, E., Tallet, A., Pomeau, Y. and Di Menza, L., *Phys. Rev. E* **66** (2002) 026203.

CHAPTER 5

GLASSY DYNAMICS AT THE EDGE OF CHAOS

A. Robledo

Instituto de Física, Universidad Nacional Autónoma de México, Apartado Postal 20-364, México 01000 D.F., Mexico

We describe our finding that the dynamics at the noise-perturbed edge of chaos in logistic maps is analogous to that observed in supercooled liquids close to vitrification. The three major features of glassy dynamics in structural glass formers, two-step relaxation, aging, and a relationship between relaxation time and configurational entropy, are displayed by orbits with vanishing Lyapunov exponent. The known properties in control-parameter space of the noise-induced bifurcation gap play a central role in determining the characteristics of dynamical relaxation at the chaos threshold. Time evolution is obtained from the Feigenbaum RG transformation.

1. Introduction

In spite of the vast knowledge and understanding gathered together on the dynamics of glass formation in supercooled liquids, this peculiar condensed matter phenomenon continues to attract much interest [1]. This is so because there remain basic unanswered questions that are both intriguing and testing [1]. A very pronounced slowing down of relaxation processes is the foremost expression of the approach to the glass transition [1, 2], and this is generally interpreted as a progressively more imperfect realization of phase space mixing. Because of this extreme circumstance, an important theoretical question is to find out whether under conditions of ergodicity breakdown the Boltzmann–Gibbs (BG) statistical mechanics remains capable of describing stationary states in the immediate vicinity of glass formation. This concern and also that about the possible applicability of generalizations of the BG statistics, such as the nonextensive statistics [3, 4], to glass formation is a subject of current attention.

Recently [6], we have proved that the dynamics of the well-known logistic map at the edge of chaos affected by noise of small amplitude is analogous to that of supercooled liquids close to vitrification. We have found that this simple dynamical system exhibits two-step relaxation, aging, and a relationship between relaxation time and configurational entropy, the three main (phenomenologically established) characteristics of glassy dynamics in structural glass formers [1]. The basic ingredient of ergodicity failure is obtained for orbits at the onset of chaos in the limit towards vanishing noise amplitude. Our study supports the idea of a degree of universality underlying the phenomenon of vitrification, and points out that it is present in different classes of systems, including some with no explicit consideration of their molecular structure. The map has only one degree of freedom but the consideration of external noise could be taken to be the effect of many other systems coupled to it, like in the so-called coupled map lattices [5]. Here we recount and attempt to improve in some respects our original discussion [6].

Our interest is to study a system that is gradually forced into a nonergodic state by reducing its capacity to traverse regions of its phase space that are space filling, up to a point at which it is only possible to move within a fractal (or multifractal) subset of this space. This situation is generated in the logistic map with additive external noise:

$$x_{t+1} = f_\mu(x_t) = 1 - \mu x_t^2 + \chi_t \sigma, \quad -1 \leq x_t \leq 1, \;\; 0 \leq \mu \leq 2, \tag{1}$$

where χ_t is Gaussian-distributed with average $\langle \chi_t \chi_{t'} \rangle = \delta_{t.t'}$, and σ measures the noise intensity [16, 17]. Notice that the formula (1) can be written as a discrete form for a Langevin equation.

2. Dynamics of Glass Formation

We recall the main dynamical properties displayed by supercooled liquids on approach to glass formation. One is the growth of a plateau and for that reason a two-step process of relaxation, as presented by the time evolution of correlations, e.g., the intermediate scattering function F_k [1, 2]. This consists of a primary power-law decay in time t (so-called β relaxation) that leads into the plateau, the duration t_x of which diverges also as a power-law of the difference $T - T_g$ as the temperature T decreases to a critical value T_g. After t_x there is a secondary power law decay (so-called α relaxation) that leads to a conventional equilibrium state [1, 2]. This behavior is shown by molecular dynamics simulations [7] and it is successfully reproduced by mode coupling (MC) theory [8]. A second important dynamic property of glasses is the loss of time translation invariance, a characteristic known as aging [9], that is due to the fact that properties of glasses depend on the procedure by which they are obtained. Remarkably, the time fall-off of relaxation functions and correlations display a scaling dependence on the ratio t/t_w where t_w is a waiting time. A third notable property is that the experimentally observed relaxation behavior of supercooled liquids is effectively described, via reasonable heat capacity assumptions [1], by the so-called Adam–Gibbs equation:

$$t_x = A \exp(B/TS_c), \tag{2}$$

where the relaxation time t_x can be identified with the viscosity or the inverse of the difussivity, and the configurational entropy S_c is related to the number of minima of the fluid's potential energy surface (and A and B are constants) [15]. Equation (2) implies that the reason for viscous slow-down in supercooled liquids is a progressive reduction in the number of configurations that the system is capable of sampling as $T - T_g \to 0$. A first principles derivation of this equation has not been developed at present. As the counterpart to the Adam–Gibbs formula, we show below that our one-dimensional map model for glassy dynamics exhibits a relationship between the plateau duration t_x, and the entropy S_c for the state that comprises the largest number of (iterate positions) bands allowed by the bifurcation gap — the noise-induced cutoff in the period-doubling cascade [16]. This entropy is obtained from the probability of chaotic band occupancy at position x.

Our results suggest that the properties known to be basic of glassy dynamics [1] are also likely to manifest in completely different physical problems, such as in nonlinear dynamics, e.g., the mentioned coupled map lattices [10], in critical dynamics [11], and other fields. This hints that new predictions might be encountered in the studies, experimental or otherwise, of slow dynamics displayed by systems other than liquids close to the glass transition. In relation to this, aspects of glassy dynamics have been observed in metastable quasistationary states in microcanonical Hamiltonian systems of N classical rotors with homogeneous long-ranged interactions. For special types of initial conditions it has been found that both two-step relaxation [12], where the length of the metastable plateau diverges with infinite size $N \to \infty$, and aging [13, 14] are present in these systems. In our simpler one-dimensional dissipative map, the amplitude σ plays a role parallel to $T - T_g$ in the supercooled liquid or $1/N$ in the system of rotors.

3. Dynamics in the Absence of Noise

As is well known [16, 18], in the absence of noise $\sigma = 0$ the Feigenbaum attractor at $\mu = \mu_c(0) = 1.40115\ldots$ is the accumulation point of both the period doubling and the chaotic band splitting sequences of transitions and it marks the threshold between periodic and chaotic orbits. The locations of period doublings, at $\mu = \mu_n < \mu_c(0)$, and band splittings, at $\mu = \hat{\mu}_n > \mu_c(0)$, obey, for large n, the power laws $\mu_n - \mu_c(0) \sim \delta^{-n}$ and $\mu_c(0) - \hat{\mu}_n \sim \delta^{-n}$, where $\delta = 0.46692\ldots$ is one of the two Feigenbaum's universal constants. The second, $\alpha = 2.50290\ldots$ measures the power-law period-doubling spreading of iterate positions. All the trajectories with $\mu_c(0)$ and initial condition $-1 \leq x_{in} \leq 1$ fall, after a (power-law) transient, into the attractor set of positions with fractal dimension $d_f = 0.5338\ldots$ Therefore, these trajectories represent nonergodic states, as $t \to \infty$ only a Cantor set of positions is accessible within the entire phase space $-1 \leq x \leq 1$. For $\sigma > 0$ the noise fluctuations smear the sharp features of the periodic attractors as these broaden into bands similar to those in the chaotic attractors, but there is still a sharp transition to chaos at $\mu_c(\sigma)$, where the Lyapunov exponent changes sign. The period doubling of bands ends at a finite value $2^{N(\sigma)}$ as the edge of chaos transition is approached and then decreases in reverse fashion at the other side of the transition. The broadening of orbits with number of periods or bands smaller than $2^{N(\sigma)}$ and the removal of orbits of periods or bands of number larger than $2^{N(\sigma)}$ in the infinite cascades introduces a bifurcation gap with scaling features [16, 17] that we shall use below. When $\sigma > 0$ the trajectories visit sequentially a set of 2^n disjoint bands or segments leading to a cycle, but the behavior inside each band is completely chaotic. These trajectories represent ergodic states as the accessible positions have a fractal dimension equal to the dimension of phase space. Thus the elimination of fluctuations in the limit $\sigma \to 0$ leads to an ergodic to nonergodic transition in the map and we contrast its properties with those known for the molecular arrest occurring in a liquid as $T \to T_g$.

The dynamics of iterates for the logistic map at the onset of chaos $\mu_c(0)$ has recently been analyzed in detail [19]. It was found that the trajectory with initial condition $x_{in} = 0$ (see Fig. 1) maps out the Feigenbaum attractor in such a way that (the absolute values of) succeeding (time-shifted $\tau = t+1$) positions x_τ form subsequences with a common power-law decay of the form $\tau^{-1/1-q}$ with $q = 1 - \ln 2/\ln\alpha \simeq 0.24449$. That is, the *entire* attractor can be decomposed into position subsequences generated by the time subsequences $\tau = (2k+1)2^n$, each obtained by running over

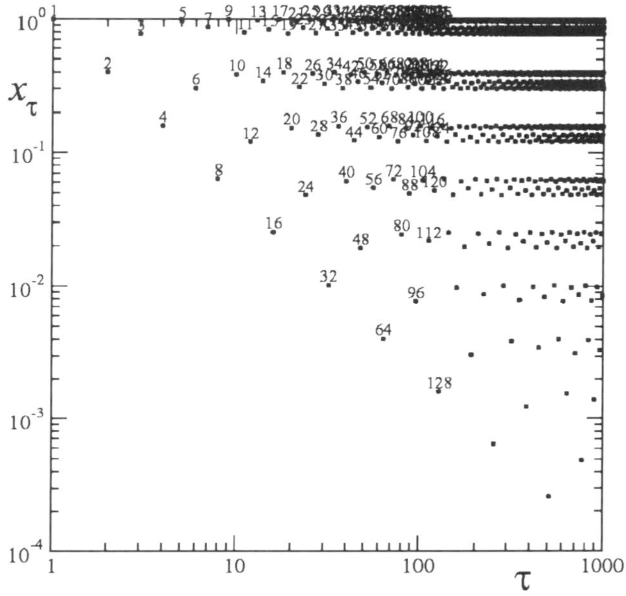

Fig. 1. Absolute values of positions in logarithmic scales of the first 1000 iterations τ for a trajectory of the logistic map at the onset of chaos $\mu_c(0)$ with initial condition $x_{in} = 0$. The numbers correspond to iteration times. The power-law decay of the time subsequences described in the text can be clearly appreciated.

$n = 0, 1, 2, \ldots$ for a fixed value of $k = 0, 1, 2, \ldots$ Noticeably, the positions in these subsequences can be obtained from those belonging to the "superstable" periodic orbits of lengths 2^n, i.e., the 2^n-cycles that contain the point $x = 0$ at $\bar{\mu}_n < \mu_c(0)$ [16]. Specifically, the positions for the main subsequence $k = 0$, that constitutes the lower bound of the entire trajectory (see Fig. 1), were identified to be $x_{2^n} = d_n = \alpha^{-n}$, where $d_n \equiv \left| f_{\bar{\mu}_n}^{(2^{n-1})}(0) \right|$ is the "nth diameter" defined at the 2^n-supercycle [16]. The main subsequence can be expressed as

$$x_t = \exp_{2-q}(-\lambda_q t) \tag{3}$$

with $\lambda_q = \ln \alpha / \ln 2$, and where

$$\exp_q(x) \equiv [1 - (q-1)x]^{1/1-q} \tag{4}$$

is the q-exponential function. Interestingly, this analytical result for x_t can be seen to satisfy the dynamical fixed-point relation, $h(t) = \alpha h(h(t/\alpha))$ with $\alpha = 2^{1/(1-q)}$ [19]. Further, the sensitivity to initial conditions

$\xi_t \equiv dx_t/dx_{in}$ obeys the closely related form

$$\xi_t = \exp_q(\lambda_q t), \quad q = 1 - \ln 2/\ln \alpha \simeq 0.24449, \tag{5}$$

where the amplitude λ_q can be identified as the q-generalized Lyapunov exponent in nonextensive statistics [19]. These properties follow from the use of $x_{in} = 0$ in the scaling relation

$$x_\tau = \left|g^{(\tau)}(x_{in})\right| = \tau^{-1/1-q}\left|g(\tau^{1/1-q}x_{in})\right|, \tag{6}$$

which in turn is obtained from the $n \to \infty$ convergence of the 2^nth map composition to $(-\alpha)^{-n}g(\alpha^n x)$ with $\alpha = 2^{1/(1-q)}$. When $x_{in} = 0$, one obtains in general [19]

$$x_\tau = \left|g^{(2k+1)}(0)g^{(2^{n-1})}(0)\right| = \left|g^{(2k+1)}(0)\right|\alpha^{-n}. \tag{7}$$

4. Dynamics in the Presence of Noise

When the noise is turned on (σ always small) the 2^nth map composition converges instead to

$$(-\alpha)^{-n}[g(\alpha^n x) + \chi\sigma\kappa^n G_\Lambda(\alpha^n x)], \tag{8}$$

where κ a constant whose numerically determined [20, 21] value $\kappa \simeq 6.619$ is well approximated by $\nu = 2\sqrt{2}\alpha(1 + 1/\alpha^2)^{-1/2}$, the ratio of the intensity of successive subharmonics in the map power spectrum [16, 21]. The connection between κ and the σ-independent ν stems from the necessary coincidence of two ratios, that of noise levels causing band-merging transitions for successive 2^n and 2^{n+1} periods and that of spectral peaks at the corresponding parameter values μ_n and μ_{n+1} [16, 21]. Following the same procedure as above we see that the orbits x_τ at $\mu_c(\sigma)$ satisfy, in place of Eq. (6), the relation

$$x_\tau = \tau^{-1/1-q}\left|g(\tau^{1/1-q}x) + \chi\sigma\tau^{1/1-r}G_\Lambda(\tau^{1/1-q}x)\right|, \tag{9}$$

where $G_\Lambda(x)$ is the first order perturbation eigenfunction, and where $r = 1 - \ln 2/\ln \kappa \simeq 0.6332$. So that use of $x_{in} = 0$ yields

$$x_\tau = \tau^{-1/1-q}\left|1 + \chi\sigma\tau^{1/1-r}\right| \tag{10}$$

or

$$x_t = \exp_{2-q}(-\lambda_q t)\left[1 + \chi\sigma \exp_r(\lambda_r t)\right], \tag{11}$$

where $t = \tau - 1$ and $\lambda_r = \ln \kappa/\ln 2$.

At each noise level σ there is a "crossover" or "relaxation" time $t_x = \tau_x - 1$ when the fluctuations start suppressing the fine structure imprinted by the attractor on the orbits with $x_{in} = 0$. This time is given by $\tau_x = \sigma^{r-1}$, the time when the fluctuation term in the perturbation expression for x_τ becomes σ-independent and so unrestrained, i.e.,

$$x_{\tau_x} = \tau_x^{-1/1-q}|1 + \chi|. \tag{12}$$

Thus, there are two regimes for time evolution at $\mu_c(\sigma)$. When $\tau < \tau_x$ the fluctuations are smaller than the distances between adjacent subsequence positions of the noiseless orbit at $\mu_c(0)$, and the iterate positions in the presence of noise fall within small non-overlapping bands each around the $\sigma = 0$ position for that τ. In this regime the dynamics follows in effect the same subsequence pattern as in the noiseless case. When $\tau \sim \tau_x$ the width of the fluctuation-generated band visited at time $\tau_x = 2^N$ matches the distance between two consecutive diameters, $d_N - d_{N+1}$ where $N \sim -\ln\sigma/\ln\kappa$, and this signals a cutoff in the advance through the position subsequences. At longer times $\tau > \tau_x$ the orbits are unable to stick to the fine period-doubling structure of the attractor. In this second regime the iterate follows an increasingly chaotic trajectory as bands merge progressively. This is the dynamical image — observed along the *time evolution* for the orbits *of a single state* $\mu_c(\sigma)$ — of the static bifurcation gap first described in the map space of position x and control parameter μ [17, 20, 21].

5. Parallels with Glassy Dynamics

In establishing parallels with glassy dynamics in supercooled liquids, it is illustrative to define an "energy landscape" for the map as being composed by an infinite number of "wells" whose equal-valued minima coincide with the points of the attractor on the interval $[-1, 1]$. The widths of the wells increase as a fictitious "energy parameter" U increases and the wells merge by pairs at values U_N such that within the range $U_{N+1} < U \leq U_N$ the landscape is composed of a set of 2^N bands of widths $w_m(U)$, $m = 1, \ldots, 2^N$. This "picture" of an energy landscape resembles the chaotic band-merging cascade in the well-known (x, μ) bifurcation diagram [16]. The landscape is sampled at noise level σ by orbits that visit points within the set of 2^N bands of widths $w_m(U) \sim \sigma$, and, as we have seen, this takes place in time in the same way that period doubling and band merging proceeds in the presence of a bifurcation gap when the control parameter is run through the interval $0 \leq \mu \leq 2$. That is, the trajectories starting at $x_{in} = 0$ duplicate the number of visited bands at times $\tau = 2^n$, $n = 1, \ldots, N$, the bifurcation gap is reached at $\tau_x = 2^N$, after which the orbits fall within bands that merge by pairs at times $\tau = 2^{N+n}$, $n = 1, \ldots, N$. The sensitivity to initial conditions grows as $\xi_t = \exp_q(\lambda_q t)$ ($q = 1 - \ln 2/\ln \alpha < 1$) for $t < t_x$, but for $t > t_x$ the fluctuations dominate and ξ_t grows exponentially as the trajectory has become chaotic and so one anticipates an exponential ξ_t (or $q = 1$). We interpret this behavior to be the dynamical system analog of the α relaxation in supercooled fluids. The plateau duration $t_x \to \infty$ as $\sigma \to 0$. Additionally, trajectories with initial conditions x_{in} not belonging to the attractor exhibit an initial relaxation stretch towards the plateau as the orbit falls into the attractor. This appears as the analog of the β relaxation in supercooled liquids.

Next, we determine the entropy of the orbits starting at $x_{in} = 0$ as they enter the bifurcation gap at $t_x(\sigma)$ when the maximum number 2^N of bands allowed by the fluctuations is reached. The entropy $S_c(\mu_c(\sigma))$ associated to the 2^N bands at $\mu_c(\sigma)$ has the form

$$S_c(\mu_c(\sigma)) = 2^N \sigma s, \tag{13}$$

since each of the 2^N bands contributes with an entropy σs, where

$$s = -\int_{-1}^{1} p(\chi) \ln p(\chi)\, d\chi \tag{14}$$

and where $p(\chi)$ is the distribution for the noise random variable. In terms of t_x, given that $2^N = 1 + t_x$ and $\sigma = (1 + t_x)^{-1/1-r}$, one has

$$S_c(\mu_c, t_x)/s = (1 + t_x)^{-r/1-r} \tag{15}$$

or, conversely,

$$t_x = (s/S_c)^{(1-r)/r}. \tag{16}$$

Since $t_x \simeq \sigma^{r-1}$, $r - 1 \simeq -0.3668$ and $(1 - r)/r \simeq 0.5792$ then $t_x \to \infty$ and $S_c \to 0$ as $\sigma \to 0$, i.e., the relaxation time diverges as the "landscape" entropy vanishes. We interpret this relationship between t_x and the entropy S_c to be the dynamical system analog of the Adam–Gibbs formula for a supercooled liquid. Notice that Eq. (16) is a power-law in S_c^{-1} while for structural glasses it is an exponential in S_c^{-1} [1]. This difference is significant as it indicates how the superposition of molecular structure and dynamics upon the bare ergodicity breakdown phenomenon described by the map modifies the vitrification properties.

Last, we examine the aging scaling property of the trajectories x_t at $\mu_c(\sigma)$. The case $\sigma = 0$ is more readily appraised because this property is, essentially, built into the same position subsequences $x_\tau = \left|g^{(\tau)}(0)\right|$,

$\tau = (2k+1)2^n$, $k, n = 0, 1, \ldots$ that we have been using all along. These subsequences are relevant for the description of trajectories that are at first held at a given attractor position for a waiting period of time t_w and then released to the normal iterative procedure. We chose the holding positions to be any of those along the top band shown in Fig. 1 for a waiting time $t_w = 2k+1$, $k = 0, 1, \ldots$. Notice that, as shown in Fig. 1, for the $x_{in} = 0$ orbit these positions are visited at odd iteration times. The lower-bound positions for these trajectories are given by those of the subsequences at times $(2k+1)2^n$ (see Fig. 1). Writing τ as $\tau = t_w + t$ we have that $t/t_w = 2^n - 1$ and

$$x_{t+t_w} = g^{(t_w)}(0)g^{(t/t_w)}(0) \tag{17}$$

or

$$x_{t+t_w} = g^{(t_w)}(0)\exp_q(-\lambda_q t/t_w). \tag{18}$$

This property is gradually modified when noise is turned on. The presence of a bifurcation gap limits its range of validity to total times $t_w + t < t_x(\sigma)$ and so progressively disappears as σ is increased.

6. Discussion

Thus, the dynamics of noise-perturbed logistic maps at the chaos threshold exhibit the most prominent features of glassy dynamics in supercooled liquids. Specifically our results are: (i) The two-step relaxation occurring when $\sigma \to 0$ was determined in terms of the bifurcation gap properties, in particular, the plateau duration is given by the power-law $t_x(\sigma) \sim \sigma^{r-1}$, where $r \simeq 0.6332$ or $r - 1 \simeq -0.3668$. (ii) The map equivalent of the Adam–Gibbs law was obtained as a power-law relation $t_x \sim S_c^{-\zeta}$, $\zeta = (1-r)/r \simeq 0.5792$, between $t_x(\sigma)$ and the entropy $S_c(\sigma)$ associated to the noise broadening of chaotic bands. (iii) The trajectories at $\mu_c(\sigma \to 0)$ were shown to obey a scaling property, characteristic of aging in glassy dynamics, of the form $x_{t+t_w} = h(t_w)h(t/t_w)$, where t_w is a waiting time. These properties were determined from the quasi-stationary trajectories followed by iterates at $\mu_c(\sigma)$, and these in turn were obtained via the fixed-point map solution $g(x)$ and the first noise perturbation eigenfunction $G_\Lambda(x)$ of the RG doubling transformation consisting of functional composition and rescaling, $\mathbf{R}f(x) \equiv \alpha f(f(x/\alpha))$. Positions for time subsequences within these trajectories are expressed analytically in terms of the q-exponential function.

The existence of this analogy cannot be considered accidental since the limit of vanishing noise amplitude $\sigma \to 0$ (the counterpart of the limit

$T - T_g \to 0$ in the supercooled liquid) entails loss of ergodicity. The incidence of these properties in such simple dynamical systems, with only a few degrees of freedom and no reference to molecular interactions, suggests a universal mechanism underlying the dynamics of glass formation. As definitely proved [19], the dynamics of deterministic unimodal maps at the edge of chaos is a genuine example of the pertinence of nonextensive statistics in describing states with vanishing ordinary Lyapunov exponent. Here we have shown that this nonergodic state corresponds to the limiting state, $\sigma \to 0$, $t_x \to \infty$, for a family of small σ noisy states with glassy properties, that are noticeably described for $t < t_x$ via the q-exponentials of the nonextensive formalism [19]. The fact that these features transform into the usual BG exponential behavior for $t > t_x$ provides a long-awaited opportunity for investigating the crossover from the ordinary BG to the nonextensive statistics in the physical circumstance of loss of mixing and ergodic properties.

It has been suggested on several occasions [4, 22] that the setting in which nonextensive statistics appears to emerge is linked to the incidence of nonuniform convergence, such as that involving the thermodynamic $N \to \infty$ and very large time $t \to \infty$ limits. For example, in the rotor problem mentioned above — for specific choices of initial conditions — if $N \to \infty$ is taken before $t \to \infty$ the anomalous metastable states with noncanonical properties appear to be the only observable stationary states, whereas if $t \to \infty$ is taken before $N \to \infty$ the usual BG equilibrium states are obtained. Here it is clear that a similar situation takes place, that is, if $\sigma \to 0$ is taken before $t \to \infty$, a nonergodic orbit confined to the Feigenbaum attractor and with fully developed glassy features is obtained, whereas if $t \to \infty$ is taken before $\sigma \to 0$ a typical $q = 1$ chaotic orbit is observed.

The point of view that our study offers is that the observed slow dynamics in a given system can be seen to be composed of the ideal glassy features stemming from ergodicity breakdown and other superimposed system-dependent features. The actual differences to be found between supercooled-liquid dynamics (from experimental or from fluid model calculations) and the ideal map dynamics would then be attributed to the presence of molecular structure and other effects. Finally, it is worth mentioning that while the properties displayed by the map capture in a qualitative, heuristic way the phenomenological issues of vitrification, they are obtained in a quantitative and rigorous manner as the map is concerned. Our map setup is a rarely available "laboratory" where every aspect of glassy dynamics can be studied analytically.

Acknowledgments

The author thanks Fulvio Baldovin for contributing the figure. Work partially supported by CONACyT Grant No. P-40530-F.

References

[1] For a recent review see, De Benedetti, P.G. and Stillinger, F.H., *Nature* **410** (2001) 267.
[2] De Benedetti, P.G., *Metastable Liquids. Concepts and Principles* (Princeton University Press, Princeton, 1996).
[3] Tsallis, C., *J. Stat. Phys.* **52** (1988) 479.
[4] For a recent review see, Tsallis, C., Rapisarda, A., Latora, V. and Baldovin, F., in *Dynamics and Thermodynamics of Systems with Long-Range Interactions*, eds. Ruffo, S., Arimondo, E. and Wilkens, M., Lecture Notes in Physics, Vol. 602 (Springer, Berlin, 2002) p. 140. See http://tsallis.cat.cbpf.br/biblio.htm for full bibliography.
[5] Kaneko, K., *Chaos* **2** (1992) 279.
[6] Robledo, A., *Phys. Lett. A* **328** (2004) 467.
[7] Kob, W. and Andersen, H.C., *Phys. Rev. E* **51** (1995) 4626.
[8] Götze, W. and Sjögren, L., *Rep. Prog. Phys.* **55** (1992) 241.
[9] See, for example, Bouchaud, J.P., Cugliandolo, L.F., Kurchan, J. and Mezard, M., in *Spin Glasses and Random Fields*, ed. Young, A.P. (World Scientific, Singapore, 1998).
[10] Simdyankin, S.I., Mousseau, N. and Hunt, E.R., *Phys. Rev. E* **66** (2002) 066205.
[11] Berthier, L. and Holdsworth, P.C.W., *Europhys. Lett.* **58**(1) (2002) 35–41.
[12] Latora, V., Rapisarda, A. and Tsallis, C., *Phys. Rev. E* **64** (2001) 056134.
[13] Montemurro, M.A., Tamarit, F. and Anteneodo, C., *Phys. Rev. E* **67** (2003) 031106.
[14] Pluchino, A., Latora, V. and Rapisarda, A., cond-mat/0303081; cond-mat/0306374.
[15] Adam, G. and Gibbs, J.H., *J. Chem. Phys.* **43** (1965) 139.
[16] See, for example, Schuster, H.G., *Deterministic Chaos. An Introduction*, 2nd Revised Edition (VCH Publishers, Weinheim, 1988).
[17] Crutchfield, J.P., Farmer, J.D. and Huberman, B.A., *Phys. Rep.* **92** (1982) 45.
[18] Beck, C. and Schlogl, F., *Thermodynamics of Chaotic Systems* (Cambridge University Press, UK, 1993).
[19] Baldovin, F. and Robledo, A., *Phys. Rev. E* **66** (2002) 045104(R), *Phys. Rev. E* **69** (2004) 045202(R).
[20] Crutchfield, J., Nauenberg, M. and Rudnick, J., *Phys. Rev. Lett.* **46** (1981) 933.
[21] Shraiman, B., Wayne, C.E. and Martin, P.C., *Phys. Rev. Lett.* **46** (1981) 935.
[22] Tsallis, C., in *Nonextensive Statistical Mechanics and Its Applications*, eds. Abe, S. and Okamoto, Y., Lecture Notes in Physics, Vol. 560 (Springer, Berlin, 2001) p. 3.

PART 2

STRUCTURES AND GLASSES

CHAPTER 6

FLEXIBILITY IN BIOMOLECULES

M.F. Thorpe*, Mykyta Chubynsky, Brandon Hespenheide and Scott Menor
Physics & Astronomy Department, Arizona State University,
Tempe, AZ 85287, USA
**mft@asu.edu*

Donald J. Jacobs
Department of Physics & Astronomy, California State University Northridge,
Northridge, CA 91330, USA

Leslie A. Kuhn and Maria I. Zavodszky
Department of Biochemistry & Molecular Biology, Michigan State University,
East Lansing, MI 48824, USA

Ming Lei
Department of Biochemistry, Brandeis University,
Waltham, MA 02454, USA

A.J. Rader
Center for Computational Biology & Bioinformatics,
University of Pittsburgh, PA 15261, USA

Walter Whiteley
Department of Mathematics & Statistics,
York University, Toronto, Canada

In this chapter we review recent theoretical and computational work on the flexibility of biomolecules. This approach uses constraint theory and includes all the constraints in a biomolecule that are important at room temperature. A rigid region decomposition determines the rigid regions (both stressed and unstressed) and the flexible regions that separate them. Enzymes usually have a rigid core for stability and flexible regions for functionality. The rigid region decomposition can be used as input to for a Monte Carlo dynamics in which the flexible regions are allowed to move, consistent with the constraints. Results are illustrated with the proteins HIV protease and barnase.

1. Introduction

The relationship between the structural flexibility and dynamics of biomolecules and function is one of the key areas of research in modern biological science. While structural flexibility has long been known to play an essential role in function, experimental and also computational data for many biomolecules has been slow in coming due to the difficulty in observing the motion over long time periods, especially for large systems. Drawing upon a range of disciplines, including biology, chemistry, physics, mathematics, and computer science, rigidity theory now offers a means to predict the flexibility inherent within a biomolecule, given only a single, static, three-dimensional structure. Knowledge of the rigid and flexible regions of a biomolecule can provide insight into its function, allow detailed exploration of the ensemble of conformations available for a given state, and provide a means to predict changes in structural flexibility as local environmental conditions such as temperature and pH change.

Advances in mathematical rigidity theory and computational algorithms, together with a representation of molecular forces as mathematical constraints, have resulted in the development of two programs, *FIRST* (Floppy Inclusions and Rigid Substructure Topography) and *ROCK* (Rigidity Optimized Conformational Kinetics). The *FIRST* software can decompose a static protein structure into rigid and flexible regions, and track changes in these regions during simulated thermal denaturation. The program *ROCK* extends the results of *FIRST* by exploring the ensemble of conformations accessible to the flexible regions in a structure, keeping the rigid regions stationary. Additionally, *ROCK* can be used in a directed dynamics mode to identify a conformational pathway between two distinct, known structures of a single biomolecule. In the next two sections, we give examples of the results obtained by this approach.

2. Constraint Theory

For the last 30 years, there has been an evolving geometric and combinatorial theory of structural rigidity [1–3]. This work builds upon and extends a body of work spanning the last century, including that of

James Clerk Maxwell [4], and a host of engineers studying the statics (and first-order kinematics) of bar and joint frameworks. Within this setting, it is useful to define a framework as a set of edges (E) and vertices (V) that define a simple graph (no loops or multiple edges) G = (V, E) and a configuration $\boldsymbol{p}$ of points for the vertices, together written as G($\boldsymbol{p}$). The edges of the graph represent distances between points that are constrained to remain constant in any transformation or motion of the framework.

A more abstract combinatorial theory of generic rigidity describes the rigidity properties of a given graph for "almost all" configurations $\boldsymbol{p}$. Choosing a configuration at random will give the necessary generic behavior with probability 1, and these graphs, which lack any special symmetries, are classified as generic. For realizations of a graph in the plane, there is a simple combinatorial criterion for identifying minimal graphs which are generically rigid. Known in the last century, but first proven by Laman [2] in 1970, *a generic network in 2D with V sites and E bonds (defining a graph) does not have a redundant bond if and only if no subset of the network containing v sites and e bonds defining a subgraph violates* $e \leq 2v - 3$. This criterion, in turn, leads to fast (almost linear) time combinatorial algorithms for decomposing a given graph into rigid components as well as predicting the degrees of freedom (dimension of the space of non-trivial motions) for the graph and its subgraphs. A clear implementation of this algorithm in 2D was suggested by Hendrickson [5], and subsequently developed into the *2D pebble game* algorithm by Thorpe and Jacobs [6]. Such a combinatorial algorithm for decomposing a graph into rigid and non-rigid subgraphs has several important advantages. It is fast and it is stable compared to the slower numerical evaluations of the rank of the rigidity matrices.

The underlying projective and combinatorial theory for representing molecular frameworks extends to a wider variety of structures with distance constraints as edges. One extension studies frameworks with larger "bodies" and hinges (removing up to five degrees of freedom between bodies) as constraints, known as body-bar graphs [7–9]. There are several surprises in this extension. The biggest is that the theory of body-bar frameworks does have a good combinatorial theory leading to extensions of the *pebble game* in 3D. These are fast polynomial time algorithms for decomposing a generic 3D graph into rigid components, as well as the total degrees of freedom for non-rigid pieces. Essentially, this extension searches for six edge-disjoint spanning trees in a modified graph with vertices for the bodies and five edges for each hinge (corresponding to the five constraints imposed by the hinge). At a geometric level there is a complete first-order projective theory for these structures. (To date, this full blown projective theory has only been sketched in the literature [10], but the theory is well developed and should receive a proper exposition soon.)

At the combinatorial level, the *Molecular Framework Conjecture* of Tay and Whiteley [8, 10, 12] is central to all our studies. There is overwhelming evidence for this conjecture, including comparing the combinatorial predictions and the actual rank computations for dynamical matrices. The conjecture, and the associated *3D pebble game*, has been verified explicitly for all structures studied with up to ~700 atoms, using costly matrix diagonalization techniques as a standard for comparison. The *pebble game* is an integer algorithm which is why it is so fast — handling a typical protein in less a minute in real time. These results use extensions of the projective and combinatorial techniques in [7, 8] and all evidence points to the algorithms used in the *3D pebble game* being exact [13, 14]. It would be irresponsible and also unnecessary to wait for a strict mathematical proof before proceeding with applications.

3. The Pebble Game and *FIRST*

The *3D pebble game* algorithm for frameworks is embodied in the *FIRST* software for measuring rigidity in biomolecular structures [15] and an example is shown in Fig. 1. The fundamental step in the application is how to represent the microscopic forces in a molecule as distance constraints in a body-bar graph. For example, to fully model proteins, it is important to represent the prevalent and structurally crucial non-covalent interactions, in addition to the covalent bonds and angles. These non-covalent interactions and additional bond-rotational constraints include hydrogen bonds, salt bridges, hydrophobic tethers, and double bonds. With these representations, we have performed flexibility analyses for many proteins from different structural classes in their functional native state, including inter-domain hinge motions (lysine-arginine-ornithine binding protein and dihydrofolate reductase), loop or flap motions (HIV protease, cytochrome *c*, dihydrofolate reductase), and grip-like motions involving the concerted curling of multiple segments of the protein (adenylate kinase) [1, 15–17]. These results indicate that the set of covalent and non-covalent interactions that we model as distance constraints is sufficient for reproducing the experimentally observed flexibility in these protein structures. Such information on native-state flexibility can be valuable, for example when studying enzyme mechanisms involving motion in parts of the protein.

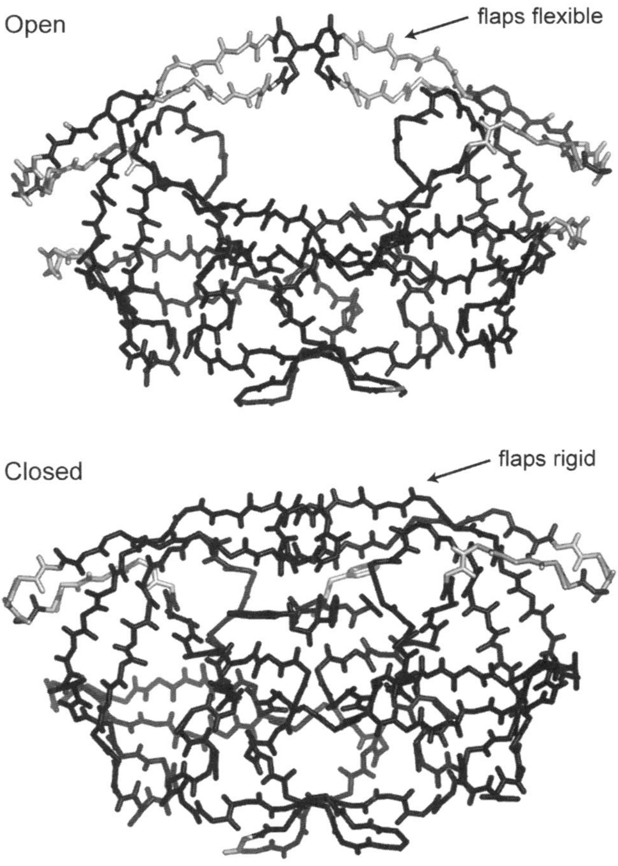

Fig. 1. Showing a rigid region decomposition. The protein HIV protease is a dimer that consists of two polypeptide chains, shown on the left and right sides of each panel above, each containing 99 amino acids, and is an important part of the HIV virus. The open form shows the protease in the native state where it functions as a chemical scissors, whereas this function is inhibited in the closed form where the flaps are pinned against the inhibitor [11]. The black regions are rigid.

An accurate methodology for computing native-state flexibility in proteins has an interesting extension to the prediction of protein folding pathways. This extension is based on the concept that as a protein folds, specific non-covalent bonds form, and remain formed throughout the remainder of the folding reaction. This suggests that the network of bonds in a native-state protein contains sub-networks corresponding to substructures formed along the folding pathway. Conversely, as a protein *un*folds, one might expect these early formed substructures to remain structurally stable the longest. We have simulated the effect of thermally induced unfolding on a protein by assigning an energy, or better a free energy change, to every non-covalent bond and then removing them from the flexibility calculation in order of energy, or free-energy. Changes in structural rigidity are tracked as the bonds are removed. The results of this simulation, referred to as *hydrogen bond dilution*, can be displayed graphically by mapping the rigidity results for the protein main chain onto a 1D line (sequence). The results of hydrogen bond dilution for barnase are shown in Fig. 2, where the shaded bars represent rigid main-chain bonds, and different shades indicate mutually exclusive rigid regions, often referred to as rigid clusters. The thin black line represents flexible main-chain bonds. This figure shows onset of flexibility as the protein is denatured, and also shows how the size of the

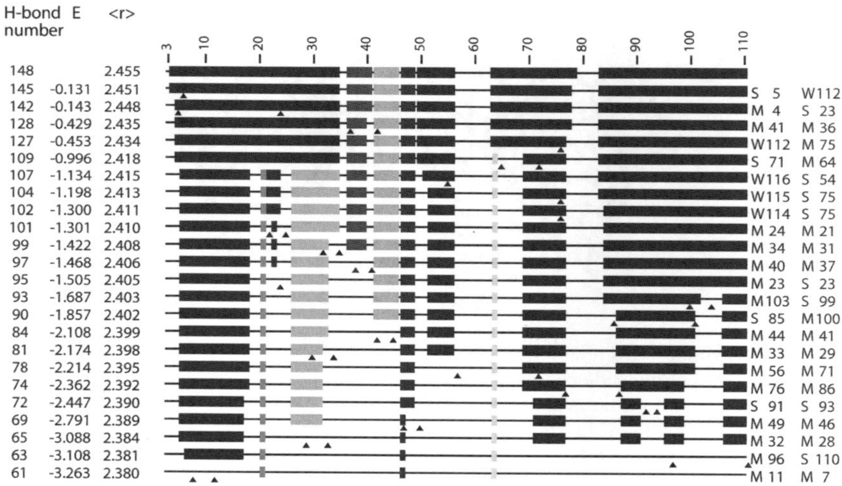

Fig. 2. Hydrogen bond dilution results for barnase. Each line represents *FIRST* flexibility results for the main-chain bonds of barnase for a given concentration of hydrogen bonds (the total hydrogen bonds in the protein for a given line is listed in the first column on the right). The shaded bars indicate rigid bonds, and different shades represent different rigid clusters. Flexible bonds are shown as a thin black line. The top line represents the "native state" of the protein, and depicts a largely rigid structure consisting of a predominant rigid cluster. As stronger hydrogen bonds are removed from the structure (the energies of a given bond are listed in the second column on the left), the protein becomes more and more flexible. A key feature for this protein is the presence of a rigid core that persists until the protein becomes completely flexible.

largest rigid cluster, decreases until a point at which it fragments into small rigid clusters and flexible bonds. This result represents a plausible unfolding pathway for barnase, and as such, can be verified by comparison to experiment.

For barnase, and many other proteins, we have shown that those secondary structures that remain mutually rigid the longest during our denaturation simulation correspond well to the secondary structures that form the folding core [18, 19]. Experimentally, the folding core is identified by using a combination of hydrogen-deuterium exchange and 2D NMR [20]. Such results of *FIRST* demonstrate that given only a single static protein structure, it is possible to obtain information about a dynamic process such as folding and predict the most stable regions of the protein. However, given that *FIRST* can also identify the flexible regions of a protein, it is possible to explore alternative conformations of a protein by moving the flexible regions in such a way that is consistent with the internal bond lengths and angles.

There have been significant studies in computational geometry using a variety of techniques, including extensions of first-order motions, to consider paths followed by flexible linkages. Some of this work has focused on small linkages connected with motion planning in robotics (or similar scaled biological problems such as the necks of birds). Other recent studies explore the 3D motions for a variety of structures, which include the equivalent of body-bar frameworks, or linkages for polymers and related structures [21]. These flexible linkage representations include polygonal chains, which are bonds linked end-to-end to form a linear polymer, and can include bond-coordination angle constraints coupled with dihedral rotations. In many cases, rather than generating algorithms for solving these problems in full generality, the complexity of the problems are confirmed, many of which are probably NP hard. Within the larger space of possible motions permitted by the basic constraints of bond lengths and angles, biochemists have used simulations that select appropriate paths by following an energy landscape to "steer" the dynamic motion. These molecular dynamics simulations require potential-energy functions to apply classical physics equations of motion to the atoms in a molecule. While molecular dynamics remains the state-of-the-art for examining the details of molecular motion, the computational complexity involved and short time step limits its usefulness to examine the low-frequency modes associated with the large-scale motions of a biomolecule; motion that typically occurs in the microsecond to second regime.

The *3D pebble game* [6, 22, 23] algorithm, as currently implemented in *FIRST*, can identify all the structurally rigid and flexible regions within a protein of hundreds of residues in a few seconds. Computationally, *FIRST* generates a directed graph of the covalent and non-covalent constraints present within the protein. This graph corresponds to a body-bar network of distance constraints that represent the physical forces present in the protein, and *FIRST* determines for every bond in this network whether it is part of a rigid cluster or an under-constrained region composed of flexible bonds. In addition, the number of independent degrees of bond-rotational freedom, or floppy modes, associated with each under-constrained region is also determined. The concept of an independent flexible bond is important, as it is often found that the rotatable bonds in a flexible region are coupled to each other. For example, a ring of seven bonds is flexible, but contains only a single independent degree of freedom (DOF). Using *FIRST*, it is possible to determine whether a given flexible bond is independent, or part of a larger group of flexible bonds referred to as a *collective motion*. The conformational space available to the set of independent flexible bonds and collective motions can then be explored using the program *ROCK*.

The original *3D pebble game* algorithm as encoded in *FIRST* allows for only two types of bonds to be modeled, a rotatable bond or a non-rotatable bond, corresponding to a constraint with five bars or six bars respectively, within the body-bar formalism. This representation is sufficient for representing covalent bond networks, but has led to complications when attempting to model weaker, less-specific interactions such as those associated with the hydrophobic effect. This was rectified by replacing a single 5-bar constraint with a series of 5-bar constraints, which is explicitly included in the calculation through the use of pseudo-atoms. This modeling scheme reduces the impact of an interaction on network rigidity because each pseudo-atom introduces one more DOF to the system than the associated constraints remove (it is important to note that a series of pseudo-atoms still represents a single, real, microscopic interaction such as a hydrogen bond). Figure 3 depicts the impact a chain of pseudo-atoms has on the rigidity of a structure. For example, a single 5-bar constraint between a pair of atoms, as is used to represent a covalent bond, removes 5 degrees of freedom from the network (1 DOF is removed for each bar in the network). In contrast, a chain of 4 pseudo-atoms, shown at the bottom of Fig. 3, removes only 1 DOF [the total loss of DOF comes from (4 pseudo-atoms $\times$ 6 DOF per pseudo-atom) minus (5 constraints $\times$ 5 bars per constraint) $= -1$ DOF]. Interestingly, a chain of 5 pseudo-atoms between atoms removes zero DOF

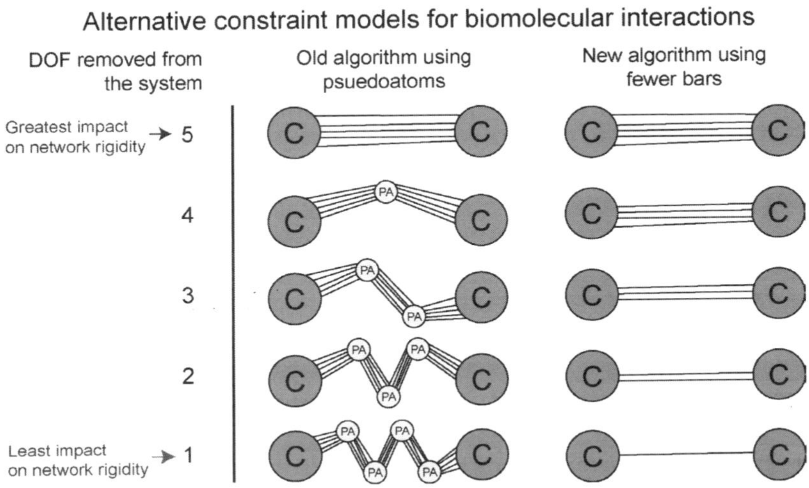

Fig. 3. Modeling the effect of a carbon–carbon constraint on network rigidity. The more DOF a constraint or series of constraints removes, the more rigid the structure will be. Previously, pseudo-atoms (spheres) were included in a structure to attenuate the number of DOF removed by a bond between two atoms because the algorithm required 5 bars for every constraint. Non-intuitively, the higher the number of pseudo-atoms used, the smaller the affect on network rigidity. The new pebble game algorithm has been modified to allow 1–6 bars. Now, the number of bars directly corresponds to the number of DOF removed from the system. Not only does this have a clear physical interpretation, it is computationally less complex, reducing the run time for *FIRST*.

from the system, and has absolutely no affect on the network rigidity. By taking advantage of the 4 different sized chains of pseudo-atoms shown in Fig. 3, we can represent a range of microscopic forces, the physical interpretation being that a constraint removing the least number of DOF from a system has the least rigidifying effect on the structure.

While the results of the *3D pebble game* are not affected by the inclusion of pseudo-atoms, they have no physical interpretation. Additionally, because they are explicitly included as part of the input structure, they increase the size of the system and hence the computation time.

We have recently eliminated the need for pseudo-atoms due to a new implementation of the *3D pebble game* algorithm. This new algorithm allows for any number of bars between 1 and 6 to be placed between a pair of atoms, and this number directly corresponds to the degrees of freedom removed from the system. The mapping between a chain of pseudo-atoms and the new "fewer-bars" representation is shown in Fig. 3. Not only is the new implementation faster, but it is more intuitive. While *more* pseudo-atoms implies *less* affect on the rigidity, in the new algorithm the number of bars used to model a microscopic force is directly proportional to the effect on the network rigidity. The equivalence between a chain of pseudo-atoms and fewer bars has been demonstrated both mathematically, and in practice.

We have taken advantage of the new *3D pebble game* algorithm to model the effect of hydrophobicity in protein structures. Hydrophobic interactions are known to contribute significantly to protein stability and are generally believed to be critical in driving the protein folding process [24]. Hydrophobicity reflects the tendency of the system to optimize entropy by folding hydrophobic groups that cannot form hydrogen bonds to the interior of the protein. This allows water molecules to associate randomly with each other and with polar groups on the protein surface, rather than being forced to become ordered relative to each other when presented with a hydrophobic surface of atoms. We model this tendency for hydrophobic atoms, principally carbon and sulfur atoms within proteins, to remain relatively near one another rather than unfolding to interact with the solvent. These hydrophobic contacts can be thought of as slippery, loosely constraining the local motion.

Hydrophobic interactions are identified geometrically between a pair of carbon and/or sulfur atoms [18]. The resulting constraint, which we refer to as a hydrophobic tether, is modeled with 2 bars. The decision to use 2 bars was the result of computing native-state flexibility and folding-core data in many proteins with an exhaustive sampling of geometric criteria and 4-, 3-, 2-, and 1-bar constraint representations. The 2-bar representation gave the best correlation to experiment. A physical interpretation of this 2-bar model is that it restricts the maximum distance between the two hydrophobic atoms, while allowing them to slide with respect to one another.

The most important of the microscopic forces that we model as constraints are hydrogen bonds and salt bridges. These non-covalent bonds are identified according to geometrical rules [16]. For PDB entries lacking polar hydrogen atom positions, the *What If* software package [25] is used to define hydrogen atom positions optimal for hydrogen bonding. Water molecules present in the input file are included in the analysis. In our recent research, we have only included water molecules if they are entirely buried within the structure, determined by the software *PRO_ACT* [26], as there is no mechanism in *FIRST* to identify these waters. Bonds between the protein and ligands, including metals and other ions, are treated as covalent bonds if so specified in the PDB file (or if they are within covalent bonding distance); otherwise, their polar and hydrophobic atoms are subject to the same rules as protein atoms for determining non-covalent interactions with the protein. Each potential hydrogen bond is assigned an energy using a modified Mayo potential, which evaluates the favorability of the bond based on a combination of distance and angular functions. We have modified the potential by strengthening the angular dependence on the donor–H–acceptor angle, so that it must be $\geq 120°$ for the bond to receive a favorable (negative) energy [27]. This avoids including non-physical H-bonds with angles near 90° (e.g., between C=O(i) and NH(i+3), rather than NH(i+4), in α-helices).

4. Dynamics Using *ROCK*

While *FIRST* analysis can quickly identify flexibility in a biomolecule, one drawback is that atoms do not actually move — the analysis is based on statics, and so shows potential or virtual motion, rather than actual motion. While the atoms in the flexible regions have the possibility to move in various collective motions, *FIRST* does not give the amplitude of the motion. Some motions are restrained from having large amplitude by the constraints within a single flexible region, and the motion may also be restricted by collisions with adjacent regions.

In order to clarify these points, the program *ROCK* (Rigidity Optimized Conformational Kinetics) has been developed. *ROCK* [28, 29] uses *FIRST* as input and then makes Monte Carlo moves within the individual flexible regions, while maintaining ring closure. A typical flexible region, containing say 100 atoms, will have very many rings, involving both covalent and non-covalent interactions. This is because rings consist of hydrogen bonds and hydrophobic tethers as well as covalent bonds, and so the rings themselves form a dense interlocking network. In addition to maintaining ring closure, the Monte Carlo moves must respect the hard spheres associated with the van der Waals radii, and also the Ramachandran constraints on the main-chain dihedral angles. After each attempted move in *ROCK*, the motions of adjacent flexible regions are checked for collisions. The conformations produced by *ROCK* show the diffusive motion of the protein (http://www.pa.msu.edu/~lei/Research/*ROCK*/Proteins/HIV-1/HIV-1.html). Thus the technique is complimentary to molecular dynamics which is good for time scales of up to a millisecond for small proteins. *ROCK* does not use a potential, and so does not give as accurate a picture as MD. However, by effectively freezing out the high frequency motions,

(a)

(b)

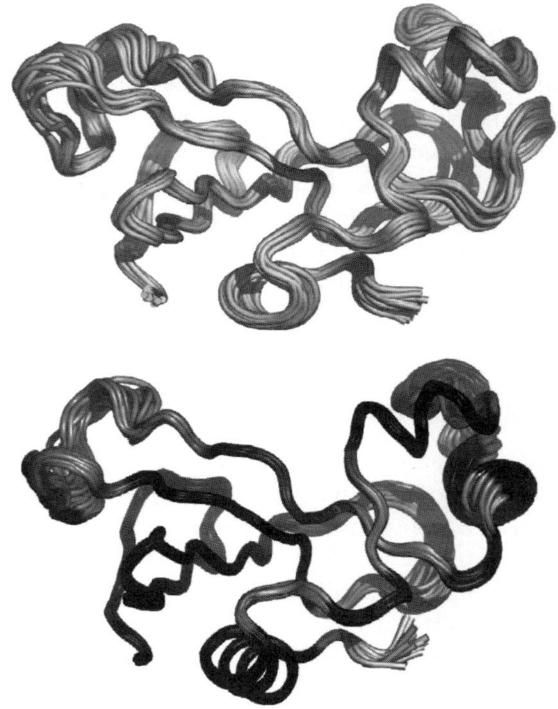

Fig. 4. Flexibility in the native-state structure of barnase. The solid tubes represent the main chain of the protein. (a) Result of *ROCK* dynamics using the X-ray structure **(1A2P)** of barnase [32]. The rigid regions are determined to be rigid from *FIRST* analysis, and as such, are kept fixed. The flexible regions are used to generate the 20 alternative conformations created by *ROCK*, which are shown superimposed. Interestingly, the rigid helix on the right is flanked by two flexible regions, and *ROCK* explores a range of motion in which this helix moves as a rigid unit. (b) Superposition of 20 conformers of barnase **(1BNR)** as determined by NMR spectroscopy [31]. The dynamic nature of NMR experiments results in many possible structures that fit the observed data. Those regions of the structure that overlap well imply little difference in the local conformation among the 20 structures. In contrast, those regions of the structure that can adopt many different conformations, as shown in the upper left portion of the panel, indicate structural flexibility. A comparison between the predicted structural flexibility in panel (a) and the experimentally observed flexibility shown in panel (b) reveals a good correspondence. The flexible regions in (a) correspond well to the thick regions in panel (b), and the rigid regions in (a) align with the thin regions in panel (b) [29].

which have little to do directly with biological function, the large diffusive motions can be visualized, giving the researcher a good sense of the possible motions. As an application, *ROCK* can be used to study directed dynamics by driving a protein from a known initial state towards a known

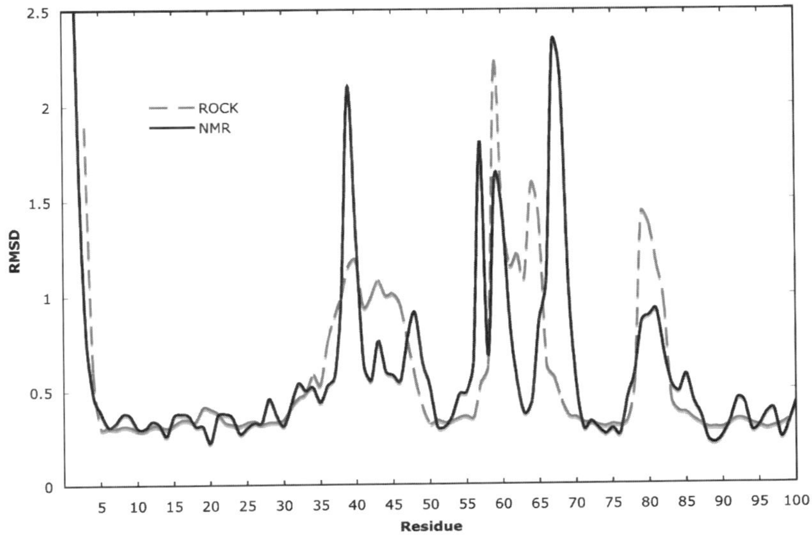

Fig. 5. Showing a comparison for the two sets of aligned conformations shown in Fig. 4 for barnase. The quantity RMSD is the root mean square deviation from the average structure in each case, measured in Angstroms, and plotted against residue number. The NMR structure (1BNR) [31] and X-ray structure (1A2P) [32] are from high resolution experiments. A background has been added to the *ROCK* results of an RMSD of 0.3 Å which represents the high frequency motions, which are explicitly excluded from *ROCK*. To get the result displayed as *ROCK* above, the square deviations have been added [29].

target state, such as the ligand bound and unbound conformations of an enzyme. This allows the possible pathway(s) to be determined, through a set of conformationally allowed intermediate steps, which can be amplified later by more detailed optimization methods using realistic potentials.

ROCK can produce various conformations from a single static structure. This is useful when trying to find the flexible regions from X-ray crystallographic data when NMR data is not available. Using B-values alone is not very informative as no correlation information is obtained from the B-values which associate an amplitude with single atoms. Figure 4 shows a comparison of the flexibility of barnase using a single X-ray crystallographic structure (PDB: 1A2P) and *ROCK*, together with NMR data (PDB: 1BNR). The NMR figure is a superposition of the best 20 fits and contains sufficient measured constraints that the "spread" represents the flexibility of the protein.

It is important to ascertain such flexibility in order to understand function. The set of conformational states determined by *ROCK* is also useful in docking studies [29] which are facilitated by having a partially flexible interface because this adds a favorable entropic contribution to the free energy gain associated with the docking. Such studies are important in the early stages of finding drug candidates.

5. Conclusions

In this brief review, we have outlined various methods that can be used to study flexibility and the associated motion in biomolecules. Much of the work here has been previously published, and so is only summarized. Full details can be found in the publications cited involving the authors of this paper in various combinations. The software used in this paper to study flexibility, including *FIRST* and *ROCK* can be found on the *Flex*web site at *Flex*web.asu.edu.

Acknowledgments

This work has been inspired by the approaches to science used by Sir Roger Elliott during a career in which he introduced many new methods and insights into the study of disorder in non-crystalline systems. Such approaches have propagated through his many graduate students, like Mike Thorpe, and now through the students of those students (grand-students) and recently some great-grand-students. The work described here is a natural evolution of the ideas and approaches that Roger has pioneered. It has been an honor to be asked to include this work in this volume that celebrates Roger's seventy-fifth birthday. This work was supported by the National Institutes of Health under Grant No. R01 GM 67249-01.

References

[1] Jacobs, D.J., Rader, A.J., Kuhn, L.A. and Thorpe, M.F., Protein flexibility predictions using graph theory, *Proteins* **44** (2001) 150–165.
[2] Laman, G., On graphs and rigidity of plane skeletal structures, *J. Eng. Mathematics* **4** (1970) 331–340.
[3] Graver, J., Servatius, B. and Servatius, H., Graduate Studies in Mathematics, *Am. Math. Soc.* (1993).
[4] Maxwell, J.C., On reciprocal figures and diagrams of forces, *Phil. Mag.* **27** (1864) 250–261.
[5] Hendrickson, B., *Siam J. Comput.* **21** (1992) 65.
[6] Jacobs, D.J. and Thorpe, M.F., Generic rigidity percolation in two dimensions, *Phys. Rev. E.* **53** (1996) 3682–3693.
[7] Whiteley, W., The union of matroids and the rigidity of frameworks, *SIAM J. Disc. Math.* **1** (1988) 237–255.

[8] Tay, T.-S. and Whiteley, W., Recent progress in the rigidity of frameworks, *Struct. Topology* **9** (1984) 31–38.

[9] White, N. and Whiteley, W., The algebraic geometry of motions of bar and body frameworks, *SIAM J. Alg. Disc. Math.* **8** (1987) 1–32.

[10] Crapo, H. and Whiteley, W., Statics of frameworks and motions of panel structures: A projective geometric introduction, *Struct. Topology* **6** (1982) 43–82.

[11] Thorpe, M.F., Protein folding, HIV and drug design, *Physics and Technology Forefronts*, APS News (February, 2003).

[12] Whiteley, W., Infinitesimal motions of a bipartite framework, *Pac. J. Math.* **110** (1984) 233–255.

[13] Jacobs, D.J., Generic rigidity in three-dimensional bond-bending networks, *J. Phys. A* **31** (1998) 6653–6668.

[14] Whiteley, W., Rigidity of molecular structures: Generic and geometric analysis, in *Rigidity Theory and Applications*, eds. Thorpe, M.F. and Duxbury, P.M. (Academic/Kluwer, 1999), pp. 21–46.

[15] Jacobs, D.J., Kuhn, L.A. and Thorpe, M.F., Flexible and rigid regions in proteins, in *Rigidity Theory and Applications*, eds. Thorpe, M.F. and Duxbury, P.M. (Academic/Kluwer, 1999), pp. 357–384.

[16] Thorpe, M.F., Hespenheide, B.M., Yang, Y. and Kuhn, L.A., Flexibility and critical hydrogen bonds in cytochrome c, in *Pac. Symp. Biocomput.* (2000) 191–202.

[17] Thorpe, M.F., Lei, M., Rader, A.J., Jacobs, D.J. and Kuhn, L.A., Protein flexibility and dynamics using constraint theory, *J. Mol. Graph. Model.* **19** (2001) 60–69.

[18] Hespenheide, B.M., Rader, A.J., Thorpe, M.F. and Kuhn, L.A., Identifying protein folding cores from the evolution of flexible regions during unfolding, *J. Mol. Graph. Model.* **21** (2002) 195–207.

[19] Rader, A.J. and Bahar, I., Folding core predictions from network models of proteins, *Polymer* (2004) 655–668.

[20] Li, R. and Woodward, C., The hydrogen exchange core and protein folding, *Protein Sci.* **8** (1999) 1571–1590.

[21] Soss, M. and Toussaint, G.T., Geometric and computational aspects of polymer reconfiguration, *J. Math. Chem.* **27** (2000) 303–318.

[22] Thorpe, M.F., Jacobs, D.J., Chubynsky, M.V. and Rader, A.J., Generic rigidity of network glasses, in *Rigidity Theory and Applications*, eds. Thorpe, M.F. and Duxbury, P.M. (Kluwer Academic, 1999), pp. 239–277.

[23] Jacobs, D.J. and Thorpe, M.F., Generic rigidity percolation: The pebble game, *Phys. Rev. Lett.* **75** (1995) 4051–4054.

[24] Dill, K.A., Dominant forces in protein folding, *Biochemistry* **29** (1990) 7133–7155.

[25] Vriend, G. WHAT IF: A molecular modeling and drug design program, *J. Mol. Graph.* **8** (1990) 29, 52–56.

[26] Williams, M.A., Goodfellow, J.M. and Thornton, J.M., Buried waters and internal cavities in monomeric proteins, *Protein Sci.* **3** (1994) 1224–1235.

[27] Rader, A.J., Hespenheide, B.M., Kuhn, L.A. and Thorpe, M.F., Protein unfolding: Rigidity lost, in *Proc. Natl. Acad. Sci. USA* **99** (2002) 3540–3545.
[28] Lei M., Kuhn, L.A., Zavodszky, M.I. and Thorpe, M.F., Sampling protein conformations and pathways, *J. Comput. Chem.* **25** (2004) 1133–1148.
[29] Zavodszky, M.I., Lei Ming, Thorpe, M.F., Day, A.R. and Kuhn, L.A., Modeling correlated main-chain motions in proteins for flexible molecular recognition, accepted by *Proteins: Structure, Function and Genetics* (2004).
[30] Menor, S. and Thorpe, M.F., unpublished (2004).
[31] Bycroft M., Ludvigsen S., Fersht, A.R. and Poulsen, F.M., Determination of the three-dimensional solution structure of barnase using nuclear magnetic resonance spectroscopy, *Biochem.* **30** (1991) 8697.
[32] Mauguen, Y., Hartley, R.W., Dodson, E.J., Dodson, G.G., Bricogne, G., Chothia, C. and Jack, A., Molecular structure of a new family of ribonucleases, *Nature* **297** (1982) 162.

CHAPTER 7

LATTICE DYNAMICS OF CARBON NANOTUBES

Valentin N. Popov

Faculty of Physics, University of Sofia,
5 James Bourchier Blvd. BG-1164 Sofia, Bulgaria

Minko Balkanski

Université Pierre et Marie Curie,
Tour 13, 4 etage, Boîte 256, 4 Place Jussieu,
F-75252 Paris Cedex 05, France

The theory of the lattice dynamics of carbon nanotubes in terms of force constants is presented. The screw symmetry of the nanotubes is taken into account explicitly, which has computational advantages. It is shown that the theory is free from the drawbacks of previous studies in that it correctly predicts the existence of four acoustic branches. The longitudinal and twist branches are linear in the wave vector at the origin while the transverse acoustic branches are quadratic with the wave vector. This behaviour has crucial consequences for the elastic and thermal properties of the nanotubes. The effect of bundling of the nanotubes on their vibrational, elastic and thermal properties is studied in more detail. Special attention is paid to the breathing-like modes of nanotube systems because of their importance for characterization purposes.

1. Introduction

Multiwalled carbon nanotubes (MWNTs) were first discovered as a by-product of fullerene production in an electric arc between two graphite rods [1]. Introducing transition metals in the graphite electrodes of the electric arc made it possible to produce single-walled carbon nanotubes (SWNTs) [2, 3]. Soon after that, a high yield of nanotubes, forming bundles of parallel nanotubes, was obtained by laser ablation of a graphite target [4, 5]. In this case, the X-ray diffraction (XRD) measurements yielded mean diameters of the tubes of 13.8 ± 0.2 Å. Transmission electron microscope (TEM) measurements of similar samples [6] revealed that the tube diameter distribution was consistent with the range of armchair tubes (8, 8), (9, 9), (10, 10), and (11, 11). Later, scanning tunnelling microscope (STM) studies on ropes of nanotubes showed that nanotubes with a wide range of chiral angles were present [7, 8].

The field of carbon nanotubes has experienced an explosive development in the recent years. Many technological applications have been proposed. Their realization depends on good preparation methods and the possibility of precise characterization of the synthesized samples. Since the direct measurement of the structural parameters is a rather time-consuming and expensive procedure, it is highly desirable to use for this purpose alternative methods such as light scattering (LS) spectroscopy. The phonon frequencies determined by LS are directly related to the structure and the binding forces in the lattice. This is, therefore, an excellent non-destructive and inexpensive method, and rapidly yields useful results.

Up to now, a few calculations of the phonon dispersion in nanotubes have been accomplished either by the ZF method with correction of the dynamical matrix in order to obtain the two acoustic branches [9] or by using a simple force-constant model with a modification of the force constants in order to fulfil the rotational sum rule and to obtain the torsional acoustic branch [10]. In the latter work it has been shown however, that the four acoustic modes have non-zero slope at the origin. Recently, tight-binding [11] and *ab initio* [12] phonon dispersions of a few nanotubes that are free from the above-mentioned deficiencies have been published.

We developed [13] a model of the lattice dynamics of carbon nanotubes for the purpose of precise and rapid comparison with experimentally measured phonon frequencies, which should allow the determination of the exact structure of the material. This model takes into account the screw symmetry of the nanotubes, which results in a reduction of the size of the dynamical matrix to six for all tube types and allows one to study the vibrations of, practically, all observable nanotubes. Our model differs from the zone-folding (ZF) method that is also based on a dynamical matrix of size six but entirely ignores the tube curvature. Here, the calculation of the phonons is accomplished within a valence force field (VFF) model with parameters taken over from graphite. The light scattering intensity is estimated using a bond-polarization model. Special attention is paid to the investigation of the radial breathing mode because of the importance of this mode for the characterization of nanotube samples on the basis of light-scattering data. The results of the lattice dynamical investigation are further used in the study of the elastic and thermal properties of various nanotube systems.

This paper is organized as follows. The lattice dynamical model is presented in Sec. 2 followed by results of the calculation of the phonon dispersion of nanotubes (Sec. 3) and the radial breathing mode of isolated SWNTs (Sec. 4), bundles of SWNTs (Sec. 5) and MWNTs (Sec. 6). The elastic properties of isolated and bundled nanotubes are studied in Sec. 7. The theoretical results for the specific heat of nanotube systems are given in Sec. 8. The paper ends with conclusions (Sec. 9).

2. Lattice Dynamical Model

A nanotube can be viewed as constructed by rolling up of a graphite sheet (graphene) into a seamless cylinder leading to coincidence of the lattice point O at the origin and another one A defined by the chiral vector $\mathbf{C}_h = (n, m)$ (see Fig. 1) [14–16]. A tube is uniquely specified by the pair of integer numbers (n, m) or by its radius and chiral angle θ. The latter is defined as the angle between the chiral vector $\mathbf{C}_h$ and the nearest zigzag of carbon–carbon bonds with values in the interval $0 \leq \theta \leq \pi/6$. The tubes are called achiral for $\theta = 0$ ("zigzag" type) and $\theta = \pi/6$ ("armchair" type),

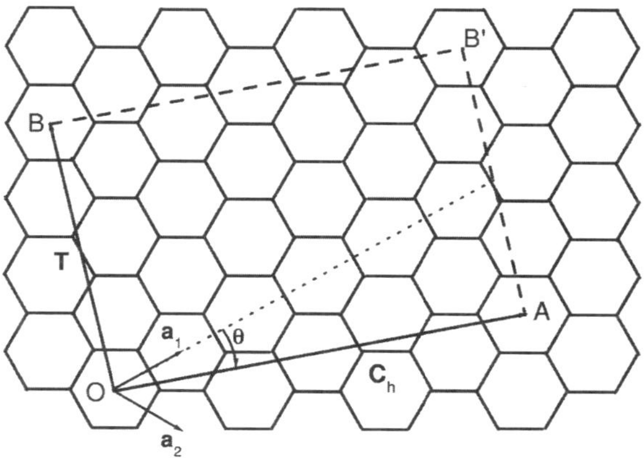

Fig. 1. The unrolled honeycomb lattice of a nanotube. The lattice points O and A define the chiral vector $\mathbf{C}_h$ and the points O and B define the translation vector $\mathbf{T}$ of the tube. The rectangle formed by the two vectors defines the unit cell of the tube. The figure corresponds to $\mathbf{C}_h = (4, 2)$ and $\mathbf{T} = (4, -5)$.

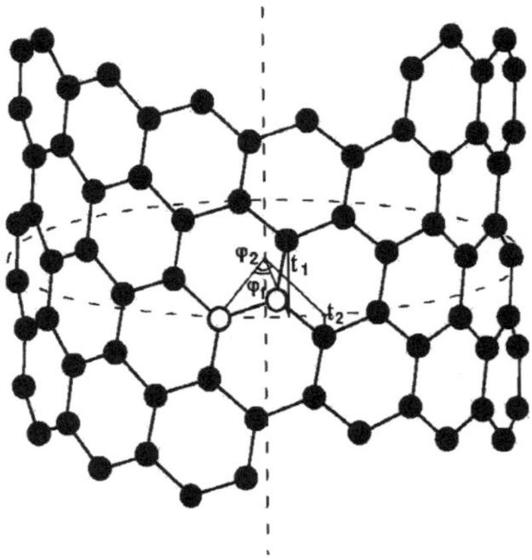

Fig. 2. A front view of a nanotube [13]. The closed dashed line is the circumference and the straight dashed line is the axis of the tube. The nanotube can be constructed by mapping of the two atoms of the unit cell (depicted by empty symbols) onto the entire cylindrical surface using two different screw operators.

and chiral for $\theta \neq 0, \pi/6$. All the carbon atoms of a tube can be reproduced by using two different screw operators [17] (see Fig. 2).

A screw operator $\{S|\mathbf{t}\}$ rotates the position vector of an atom at an angle φ about the tube axis and translates it a vector $\mathbf{t}$ along the same axis. Thus, the equilibrium position vector $\mathbf{R}(\mathbf{l}k)$ of the kth atom of the $\mathbf{l}$th atomic pair of the tube is obtained from $\mathbf{R}(k) \equiv \mathbf{R}(\mathbf{0}k)$ using two screw operators $\{S_1|\mathbf{t}_1\}$ and $\{S_2|\mathbf{t}_2\}$:

$$\mathbf{R}(\mathbf{l}k) = \{S_1|\mathbf{t}_1\}^{l_1}\{S_2|\mathbf{t}_2\}^{l_2}\mathbf{R}(k) = S_1^{l_1}S_2^{l_2}\mathbf{R}(k) + l_1\mathbf{t}_1 + l_2\mathbf{t}_2, \qquad (1)$$

where $\mathbf{l} = (l_1, l_2)$, l_1 and l_2 are integer numbers labelling the atomic pair, and $k = 1, 2$ enumerates the atoms in the pair. It is convenient to adopt the compact notation $S(\mathbf{l}) = S_1^{l_1}S_2^{l_2}$ and $\mathbf{t}(\mathbf{l}) = l_1\mathbf{t}_1 + l_2\mathbf{t}_2$, and to rewrite Eq. (1) as $\mathbf{R}(\mathbf{l}k) = S(\mathbf{l})\mathbf{R}(k) + \mathbf{t}(\mathbf{l})$.

For a small displacement $\mathbf{u}(\mathbf{l}k)$ of the atoms from their equilibrium positions, the harmonic approximation may be used for the potential energy of the tube and the resulting equations of motion are readily derived in the form

$$m_k\ddot{u}_\alpha(\mathbf{l}k) = -\sum_{\mathbf{l}'k\beta}\Phi_{\alpha\beta}(\mathbf{l}k, \mathbf{l}'k')u_\beta(\mathbf{l}'k'), \qquad (2)$$

where $\Phi_{\alpha\beta}(\mathbf{l}k, \mathbf{l}'k')$ are the force constants.

The helical symmetry of the nanotube suggests searching for a solution of the type

$$u_\alpha(\mathbf{l}k) = \frac{1}{\sqrt{m_k}} \sum_\beta S_{\alpha\beta}(\mathbf{l}) e_\beta(k|\mathbf{q}) \exp i\,(\mathbf{q}\cdot\mathbf{l} - \omega t)\,, \tag{3}$$

representing a wave with wave vector $\mathbf{q} = (q_1, q_2)$ and angular frequency $\omega(\mathbf{q})$. Substituting Eq. (3) in Eq. (2), we get a system of linear equations of the form

$$\omega^2(\mathbf{q}) e_\alpha(k\,|\mathbf{q}) = \sum_{k'\beta} D_{\alpha\beta}(kk'|\mathbf{q}) e_\beta(k'|\mathbf{q})\,, \tag{4}$$

where the dynamical matrix is defined as

$$D_{\alpha\beta}(kk'|\mathbf{q}) = \frac{1}{\sqrt{m_k m_{k'}}} \sum_{\mathbf{l}'\gamma} \Phi_{\alpha\gamma}(\mathbf{0}k, \mathbf{l}'k') S_{\gamma\beta}(\mathbf{l}') \exp i\mathbf{q}\cdot\mathbf{l}'. \tag{5}$$

The eigenfrequencies $\omega(\mathbf{q})$ are solutions of the characteristic equation

$$\|D_{\alpha\beta}(kk'|\mathbf{q}) - \omega^2(\mathbf{q})\delta_{\alpha\beta}\delta_{kk'}\| = 0. \tag{6}$$

Using the eigenvalues $\omega^2(\mathbf{q})$ one can obtain from Eq. (4) the corresponding eigenvectors $e_\alpha(k|\mathbf{q}j)$ ($j = 1, 2, \ldots, 6$). From Eq. (5) it can be proven that $D(\mathbf{q})$ is Hermitian and therefore $\omega^2(\mathbf{q}j)$ are real and $e_\alpha(k|\mathbf{q}j)$ may be chosen orthonormal.

The rotational boundary condition imposes the constraint

$$2\pi l = k_1 n + k_2 m \tag{7}$$

($l = 0, 1, \ldots, N–1$) on the $\mathbf{q}$-vector components. The theory presented above is valid for any helical structure and no translational periodicity was accounted for so far. In all nanotubes obtained by wrapping of graphene into a cylinder, such a symmetry exists and the primitive translation vector of a given tube $\mathbf{T} = (n_1,\, m_1)$ is given by

$$n_1 = (2m + n)/d_R, \quad m_1 = -(2n + m)/d_R, \tag{8}$$

where d_R is the greatest common divisor of $(2m + n)$ and $(2n + m)$. The total number N of atomic pairs is

$$N = 2\left(n^2 + nm + m^2\right)/d_R. \tag{9}$$

The translational periodicity of a tube leads to a Bloch-type of displacement field with a one-dimensional wave vector q:

$$q = n_1 q_1 + m_1 q_2. \tag{10}$$

Equations (7) and (10) allow replacing the **q**-vector components with the pair l and k. For each q the eigenvalues and the eigenvectors will be enumerated by the indices l and j. While the number of eigenvectors will be $6N$, the number of the eigenvalues can be less than $6N$ due to the degeneracy of some of the modes.

The force constants are invariant under infinitesimal translations along and perpendicular to the tube axis that leads to the translational sum rules and to three zone-centre zero-frequency modes. The infinitesimal rotation invariance condition imposed on the force constants gives rise to a rotational sum rule and to an additional zero-frequency mode [13].

Due to the explicit accounting for the screw symmetry of the tubes, the computation time for each q scales as 6^3N. This ensures a great advantage for phonon calculations of tubes with very large N in comparison to the approach that does not use the screw symmetry, where the computation time scales as $(6N)^3$. In practice, all observable nanotubes can be handled with the presented lattice dynamical model.

3. Phonon Dispersion of Single-Walled Carbon Nanotubes

The light scattering from nanotubes is normally observed under resonant conditions when the incident laser energy coincides with separations between electronic van Hove singularities. This situation is complicated by the fact that the electronic structure of nanotubes is not known with sufficient accuracy. On the other hand, the LS spectra are relatively simple because the resonant conditions are fulfilled only for certain nanotubes. The LS spectra of nanotubes exhibit mainly three bands of lines. The lowest frequency band comes from the radial breathing mode (BM) in which all atoms experience uniform radial displacements. The band around $1300\,\text{cm}^{-1}$ is due to disorder and the high-frequency band around $1590\,\text{cm}^{-1}$ arises from tangential stretching modes of the nanotubes. Among the low-frequency modes, the BM is particularly important for experimental investigations because of its high scattering cross-section and resonant behavior. The frequency of the BM is roughly inversely proportional to the tube radius R and is almost independent of the tube chirality. In Ref. 9, the value $C = 1170\,\text{cm}^{-1}\,\text{Å}$ was proposed. Accounting for the re-hybridization of the C–C bond during the vibration, a slight chirality dependence was deduced: $C(n,n) = 1180\,\text{cm}^{-1}\,\text{Å}$ and $C(n,0) = 1160\,\text{cm}^{-1}\,\text{Å}$ (*ab initio* local-density approximation in the density functional theory [12, 18]), and $C(n,n) = 1301\,\text{cm}^{-1}\,\text{Å}$ and $C(n,\,0) = 1282\,\text{cm}^{-1}\,\text{Å}$ (non-orthogonal tight-binding approach [19]).

The direct comparison of these values to the experimental data is not easy nor obvious. The situation is particularly complicated by the fact that the light scattering from carbon nanotubes is essentially resonant.

The theory of the lattice dynamics of carbon nanotubes presented in Sec. 2 can now be used to calculate the phonon dispersion curves for arbitrary q within a force-constant model. The experimental data do not provide enough information on the structural parameters of the tubes such as bond lengths and bond angles. On the other hand, the results of *ab initio* calculations of the optimized structure of nanotubes show that these quantities differ only by a few percent with those in graphite and that the carbon atoms do not lie on a cylindrical surface [20]. It is found here that such minor differences do not affect the eigenmodes much. Therefore, it was assumed that for all nanotubes the bond lengths are those of graphite and that all bond angles are equal. We note that other authors [9] have used an alternative choice of the tube structure, the one obtained by rolling graphene into a tube. The calculation of the eigenmodes in nanotubes was carried out within the VFF model with nearest-neighbor stretch, next-to-nearest-neighbor stretch, in-plane bend, out-of-plane bend, and twist interactions [21]. The values of the VFF force constants were obtained by fitting to the surface phonon dispersion curves of graphite investigated by high-resolution electron energy-loss spectroscopy. Here it was also assumed that these constants could be transferred to nanotubes without any modifications. Nevertheless, effects due to curvature of the tube will still exist because the bond angles enter the force-constant matrix explicitly.

The calculated phonon dispersion of a $(10, 10)$ nanotube [13] given in Fig. 3 shows the presence of four acoustic branches: longitudinal, torsional (or twist), and doubly degenerate ones. The first two increase linearly with the wave number and the latter one increases as the square of the wave number near the origin in agreement with the long-wavelength results from the theory of elasticity. For each q the number of phonons is equal to $6N = 120$. The phonons with $q = 0$ are atomic displacements either along the tube axis or along the tube circumference. The displacements form standing waves along the circumference with $2l$ nodes. The number l is associated with the symmetry species of the phonon. Considering the symmetry sub-group of the screw operations of the tube, C_N, the total vibrational representation of a tube splits into irreducible representations as $6A + 6B + 6E_1 + 6E_2 + \cdots$. The modes of symmetry A have $l = 0$ (no nodes), the E_l modes have $2l$ nodes, the B modes have $2(N–1)$ nodes. The phonons with $q \neq 0$ form in addition running waves along the tube axis.

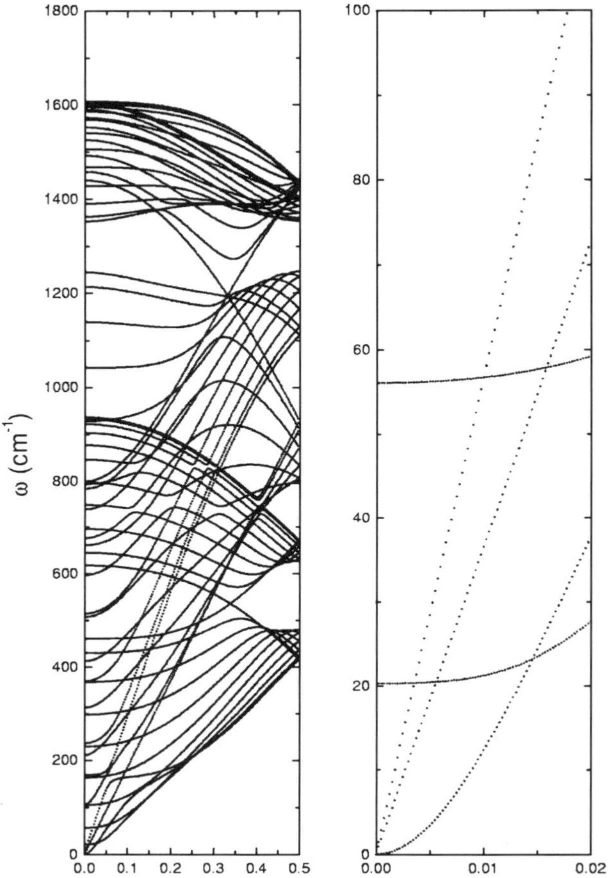

Fig. 3. Calculated phonon dispersion curves for a (10, 10) SWNT (left) and the low-energy region of the same curve containing the translational, the torsional (or twist), and the doubly degenerate transverse acoustic branches (right) [22].

The predictions of our lattice dynamical model for the tangential band deviate from the experimental data because of the underestimation of the tangential phonon branch overbending. The agreement with experiment can be improved by including force constants between more distant neighbors. On the other hand, the BM can be modelled successfully within the force-constant approach. Simplifying a tube down to a single ring of carbon atoms, it is straightforward to show that the force constant for the breathing motion of the ring is inversely proportional to the square of

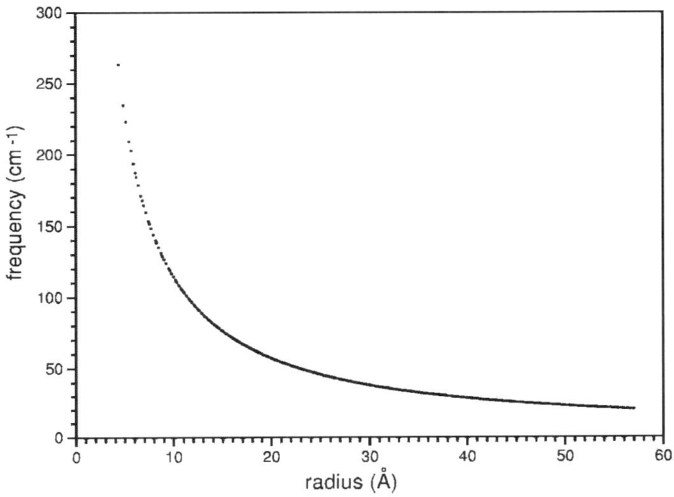

Fig. 4. Calculated dependence of the frequency of the radial breathing mode on the tube radius for all tubes in the range $4 < R < 55\,\text{Å}$. The data was fitted by the simple power law $1148/R^{1.00181}$ (in cm^{-1}) [22].

the ring radius R. Hence, the frequency of the breathing mode of the ring will depend on the radius as $1/R$. The calculations of the BM frequency within the VFF model, carried out for all nanotubes with radii in the range $4 < R < 55\,\text{Å}$, can be approximated by a simple power law C/R^n with $C = 1148\,\text{cm}^{-1}\,\text{Å}$ and $n = 1.00181$ [22] (see Fig. 4). The value of the constant C agrees well with previous estimations [9, 12, 18, 19]. However, the BM frequency was found to be chirality independent while *ab initio* calculations show that it slightly depends on the tube chirality [12, 18]. The reason for this is that although the non-planarity of the three bonds of each carbon atom enters the model explicitly, the force constants themselves are transferred from graphene to nanotubes without any corrections for bond non-planarity. The latter corrections can be done on the basis of *ab initio* or tight-binding results. Then the improved force constants can be used in the presented lattice dynamical model to yield more precise predictions for the phonons in carbon nanotubes.

4. Breathing-Like Modes in Bundles of Carbon Nanotubes

In order to gain useful information on the structure of carbon nanotubes one has to confront the theoretical calculations of the lattice dynamics giving the phonon dispersion curves and the experimental data obtained by light scattering spectroscopy. We have already mentioned some of the difficulties of this confrontation. Now we shall discuss the problems arising from the fact that the carbon nanotubes available for experimental investigations are not single isolated nanotubes but in most cases come packed into hexagonal microcrystals of several tens of individuals. The intertube interactions influence mainly the low-frequency modes and the effects of these interactions have to be taken into account.

Infinite crystals of nanotubes have been considered within tight-binding models using Lennard–Jones (LJ) potentials for the intertube interactions by several authors. Venkateswaran *et al.* [23] found an 8% upshift of the BM of the $(9,9)$ tube when bundled. Kahn and Liu [24] reported a 6% increase for a $(10,10)$ tube bundle. In a more exhaustive data set one finds [19] a 10% upshift for $(9,9)$ bundles and this shift ranges from 5% for tubes with $R = 3.45$ Å to 15% for $R = 8.21$ Å. The differences between these results stem from different parameterization for the tight-binding model and the LJ potential.

In a recent investigation Popov *et al.* [25, 26] studied the influence of packing on the vibrational properties of infinite and finite bundles of carbon nanotubes and we shall discuss some of their results. A microscopic model of the lattice dynamics was implemented in which the carbon–carbon interactions were described by a valence force field model and a bond polarization model. The interactions between carbon atoms belonging to different tubes were described by a LJ potential. In the case of infinite crystals of (n, m) tubes, the crystal is three-dimensional with a unit cell containing N_t carbon atoms. For finite bundle of N tubes, the crystal is one-dimensional with NN_t atoms in the unit cell. The dynamical matrix is constructed using force constants of the VFF type for the intratube interactions and force constants deduced from the LJ potential for the intertube interactions. The light scattering (LS) spectra are calculated within the bond polarization model [13]. In all cases, the LS intensity is averaged over all tube orientations in space.

The effect of bundling is most pronounced for the low-frequency modes. Those of them originating from displacement of tubes with respect to each other, were found to have relatively low frequencies (below $\sim 50\,\mathrm{cm}^{-1}$) and small LS intensity. The higher-frequency breathing-like modes have larger LS intensity and can serve as indicators for the presence of bundles of tubes with definite structural parameters.

The results of the calculations for infinite bundles of armchair type reveal that there are two breathing-like modes, BM(1) and BM(2), with intensive LS lines [25] (see Fig. 5). For small tube radii, the elastic force constant for radial extension is much larger than the intertube force constant and, consequently, the bundle eigenmodes are nearly unchanged with respect to these of isolated tubes. It is clear from Fig. 6 that BM(1) originates from the BM and BM(2) originates from a doubly degenerate mode,

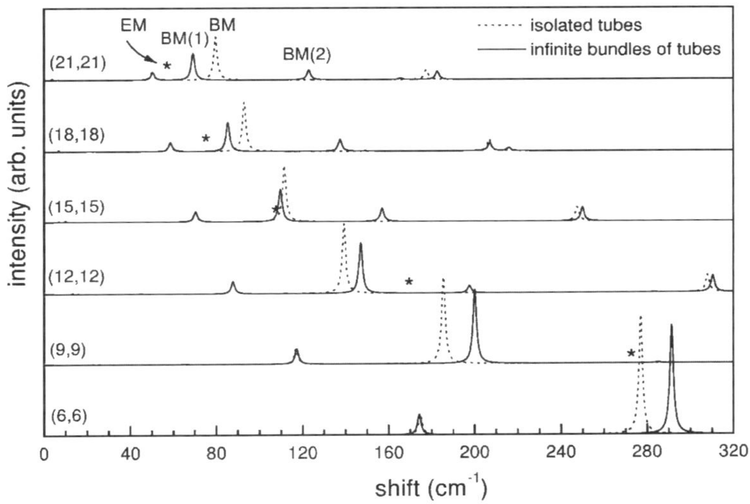

Fig. 5. The LS spectra of isolated tubes (dotted lines) and infinite bundles of tubes (solid lines) of armchair types (6, 6), (9, 9), (12, 12), (15, 15), (18, 18), and (21, 21) with radii in the interval from 4 to 16 Å [25]. The peaks originating from the BM, EM, BM(1) and BM(2) are indicated.

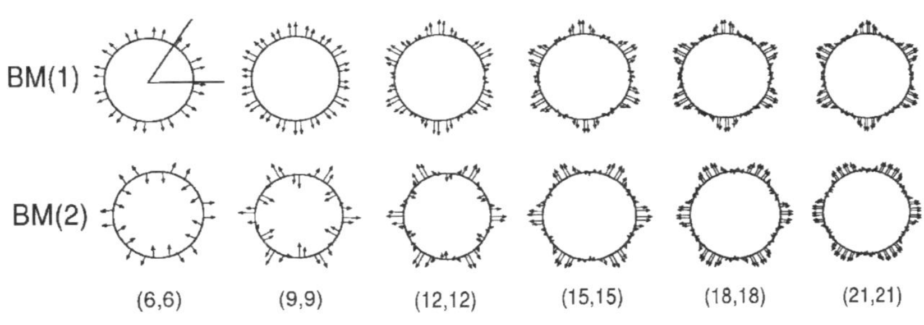

Fig. 6. Atomic displacements for the two breathing-like modes, BM(1) and BM(2), in infinite bundles of armchair tubes of the same types as in Fig. 5 [25]. The circular cross-sections of the tubes (scaled to have equal radii) and the primitive translation vectors of the regular triangular lattice are shown. The atomic displacements are depicted by arrows.

EM, in isolated tubes. With the increase of the tube radius, the vibration patterns of the two modes become more and more mixed and similar until they obtain nearly identical breathing-like shapes. The strong mixing for large radii is due to the fact that the intertube force constant is larger than the elastic one. The radius dependence of the calculated frequency of BM(1) and BM(2) is illustrated in Fig. 7 for all armchair and zigzag type tubes with radii in the range from 4 to 16 Å. The theoretical points for the breathing-like modes could not be fitted satisfactorily with single power laws C/R^n. However, reasonable fitting with such power laws could be

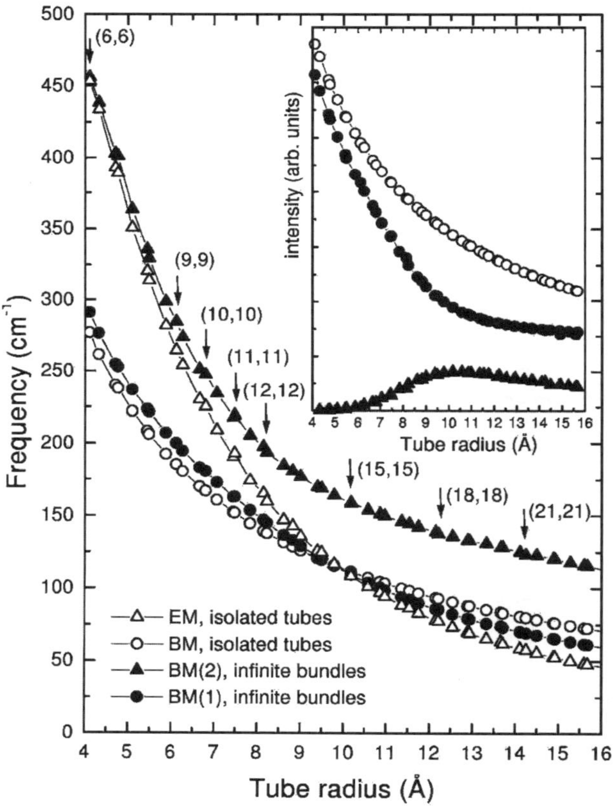

Fig. 7. Calculated frequencies of the BM and the EM in isolated tubes and originating from them BM(1) and BM(2) in infinite bundles of identical tubes for all armchair and zigzag types with radii between 4 and 16 Å [25]. This set includes all armchair tubes (n, n) with $n = 6, 7, \ldots, 21$, and all zigzag tubes $(n, 0)$ with $n = 11, 12, \ldots, 40$. The inset shows the calculated intensities of these modes versus tube radius.

obtained for two intervals separately: $4.5 < R < 8.5$ Å and $8.5 < R < 16$ Å. This result favors the idea of the existence of two regimes with a crossover at $R_c \approx 8.5$ Å, with domination of the intratube forces and intertube forces, respectively. In other words, below R_c the tubes are relatively rigid while above R_c the deformation of the tubes becomes important. The calculated LS intensity of these modes versus tube radius is shown in the inset of Fig. 7, from which it can be seen that the relative intensity of BM(2) with respect to BM(1) reaches a maximum of about 0.4 for $R \approx 10$ Å and decreases to 0.3 for $R \approx 15$ Å. The crossover between the rigid tube behavior and the deformable tube behavior is clearly demonstrated in Fig. 7.

In previous calculations of the vibrations of nanotube bundles, a second breathing-like mode has not been obtained because of the smaller diameter range [23] or because of the considered rigid tube breathing only [19]. Recent tight-binding molecular-dynamics simulations of several tube types with radii in the interval from 3.5 to 9.5 Å have revealed the existence of the second breathing-like mode in infinite bundles [27]. The estimated ratio of the intensities of the BM(2) and BM(1) lines, based only on the eigenvectors of the BM and EM, was found to increase with the radius and reach the value of about 0.9 for $R \approx 9.5$ Å while the more complete calculations here yield 0.3 for this ratio.

The results obtained for the frequency and intensity of the breathing-like modes in infinite bundles were used to explain the appearance in several experimental LS spectra of resonantly enhanced lines that are otherwise predicted to be nonresonant according to the noninteracting tubes picture and the resonance conditions [25].

We now turn to the study of finite bundles of identical tubes [26]. Figure 8 presents the calculated LS intensity of dimers of armchair tubes (n, n) with $n = 6$, 9, 10, 12, 14, and 16. In dimers, the lowering of the symmetry with respect to isolated tubes gives rise to mixing of the BM and E_l modes of the isolated tubes and to several breathing-like phonons, denoted by BM1 to BM5. For the small-radius tube $(6, 6)$ (see Fig. 9), the higher frequency mode BM2 is characterized by almost radial atomic displacement for each tube. The mode BM1 has a vibration pattern close to that of a E_4 mode of isolated tubes. For larger tubes, the greater intertube interaction leads to larger mixing between BM and E_l modes. For example, for dimers of $(10, 10)$ tubes, the atomic displacements for BM2 and BM3 are almost the same, apart from a rotation around the tube axis. This reveals a strong hybridization between the BM and the E_5 mode when the two tubes are brought together to form a dimer. Dimers of tubes $(16, 16)$ show two intense

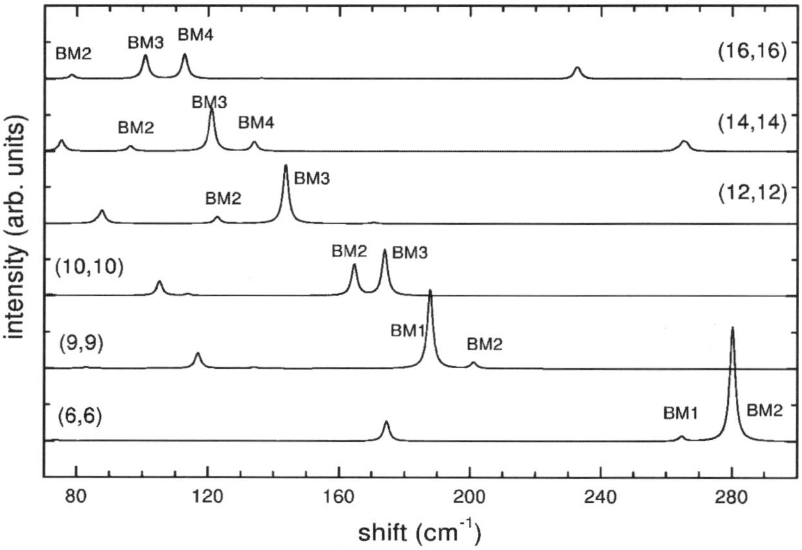

Fig. 8. LS spectra of dimers of tubes of armchair types (6, 6), (9, 9), (10, 10), (12, 12), (14, 14), and (16, 16) with radii in the interval from 4 to 11 Å [26]. The peaks due to breathing-like modes are denoted by BM1 to BM4.

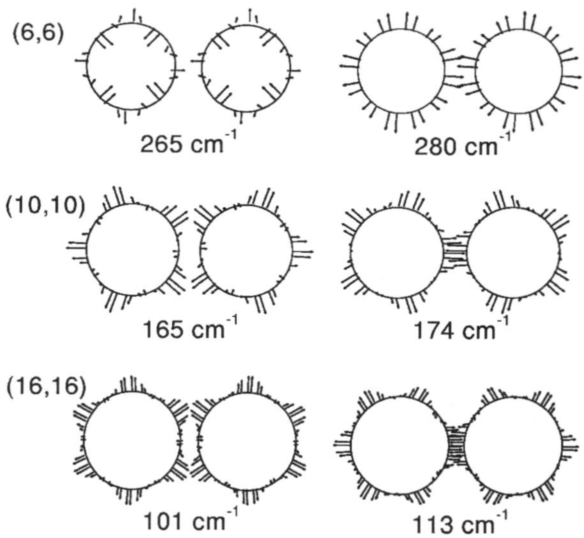

Fig. 9. Atomic displacements for the two major breathing-like modes of dimers of tubes (6, 6), (10, 10), and (16, 16) [26]. The frequencies of the modes are also given.

modes BM3 and BM4 that originate from coupling between the BM and the E_6 mode.

The radius dependence of the frequencies and intensity of the breathing-like modes BM1 to BM6 (solid symbols) is displayed in Fig. 10 in comparison to these of isolated tubes (open symbols). It is worth noting that for radii close to 7 Å, there are two breathing-like phonons with comparable LS

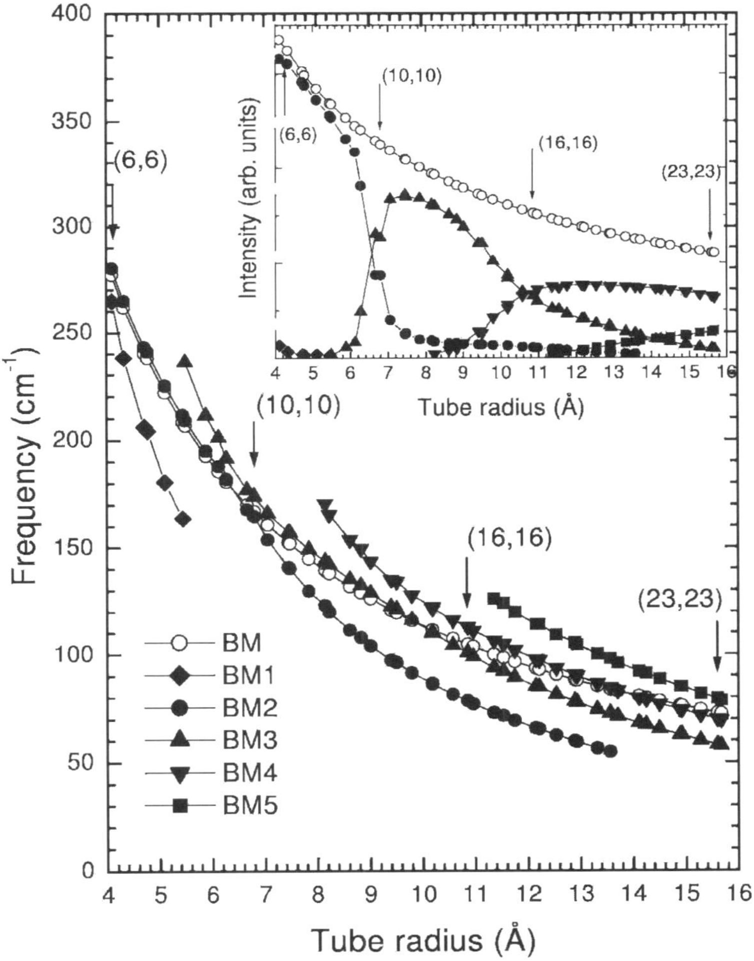

Fig. 10. Calculated frequencies of the major breathing-like modes of dimers of identical armchair and zigzag tubes in the range from 4 to 16 Å [26]. The inset shows the LS intensities of the most intense peaks originating from these modes.

intensity with a frequency separation of $10\,\mathrm{cm}^{-1}$. This separation is large enough for the peaks of BM2 and BM3 to be observed experimentally as separate peaks.

Bundles consisting of different number of identical or different nanotubes have also been studied [26]. In the latter case, there are generally different sets of breathing-like modes arising from the BM and the E_l modes of the different tubes. Therefore, many breathing-like modes are expected to exist, which will made the assignment of the low-frequency features in the LS spectra of bundles very difficult and ambiguous.

5. Breathing-Like Modes in Multiwalled Carbon Nanotubes

The synthesized carbon nanotubes often consist of many coaxial layers. These multilayer (or multiwalled) nanotubes (MWNTs) are usually found to have a certain distribution in the number and the radii of the layers. Due to this fact, the assignment of the low-frequency features in the LS spectra of MWNT samples is extremely complicated. To carry out successfully such an assignment, one needs theoretical predictions for the vibrational eigenmodes of MWNTs. As far as is known to the authors, no systematic theoretical study of the dynamical properties of MWNTs has so far been reported. From general arguments, it can be expected that the low-frequency phonons of the isolated layers will be modified significantly when the layers are stacked together in a MWNT and that the modification will be negligible for the high-frequency phonons. Again, as in the case of SWNTs, the breathing-like modes are of particular interest because intense LS peaks are expected to originate from them.

The breathing-like modes of MWNTs are calculated within a microscopic model in which a MWNT is considered as consisting of a number of coaxial SWNTs [28]. The model takes advantage of the periodicity of the nanotube so that it is restricted to commensurate layers only. The intralayer carbon–carbon interactions are described by force constants of the VFF type taken over from graphite. The structural parameters of the nanotube are not known with enough precision neither from experiment, nor from *ab initio* estimations. Therefore, it is assumed here that the bond lengths are the same as in graphene (i.e., 1.42 Å) and that all bond angles are equal. The interlayer interactions are modeled by a Lennard–Jones (LJ) potential [25, 26]. The intertube interaction energy is obtained by summing over all pairs of atoms belonging to different layers. This energy is then minimized with respect to the interlayer separations, the relative angles of rotation of the layers around the tube axis and the relative displacement of the layers along the tube axis. It is implied that the adjacent layers must be of types for which the initial radii difference is close to the optimized interlayer separation so that the change of the bond lengths is small. For the optimized tube, the dynamical matrix is calculated by use of the VFF parameters and the LJ potential. Finally, the vibrational modes of the tubes are derived as solutions of the dynamical eigenvalue problem. The LS spectra of MWNTs are calculated using a non-resonant bond-polarization model for backscattering geometry and parallel light polarization with averaging over all tube orientations in space [13]. For most of the MWNT samples the tube diameters are large enough so that the resonant scattering effect is much weaker than for SWNTs. For few-layer tubes, the resonant enhancement of some low-frequency lines may be observed.

The restrictions on the layer types limit the number of accessible tubes by the adopted lattice dynamical model. It can be noticed easily, however, that layers of armchair type $(n + 5m, n + 5m)$ for consecutive values of the integer m $(m = 1, 2, \dots)$ and for fixed values of the integer n $(n = 0, 1, 2, \dots)$ can be nested within each other to form MWNTs. The interlayer separation of the layers is $\approx$3.37 Å, which is close to that of the layers in graphite. We carried out structural optimization of such MWNTs and found that the optimal separations for two- and three-layer tubes do not deviate more that 0.1% from the initial layer separations. The corresponding difference between the frequencies of the breathing-like modes was below 1 cm^{-1}.

The calculated atomic displacements and frequencies of the breathing-like modes for two-layer tubes are shown in Fig. 11. It is seen in this figure that for the small-radius tube $(5, 5)@(10, 10)$, each of the two breathing-like modes has the characteristic features of the BM of one of the layers. This can be explained with elastic forces of the layers that are larger than the intertube ones yielding only weak mixing between the BMs of the isolated layers. For the large-radius tube $(20, 20)@(25, 25)$, the low-frequency breathing-like modes are in-phase and counter-phase collective motions of both layers which is due to large intertube forces and string hybridization between the BMs of the isolated layers. In some of the graphs in Fig. 11, the atomic displacements are not uniform radial motions rather they are strongly mixed due to lowering the symmetry of the system on bringing layers together to form a MWNT.

The results for the frequencies and the LS intensity of the breathing-like modes for many two-layer tubes are displayed in Fig. 12. The comparison

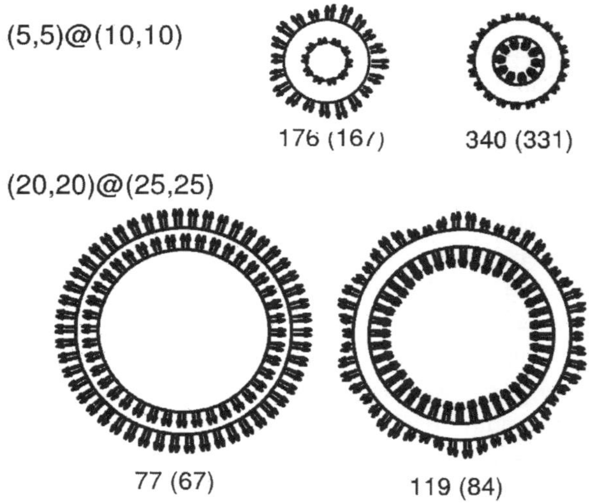

Fig. 11. Atomic displacements and frequencies for the two breathing-like modes of the two-layer tubes (5, 5)@(10, 10) and (20, 20)@(25, 25) [28]. The frequencies of the BMs of the isolated layers are given in parenthesis. The cross-sections of the two layers are given schematically by circles.

of the frequencies of the breathing-like modes (solid symbols) with the BM ones of isolated layers (empty symbols) reveals a systematic upshift of the former due to the intertube interactions. While this upshift is almost uniform of $\approx 10\,\mathrm{cm}^{-1}$ for the in-phase mode, it increases with the increase of the outer layer radius for the counter-phase mode. The dependence of the frequencies of the two breathing-like modes on the outer-layer radius R can be fitted with the power laws: $1167/R^{0.94}$ and $48127/R^{2.73} + 95.6$. The calculated LS intensity of the breathing-like modes of two-layer tubes is shown in the inset of Fig. 12. It is seen that the ratio of the intensity of the counter-phase and in-phase modes for small radius tubes is ≈ 0.5, which can be explained by the weak interlayer interactions and the volume ratio of the layers of 2. The independence of the two layers is supported by the fact that the intensity of BM of the isolated layers is nearly equal to that of the breathing-like modes. In the large-radius limit, the intensity ratio tends to zero.

The theoretical predictions for the breathing-like phonon frequency and LS intensity can be used for assignment of the low-frequency features in the LS spectra of MWNT samples. As is clear from the results above, such an assignment can be very difficult and ambiguous for samples with a

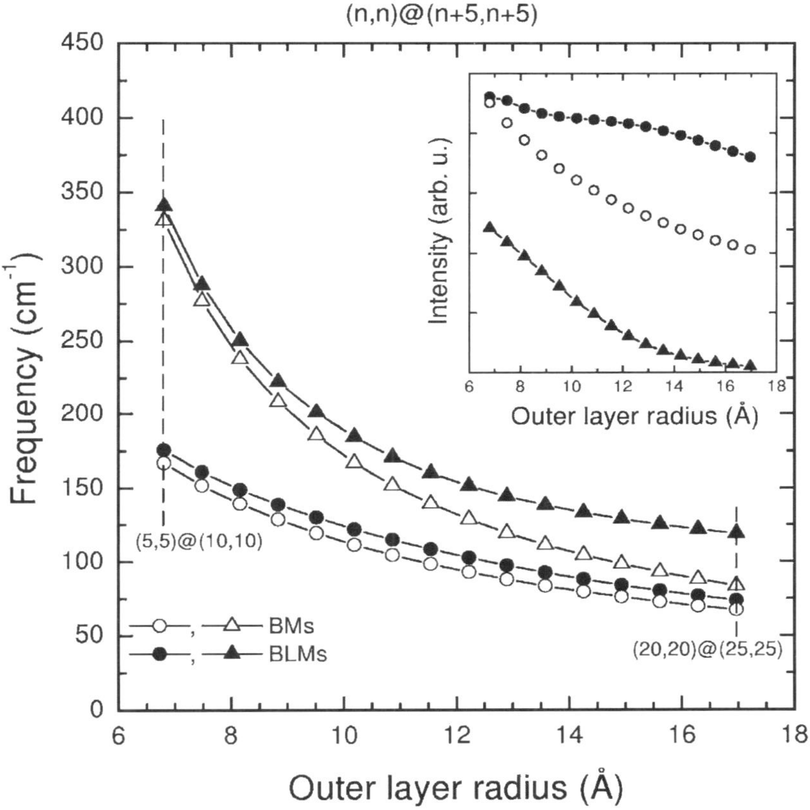

Fig. 12. Dependence of the frequencies of the breathing-like modes of two-layer tubes on the outer layer radius. The vertical dashed lines pass through the results for the tubes in Fig. 11 [28]. The inset shows the estimated LS intensity of the breathing-like modes (BLMs). The frequency and the intensity of the BMs of the isolated layers are depicted by empty symbols and those of the breathing-like modes by solid symbols.

broad distribution of the diameters of the layers and the number of layers. Recently, it has been possible to prepare double-walled nanotubes by heating of SWNTs filled with C_{60} molecules up to 1200°C to form an inner layer [29]. The measured LS spectrum of the sample shows many, well-resolved lines (see Fig. 3 in [29]) which were assigned using the $1/R$ law to BMs of the separate layers. It is interesting to check the validity of our predictions for the line positions and intensities although the observed lines are resonantly enhanced. In order to simulate the LS spectrum of the sample, first, it was taken into account that the inner layer fills maximum 2/3 of

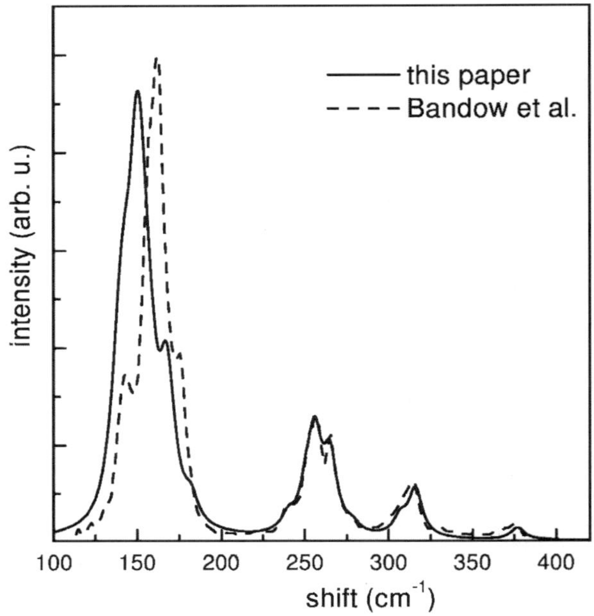

Fig. 13. The predicted LS spectrum [28] in comparison with the experimental spectrum of a double-walled nanotube sample prepared by heating of C_{60}-filled SWNTs up to 1200°C [29].

the space inside the SWNTs and that twice as many SWNTs are present as the obtained double-walled ones. Secondly, it was assumed that the lines below $\approx 200\,\mathrm{cm}^{-1}$ originate from BMs of SWNTs and in-phase breathing-like modes of double-walled tubes while the lines above $\approx 200\,\mathrm{cm}^{-1}$ are due to counter-phase breathing-like modes of double-walled tubes. Thirdly, the experimental spectrum was fitted by Lorentzians by using Fig. 12. As seen in Fig. 13 the theoretical spectrum agrees well with the experimental one apart from a missing line at about $143\,\mathrm{cm}^{-1}$. The overestimation of the separation of the in-phase and counter-phase bands can be explained by charge transfer between the layers accompanied by softening of the carbon bonds.

6. Elastic Properties of Carbon Nanotubes

Due to their specific structure, the nanotubes are expected to be as stiff as graphite along the graphene layers or even reach the stiffness of diamond. This unique mechanical property of the nanotubes combined with their light-weightiness predetermines their usage in composite materials and has motivated precise experimental measurements of their properties [30–33]. In [30], the temperature dependence of the vibration amplitude of several isolated MWNTs was analyzed in a TEM and the value of 1.8 TPa was obtained for the average Young's modulus. Later on, this technique was applied to measure Young's modulus of isolated SWNTs in the diameter range 1.0–1.5 nm and the average value 1.25–0.35/+ 0.45 TPa was derived [31]. In [32], the MWNTs were pinned to a substrate by conventional lithography and the force was measured at different distances from the pined point by atomic force microscope (ATM). The average Young's modulus for tubes with diameters from 26 to 76 nm was found to be 1.28 ± 0.59 TPa. Recently, Young's (Y) and shear (G) moduli of ropes of SWNTs were measured by suspending the ropes over the pores of a membrane and using ATM to determine directly the resulting deflection of the rope [33]. Up to now, the theoretical estimations of the elastic moduli have been accomplished by numerical second derivatives of the energy of the strained nanotubes. In the calculation of the elastic moduli of various SWNTs within a simple force-constant model [34] it was found that the moduli were insensitive to tube size and helicity and the average values of 0.97 TPa and 0.45 TPa were obtained for Y and G. In several works, molecular-dynamics simulation algorithms using the Tersoff–Brenner potential for the carbon–carbon interactions were implemented to relax the strained nanotubes and calculate their energy [14, 35, 36]. For SWNTs with diameter of 1 nm, values of 5.5 TPa [35] and 0.8 TPa [36] were obtained for Y. A non-orthogonal tight-binding scheme was applied to calculate Y of several chiral and achiral SWNTs yielding an average value of 1.24 TPa [37]. Recently, the second derivative of the strain energy with respect to the axial strain, calculated within a pseudopotential density-functional theory model for a number of SWNTs [12], was found to vary slightly with the tube type and to have the average value of 56 eV.

Here, we choose a different approach to the calculation of the elastic properties of SWNTs [22] in which analytical expressions for the elastic moduli are derived using a perturbation technique due to Born [38] within a lattice-dynamical model of nanotubes [13]. This scheme has the advantage that the moduli are consistent with the lattice dynamics of the nanotubes and that each of the moduli is obtained in one calculation step only.

In Born's perturbation technique, the dynamical matrix, the eigenvectors and eigenvalues belonging to the acoustic branches are expanded in power series in q. These expansions, substituted in the equation of motion for the translational unit cell, give rise to equations of zeroth, first and second order with respect to the perturbation parameter q.

Taking the non-trivial solution for the zeroth-order eigenvector of the form

$$e_\alpha^{(0)}(\mathbf{l}k) = \sqrt{m_k} u_\alpha, \tag{11}$$

where $\mathbf{u}$ is a constant vector, and solving zero, first and second-order equations we obtain the system of linear equations for u_α:

$$\rho v^2 u_\alpha = \sum_\beta \Lambda_{\alpha\beta} u_\beta, \tag{12}$$

where $\rho = N\Sigma_k m_k / V$ is the mass density of the tube, V is a (yet unspecified) "unit cell volume" and $v = \omega^{(1)}/q$ is the phase sound velocity. The matrix elements $\Lambda_{\alpha\beta}$ are defined by

$$\Lambda = \frac{1}{V}\left[\sum_{\mathbf{l}k\mathbf{l}'k'} \sqrt{m_k m_{k'}} D_{\alpha\beta}^{(2)}(\mathbf{l}k, \mathbf{l}'k') - \sum_{\mathbf{l}k\mathbf{l}'k'} \sum_{\mu\nu} \Gamma_{\mu\nu}(\mathbf{l}k, \mathbf{l}'k') \right.$$
$$\left. \times \sum_{\mathbf{l}''k''} \sqrt{m_{k''}}\, D_{\mu\alpha}^{(1)}(\mathbf{l}k, \mathbf{l}''k'') \sum_{\mathbf{l}'''k'''} \sqrt{m_{k'''}}\, D_{\nu\beta}^{(1)}(\mathbf{l}'k', \mathbf{l}'''k''') \right] \tag{13}$$

The matrices $D_{\alpha\beta}^{(1)}(\mathbf{l}k, \mathbf{l}'k')$ and $D_{\alpha\beta}^{(2)}(\mathbf{l}k, \mathbf{l}'k')$ are the first- and second-order dynamical matrices and the matrix $\Gamma_{\alpha\beta}(\mathbf{l}k,\mathbf{l}'k')$ is the inverse of the zeroth dynamical matrix $D_{\alpha\beta}^{(0)}(\mathbf{l}k, \mathbf{l}'k')$. This inversion cannot be performed directly because of the linear dependence of the elements of the latter. To carry out the inversion, following Born [38], we remove one row and one column of $D_{\alpha\beta}^{(0)}(\mathbf{l}k, \mathbf{l}'k')$ for each α and β from its x-y submatrix, invert the resulting matrix, and add rows and columns of zeros in the places of the removed ones.

The system of linear equations Eq. (12) has non-trivial solutions only for certain values of v that are the sound velocity of the transverse wave v_T:

$$v_T = \sqrt{\Lambda_{2,3}/\rho}, \quad v_L = \sqrt{\Lambda_1/\rho}, \tag{14}$$

where Λ_α $(\alpha = 1, 2, 3)$ are the eigenvalues of the matrix $\Lambda_{\alpha\beta}$.

Besides the non-trivial solution given in Eq. (11) the zero-order equations also have a non-trivial solution of the form

$$e_\alpha^{(0)}(\mathbf{l}k) = \sum_\nu \sqrt{m_k} \varepsilon_{\alpha z \nu} \theta_z R_\nu(\mathbf{l}k), \tag{15}$$

where, without loss of generality, $\alpha = x, y$; $\varepsilon_{\alpha\beta\gamma}$ is the Levy–Civita symbol, θ_z is an angle of rotation about z-axis. Proceeding as in [38], we obtain

$$\rho v^2 = \Lambda, \tag{16}$$

where

$$\Lambda = \frac{1}{V}\left\{\sum_{\alpha,\beta=x,y}\left[\sum_{\mathbf{l}k,\mathbf{l}'k'}\sum_{\gamma\delta}\sqrt{m_k m_{k'}}D^{(2)}_{\alpha\beta}(\mathbf{l}k,\mathbf{l}'k')\varepsilon_{\alpha z y}R_\gamma(\mathbf{l}k)\varepsilon_{\beta z\delta}R_\delta(\mathbf{l}'k')\right.\right.$$
$$-\sum_{\mathbf{l}k,\mathbf{l}'k'}\sum_{\mu\nu}\Gamma_{\mu\nu}(\mathbf{l}k,\mathbf{l}'k')\sum_{\mathbf{l}''k''\gamma}\sqrt{m_{k''}}D^{(1)}_{\mu\alpha}(\mathbf{l}k,\mathbf{l}''k'')\varepsilon_{\alpha zy}$$
$$\left.\left.\times R_\gamma(\mathbf{l}''k'')\sum_{\mathbf{l}'''k'''\delta}\sqrt{m_{k'''}}D^{(1)}_{\nu\beta}(\mathbf{l}'k',\mathbf{l}'''k''')\varepsilon_{\beta z\delta}R_\delta(\mathbf{l}'''k''')\right]\right\}. \quad (17)$$

The sound velocity of the torsional wave in the tube v_R can be deduced from Eq. (16) as

$$v_R = \sqrt{\Lambda/\rho}. \quad (18)$$

The microscopically derived sound velocities Eqs. (14) and (18) can be used to derive the Young's modulus and the shear modulus of the nanotube. For this purpose, we assume that a nanotube can be considered as an infinitely thin homogeneous cylinder with radius R and use the formula from the theory of elasticity [39]:

$$v_T^G = \sqrt{2Y/\rho}Rq = \sqrt{2}Rv_L q, \quad v_L = \sqrt{Y/\rho}, \quad v_R = \sqrt{G/\rho}. \quad (19)$$

Comparing Eqs. (14) and (18) to Eqs. (19), we identify Λ_1 and Λ as the Young's and shear moduli of the tube, respectively, and find that $\Lambda_{2,3}$ must be zero. Alternatively, Young's modulus can be determined from the expression for v_T in Eq. (19) and the transverse acoustic branches of the phonon dispersion curves.

The lattice dynamical model and the analytical expression for the sound velocities can now be applied to calculate the Young's and shear moduli of various SWNTs. Since in a force-constant model of the lattice dynamics it is not possible to carry out a real structural optimization, the structural data for the nanotubes has to be provided from the experiment or from theoretical studies.

The estimation of the elastic moduli of nanotubes requires the knowledge of the "unit cell volume" V of the tubes. There is no agreement between the different authors about the choice of the continuous model of a nanotube. Some of them consider a nanotube as a hollow cylinder with a certain wall thickness, e.g., 0.66 Å [35] or 3.4 Å [34] equal to the adjacent layer separation in graphite. Others choose a uniform cylinder with a cross-sectional area of πR^2 [36] or a prism — the unit cell in a crystalline rope of SWNT's with a cross-sectional area of $\sqrt{3}/2(2R+3.4)^2$ [33]. Recently, it was proposed to characterize the axial stiffness of a nanotube with the second derivative of the strain energy with respect to the axial strain per unit area of nanotube [37] or per atom of the tube [12]. In the latter case, the resulting quantity is equal to Young's modulus multiplied by the tube volume per atom v_a, so that it does not contain the ambiguous unit-cell volume. For this reason we adopt such a description of the elastic properties of the tubes for both axial and shear strains.

The in-plane elastic moduli, calculated for the graphene with the adopted VFF parameters, are compared to the corresponding experimental values for graphite in Table 1.

The agreement between these values is quite good and it may be expected that VFF parameters would allow for fair predictions of the elastic moduli of nanotubes as well. The Young's and shear moduli of SWNT's are calculated here using Eqs. (13) and (17) for various tube types: armchair tubes from $(3,3)$ to $(15,15)$, zigzag tubes from $(5,0)$ to $(25,0)$, and a number of chiral tubes [$(5,1)$, $(5,2)$, $(6,1)$, $(5,3)$, $(6,2)$, $(7,1)$, $(6,3)$, $(6,2)$, $(8,2)$, $(7,4)$, $(10,1)$, $(8,4)$, $(9,3)$, $(8,5)$, $(11,2)$, $(10,4)$, $(10,5)$, $(12,3)$, $(14,2)$, $(12,6)$, $(14,4)$, $(14,7)$, and (15, 6) in order of increasing tube radius]. The results for the moduli and the Poisson ratio v are displayed in Fig. 14.

We note that Y can be determined alternatively from the transverse acoustic branches of the dispersion curves fitted with a polynomial of second degree with respect to the wave number and the expression for the group sound velocity of the bending waves Eq. (19) leading to the same results as those obtained by using Eqs. (13) and (17). The results for Y presented in Fig. 14 show that for a given radius the Young's modulus for armchair tubes is slightly larger than for zigzag tubes and that for chiral tubes it has

Table 1. Experimental elastic constants (in GPa) and elastic moduli (in GPa/in eV) and Poisson ratio for graphite [40] in comparison with the calculated ones here [22]. The constant c_{66} is derived from the relation $c_{66} = (c_{11} - c_{12})/2$ for hexagonal symmetry. The last three columns contain the in-plane moduli $Y = (c_{11}^2 - c_{12}^2)/c_{11}$ and $G = c_{66}$, and Poisson's ratio $v = c_{12}/c_{11}$.

	c_{11}	c_{12}	c_{66}	Y	G	v
Exp. values [40]	1060	180	440	1029/56.43	440/24.13	0.17
Calc. values [22]	1047	219	414	1002/22.70	414/22.70	0.21

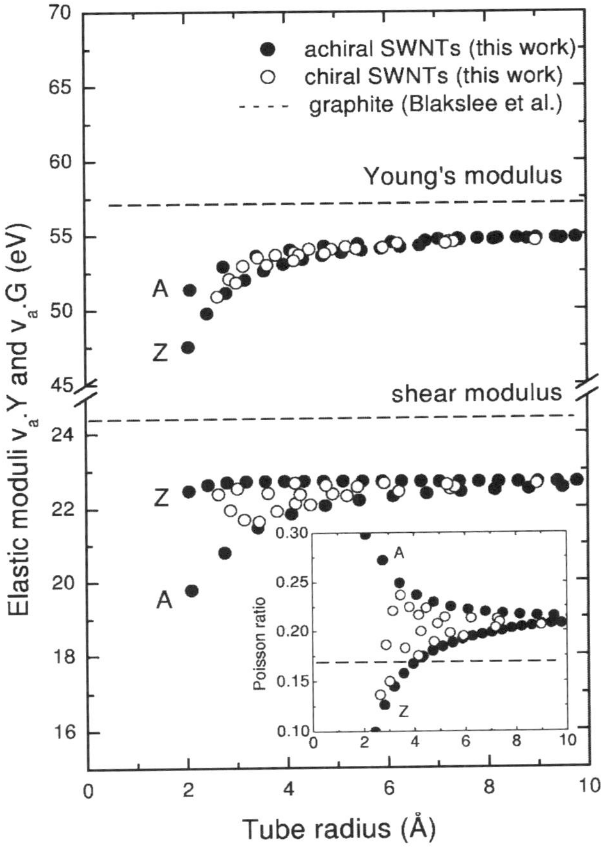

Fig. 14. Calculated Young's and shear moduli times the volume per atom of the tube v_a (in eV), and Poisson ratio estimated using the relation $v = (Y/2 - G)/G$ (inset) versus tube radius for various chiral and achiral SWNTs [22]. The letters A and Z stand for "armchair" and "zigzag," respectively.

intermediate values. As a whole, the Young's modulus is insensitive to the tube chirality and for large radii has values of about 55 eV that is about 3% smaller than the experimental one for graphite.

At small radii, the Young's modulus often tends to about 50 eV. The first molecular dynamics simulations [35] predict for the Young's modulus of a (10, 10) tube the value 59.4 eV, which differs only by few percent from our results. Recently, pseudopotential density-functional-theory calculations [12] of several SWNTs yielded an average Young's modulus of 56 eV.

The only available experimental point [31] is nearer the non-orthogonal tight-binding [37] and pseudopotential density-functional results [12] for the same tube radius but the force constant [34] results as well as those presented here are also within the experimental error of the former.

The shear modulus behaves similarly to the Young's modulus reaching values of about 23 eV for large radii but softening at small radii as shown in Fig. 14. The direct comparison of the obtained results with the experimental data for graphite (Table 1) reveals a symmetric deviation of about 6% for the shear modulus at large radii, which we attribute to the valence force field parameters of the model and to the initial assumptions. The shear moduli calculated for several tube types within the force constant model [34] appear to be insensitive to the tube radius and chirality and are about 15% higher than the ones obtained here.

Using Y and G, we can estimate the Poisson ratio, v, that is equal to the ratio of the relative radial expansion to the relative axial tube shortening, making use of the expression valid for the three-dimensional isotropic medium [30]: $v = (Y/2-G)/G$. The spread in the values of both moduli has as a consequence a spread in the values of Poisson ratio that is more prominent for small tube radii (see Fig. 14). In the limit of large radii, the Poisson ratio tends to 0.21, which is close to the experimental value for graphite (Table 1). The Poisson ratio estimated within the force constant model [34] is, in practice, a constant of 0.28 that is about 1.6 times larger than the in-plane value of graphite. A possible reason for the disagreement may be that the chosen model cannot properly describe the energy of radially strained tubes. The same behavior is exhibited by the tight-binding [37] results that range from 0.247 to 0.275. The recently calculated Poisson ratio by density functional theory model [12] varies from 0.12 to 0.19 for a number of tube types and, for a range tube radii, has values that are close to the experimental value for graphite.

In conclusion, the results for the elastic moduli and Poisson ratio estimated from analytical formula derived within a lattice-dynamical model for nanotubes, using force constants of the valence force field type are in fair agreement with the existing experimental data on graphite and nanotubes. These results compare well to the best results of more refined models — potential-based molecular dynamics, tight-binding and density-functional-theory models. The force constant model has the essential advantage to the latter models that it has a low-computational cost with respect to both computer memory and processing time. In particular the use of an analytical formula allows one to obtain the elastic moduli of a given tube in one calculation step only. Due to the large value of the Young's modulus along the tube axis, the single walled carbon nanotubes are materials of high stiffness. This property combined with their relatively small mass density makes them ideal ingredients for composites.

7. Elastic Properties of Crystals of Carbon Nanotubes

Most of the theoretical simulations of the elastic properties of nanotubes are concerned mainly with SWNTs. The experimentally observed SWNTs, however, most often form bundles of tens to hundreds of such tubes. It is also important for the technological application of the nanotubes to study the elastic properties of bulk SWNT materials. The bulk modulus of aligned SWNTs was estimated by using force constants [34] or the Tersoff potential [41] for the intratube interactions and Lennard–Jones type potentials for the intertube interactions. In both cases, a numerical second derivative of the energy of the strained tubes was used.

Here, we propose an alternative approach to the calculation of the elastic moduli and Poisson ratio of triangular close-packed crystal lattices of SWNTs [42]. First, the elastic constants of the lattice are obtained using analytical expressions [38] based on a force-constant lattice-dynamical model [13]. The main advantage of this scheme is that the elastic constants are consistent with the long-wavelength behavior of the lattice vibrations and that it allows for immediate results with reduced computational effort. Secondly, the elastic moduli and Poisson ratio are derived from the elastic constants by means of relations between them.

It has been known for several decades now that the elastic constants of crystalline solids can be estimated using a perturbation approach to the study of the long-wavelength atomic vibrations [38]. This approach is applied here to a force-constant lattice dynamical model [13], which is complemented with a Lennard–Jones type potential for the intertube atomic interactions, $V(\mathrm{r}) = 4\varepsilon[(\sigma/r)^{12} - (\sigma/r)^6]$ with parameters taken from [43]. The intertube atomic interactions are described by force constants of the valence force field type [22]. This work is based on the assumption that the tubes in the unstrained crystal are circular cylinders forming a triangular close-packed crystal lattice with lattice parameters $2R + l$, where R is the tube radius and l is the intertube separation. The assumption for rigidity of the tubes implies, in particular, that the effects of intertube van der Waals interactions on the equilibrium tube shape are ignored which, as will be shown, is justified for tubes with $R \leq 16\,\text{Å}$.

As a first step, we minimize the intertube interaction energy with respect to the intertube separation and the angle of simultaneous rotation of all tubes about their axes. The optimal intertube separation is roughly equal to 3.15 Å. Next the elastic constants are calculated using analytical expressions for the crystals consisting of achiral tubes: armchair tubes $(3n, 3n)$, $(n = 1, 2, \ldots, 13)$ or zigzag tubes $(3n, 0)$, $(n = 2, 3, \ldots, 18)$ with tube radii $2\,\text{Å} < R < 25\,\text{Å}$. We constrain ourselves to those tube types only because the tubes have six-fold symmetry and, consequently, the crystals have hexagonal symmetry with one-tube unit cells. Any crystal of tubes of other types will have a larger unit cell and the calculations will be rather time-consuming while the results will be fully predictable from those for achiral tubes.

In the case of hexagonal symmetry of the crystal lattice of single-walled carbon nanotubes, the Young's modulus and Poisson ratio are anisotropic. Let us denote the Young's moduli for directions along and perpendicular to the tube axis by $Y_{||}$ (longitudinal modulus) and $Y_{\perp}$ (transverse modulus), and the corresponding Poisson ratio by $v_{||}$ and $v_{\perp}$ (relative widening of the crystal under lateral forces). The Young's moduli, bulk modulus K and Poisson ratios are related to the elastic constants c_{ij} as shown in Table 2.

Although the calculations here show that c_{11}, c_{12}, $c_{13} \ll c_{33}$ so that simplified relations can be used (Table 2, last column), here we make use of the exact relations between the elastic constants, on the one hand, and the elastic moduli and Poisson ratio on the other. The calculated c_{11}, c_{12}, c_{13}, and c_{33} as a function of the tube radius are shown in Fig. 15. The values of a given elastic constant for both armchair and zigzag tubes lie on a smooth curve so that we might expect this to be true for tubes of arbitrary types as well. The constants c_{12} and c_{13} reach maximal values for $R \approx 6\,\text{Å}$ while c_{11} has only a kink there. Similar characteristic features can be observed in the radius dependence of the same moduli and Poisson ratios.

Table 2. Relations between the elastic moduli and Poisson ratio and the elastic constants in the case of hexagonal symmetry. The z-axis is chosen along the tube axis.

Moduli	General case	Case c_{11}, c_{12}, $c_{13} \ll c_{33}$
$Y_{||}$	$c_{33} - 2c_{13}^2/(c_{11} + c_{12})$	c_{33}
$Y_{\perp}$	$c_{13}/(c_{11} + c_{12})$	$c_{13}/(c_{11} + c_{12})$
$\nu_{||}$	$(c_{11} - c_{12})[(c_{11} + c_{12})c_{33} - 2c_{13}^2]/(c_{11}c_{33} - c_{13}^2)$	$(c_{11} - c_{12})(c_{11} + c_{12})/c_{11}$
$\nu_{\perp}$	$(c_{12}c_{33} - c_{13}^2)/(c_{11}c_{33} - c_{13}^2)$	c_{12}/c_{11}
K	$[(c_{11} + c_{12})c_{33} - 2c_{13}^2]/(c_{11} + c_{12} + 2c_{33} - 4c_{13})$	$(c_{11} + c_{12})/2$

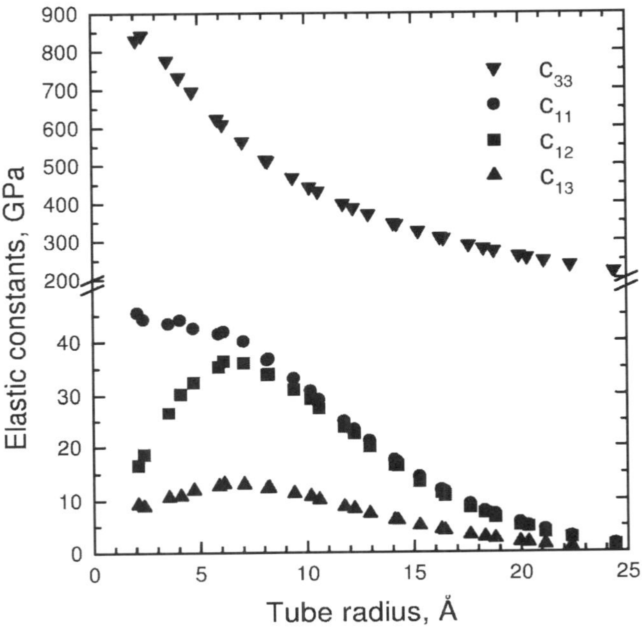

Fig. 15. Calculated elastic constants c_{11}, c_{12}, c_{13}, and c_{33} for crystals of achiral SWNTs versus tube radius R [42]. The constants c_{12} and c_{13} reach maxima at $R \approx 6$ Å while c_{11} has only a kink at this radius.

It is evident from Fig. 16(a) that the inequality $Y_{||} \gg Y_{\perp}$ holds for the calculated Young's moduli and, consequently, high mechanical anisotropy is predicted for the nanotube crystals. The longitudinal modulus $Y_{||}$ is equal to the elastic constant c_{33} within less than 1% and decreases inversely proportional to the tube radius for large radii. The transverse modulus $Y_{\perp}$ depends mainly on the elastic constants c_{11} and c_{12} and is found to decrease more rapidly with R. It is seen in Fig. 16(b) that the longitudinal Poisson ratio $v_{||}$ has values that are close to the in-plane one of graphite of 0.17. The calculated Poisson ratio $v_{\perp}$ increases steeply with the radius and becomes larger than 0.9 for $R > 6$ Å. Values of $v_{\perp}$ close to unity mean that under lateral forces the circular cross-section of the tube is deformed to an elliptic one with almost unchanged tube circumference. The change in the behavior of $Y_{\perp}$ and $v_{\perp}$ near $R \approx 6$ Å is due to the competition between the van der Waals interactions and the tube elasticity perpendicular to the tube axis, which will be discussed in more detail in the case of the bulk modulus.

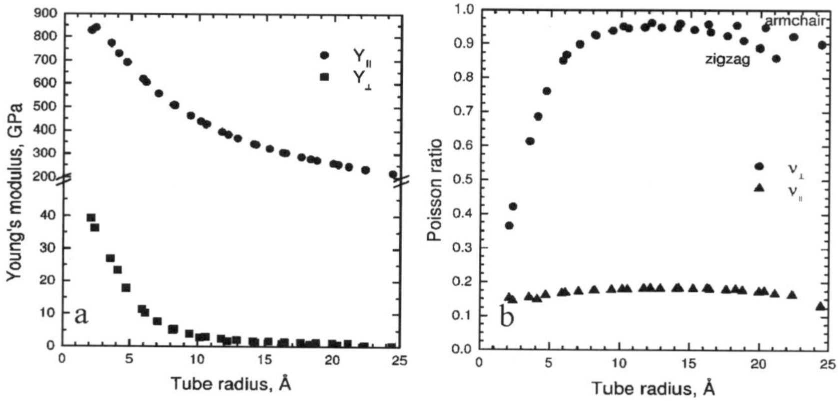

Fig. 16. Calculated (a) Young's moduli $Y_{||}$ and $Y_{\perp}$ and (b) Poisson ratios $v_{||}$ and $v_{\perp}$ for crystals of achiral SWNTs versus tube radius R [42]. Both Young's moduli decrease rapidly with the radius. For intermediate radii, $v_{||} \approx 0.17$ and $v_{\perp} \approx 0.9$.

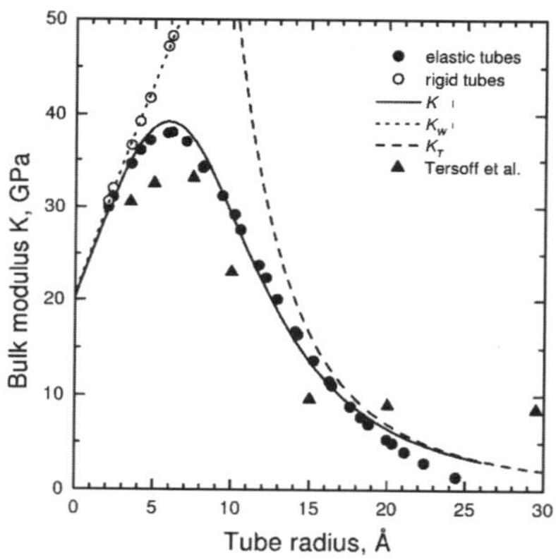

Fig. 17. Calculated bulk modulus K for crystals of achiral SWNTs versus tube radius R [42] in comparison with results from [41].

The calculated bulk modulus K (Fig. 17, solid circles) reaches a maximum at $R = 6\,\text{Å}$ and decreases to zero at large radii. This behavior of K is a consequence of the interplay of the intertube van der Waals forces and the elastic forces arising in laterally strained tubes. In the small-strain limit, these forces can be described by springs with force constants k_{W} and k_{T}, respectively, and the resulting interaction by a system of these two springs, connected in series, with a force constant $k = k_{\mathrm{W}}k_{\mathrm{T}}/(k_{\mathrm{W}} + k_{\mathrm{T}})$. Then, the strain energy per unit cell is $E = k(l - l_0)^2/2$(l_0 is the equilibrium intertube separation) and the bulk modulus is obtained in the form

$$K = v\frac{\partial^2 E}{\partial \nu^2} = \frac{\sqrt{3}}{6\tau}\frac{d^2 E}{dt^2} = \frac{\sqrt{3}}{6\tau}k, \tag{20}$$

where the unit cell volume is equal to $v = \sqrt{3}(2R + l)^2 T/2$ and T is the unit cell length. In view of Eq. (20) the bulk modulus can be split into two contributions, K_{W} and K_{T}, and expressed as

$$K = \frac{K_{\mathrm{W}} K_{\mathrm{T}}}{K_{\mathrm{W}} + K_{\mathrm{T}}}. \tag{21}$$

In order to study the effect of each of the two competing forces on the bulk modulus, we consider a crystal of rigid tubes with van der Waals interactions and calculate the moduli for the same set of tubes as above (Fig. 17, open circles). These results are fitted with a polynomial of second degree:

$$K_{\mathrm{W}} = 20.66 + 4.828R - 0.054R^2, \tag{22}$$

where R is in Å and K_{W} is in GPa (Fig. 17, dotted line).

The impact of elastic forces on K cannot be estimated in the same straightforward way. Instead, we take advantage of the result of the theory of elasticity [39] that for lateral strains of the tube due to concurrent forces, the force constant $k_{\mathrm{T}} \sim R^{-3}$ so that $K_{\mathrm{T}} \sim R^{-3}$ as well. The proportionality coefficient in the latter relation is derived by fitting Eq. (21) with K_{W} given by Eq. (22), to the results for elastic tubes (Fig. 17, solid line), which yields

$$K_{\mathrm{T}} = \frac{55509}{R^3}, \tag{23}$$

where R is in Å and K_{T} in GPa (Fig. 17, dashed line).

The degree in the radius dependence of K_{T} can be verified from the radius dependence of the tube phonon modes frequency ω, which are similar to the breathing modes but have an even number of nodes on the tube circumference. Our force-constant model yields power law $\omega \sim R^{-n}$ with $n = 1.995, 1.933, 1.883, 1.803, 1.828$ for the number of nodes equal to $4, 6, 8, 10, 12, 14, \ldots$. Such a radius dependence of ω is consistent with the power laws $k_{\mathrm{T}} \sim R^{-3}$ and $K_{\mathrm{T}} \sim R^{-3}$ used here, bearing in mind that the mass of the unit cell $m \sim R$. For radii $R > 16$ Å the lateral tube strain is not small even for lateral forces [39] and K_{T} decreases more rapidly with the tube radius nearly dropping to zero at $R \approx 25$ Å.

In a previous estimation of the bulk modulus [41] tubes were allowed relax in cylindrical shapes rather than the circular one. It was found that for tube radii larger than 15 Å, unstrained tubes flatten against each other due to van der Waals interactions and the tube cross-section deforms to a polygonized one rather than remaining circular. The polygonization increases the lateral rigidity of the lattice and yields a finite asymptotic value of K of about 10 GPa. The radius dependence of the bulk modulus is similar to that obtained here for $R \leq 16$ Å with values of K that are approximately 10% lower than our results because the intertube interactions are modeled in a different way. The onset of cross-sectional polygonization corresponds to the change of elastic behavior of the tubes observed here. Since tubes with radii larger than 16 Å are unlikely to be present in the samples synthesized in conditions reported in the literature, we conclude that this model yields as well as the one presented here, similar predictions for the elastic response of nanotube crystal under hydrostatic pressure. The bulk modulus calculated by Lu [34] is a monotonic function of the tube radius rather than displaying a maximum, which may be attributed to the use of an inadequate value of ε which is four times larger than the one obtained from the proper fitting to the graphite data and/or to the applied calculational procedure. The experimental measurements of the bulk modulus [44] carried out on samples of randomly oriented and tangled ropes with no fluid between them, yielded $K \approx 1$ GPa. Such a low value was explained by crushing and flattering of the tube cross-section to an elliptical shape under hydrostatic pressure thus providing additional mechanisms for volume reduction and decrease in tube elasticity. However, it may be expected that the experiments might yield a much larger bulk modulus if the tubes were immersed in a fluid for transmission of the hydrostatic pressure uniformly upon the surface of the nanotube ropes.

Here we present results of the calculation of the elastic constants c_{11}, c_{12}, c_{13}, and c_{33} for crystals of single-walled carbon nanotubes by means of analytical formulae derived within a lattice-dynamical model. Instead of using second derivatives of the strain energy with respect to certain strains, the Young's moduli $Y_{||}$ and $Y_{\perp}$, the corresponding Poisson ratio $v_{||}$ and $v_{\perp}$ and the bulk modulus K are expressed in terms of the obtained elastic constants. Some of the elastic constants, elastic moduli and Poisson ratios clearly exhibit three different regimes of behavior with respect to the tube radius R. In particular, the bulk modulus K is found to have a maximum value of 38 GPa for $R \approx 6$ Å — a result that may be of primary importance for future industrial applications of bulk nanotube materials.

8. Specific Heat of Carbon Nanotube Systems

The quasi-one-dimensionality of the nanotube systems has as a consequence the existence of four acoustic branches, which can result in a specific behavior of the phonon specific heat with temperature. The low-temperature specific heat of MWNTs, measured in the range from 10 to 300 K, was found to depend linearly on temperature [45]. The measured specific heat of bundles of SWNTs in the range from 1 to 200 K could not be modelled assuming linear acoustic branches [46]. The specific heat data of SWNT bundles from 2 to 300 K [47] were fitted with the theoretical curve from a two-band Debye model with linear acoustic dispersion. Recently, the specific heat of SWNT bundles was measured down to 0.1 K [48]. The data were fitted with the combination of power laws $0.043T^{0.62} + 0.035T^3$, where the first term could not be explained either with electronic, or with disorder contributions to the specific heat. The models considered for the description of the specific heat cannot provide a plausible explanation for the low-temperature (LT) dependence of the specific heat.

Here, the LT specific heat of SWNTs (isolated and bundled) and of MWNTs [49] is studied within force-constant dynamical models [13, 25, 42]. The results of the calculations of the specific heat are presented in the range below 100 K. The main contribution to the specific heat of nanotube systems is the vibrational one because the electronic one is negligible even at a few Kelvin [50]. In this study, the electronic specific heat is ignored and, therefore, the specific heat is given by

$$C(T) = k_\mathrm{B} \int \frac{(\hbar\omega/k_\mathrm{B}T)^2 \exp(\hbar\omega/k_\mathrm{B}T)}{[\exp(\hbar\omega/k_\mathrm{B}T) - 1]^2} D(\omega)\, d\omega, \qquad (24)$$

where $D(\omega)$ is the phonon density of states (PDOS). The high-temperature (or classical) limit of this expression does not depend on the particular structure of the carbon system and is equal to $3k_\mathrm{B}/m \approx 2078\,\mathrm{mJ/gK}$ with m being the atomic mass of carbon. The LT behavior of C is closely connected to the dimensionality of the system. For low enough temperatures, when the population of the lowest optical branches can be ignored, the specific heat is determined by the acoustic ones alone. If ω_o is the frequency of the lowest-energy optical phonon, then the optical phonons contribution to C can be ignored for temperatures below $T_\mathrm{o} \approx \hbar\omega_\mathrm{o}/6k_\mathrm{B}$ for which the factor multiplying $D(\omega)$ becomes smaller than 0.1 [50]. In the interval below T_o, $C(T)$ can be derived from the expression above once the acoustic-phonon dispersion is known. For the

3D system of graphite, for any of the three acoustic branches $\omega \sim q$, therefore, $D(\omega) \sim \omega^2$ and $C(T) \sim T^3$. For the 2D system of graphene, for the in-plane longitudinal acoustic (LA) and transverse acoustic (TA) phonons $\omega \sim q$, $D(\omega) \sim \omega$ and $C(T) \sim T^2$; for the out-of-plane acoustic (ZA) phonons $\omega \sim q^2$, $D(\omega) = \text{const.}$ and $C(T) \sim T$.

In the case of isolated SWNTs, within force-constant models [10, 51], tight-binding [11] and *ab initio* [12] approaches, linear q dependence of the acoustic modes frequency was obtained. It is now considered established that all four acoustic branches have linear q dependence [52]. In a systematic study of the elastic properties of isolated SWNTs [22] it is argued that the frequency of the transverse acoustic modes must be quadratic in q. This result has crucial consequences for the LT specific heat. Indeed, while for the LA and twist acoustic (TW) phonons $\omega \sim q$, $D(\omega) = \text{const.}$ and $C(T) \sim T$, for the TA phonons $\omega \sim q^2$, $D(\omega) \sim \omega^{-1/2}$ and $C(T) \sim T^{1/2}$. The applicability of these power laws depends on the value of ω_o. For example, for tubes (10, 10) $\omega_o \approx 20\,\text{cm}^{-1}$ and $T_o \approx 5$ K for tubes (10, 10). Therefore, $D(\omega)$ will show a singularity of the type $1/\sqrt{\omega}$ near $\omega = 0$.

In the case of bundles of SWNTs or MWNTs, the theoretical results for $C(T)$ of isolated SWNTs are valid. We study quantitatively the effect of bundling of SWNTs on the specific heat considering bundles of 1 to 7 tubes (9, 9). The theoretical predictions are compared to experimental data in Fig. 18. It is seen that the specific heat of isolated tubes exhibits three different regimes below $T = 100\,\text{K}$. At very low temperatures, only TA phonons are excited and $C(T) \sim T^{1/2}$ (slope 1/2 on the log–log plot). With the increase of T, the contribution of LA and TW phonons to C begins to prevail over that of the TA phonons, favoring $C \sim T$ (slope 1). Finally, above $T \approx 5\,\text{K}$, the optical phonons begin to contribute to the specific heat and its T dependence is modified again. We note that similar dependence may be expected for the thermal conductivity that is also mainly phononic and in a certain approximation is proportional to the specific heat. The $T^{1/2}$ part diminishes with the addition of tubes to the bundle due to the fact that the slope of the TA branch (i.e., the group velocity) is proportional to the radius of the bundle (see, Eq. (19)). Consequently, with the increase with the bundle lateral size, the relative contribution of the TA branches at a given temperature decreases. In the limit of infinite lateral size, the specific heat of the bundle is expected to have the behavior of a 3D system, i.e., $C \sim T^3$. It is seen in Fig. 18 that with the increase of the number of tubes in the bundle, the theoretical curve tends to the experimental data of

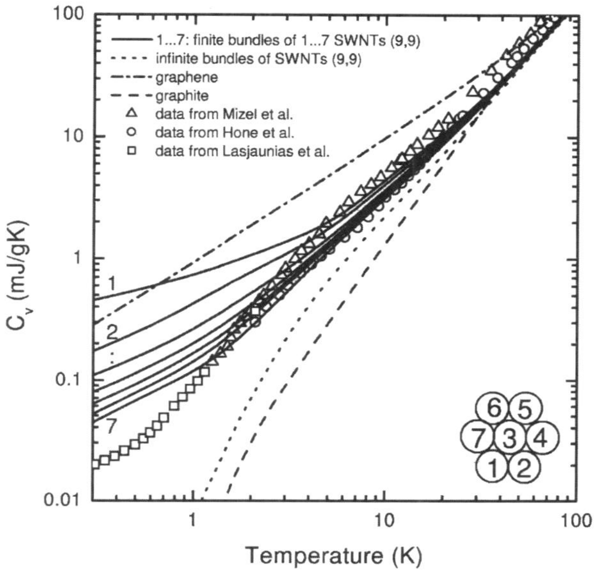

Fig. 18. Calculated specific heat of finite bundles of SWNTs of 1 to 7 tubes (9, 9), infinite bundles, graphene, and graphite [49] in comparison with available experimental data [46–48].

[46, 47]. Our predictions agree well with recent data on bundles measured down to 0.1 K and fitted with $0.043T^{0.62} + 0.035T^3$ [48]. The power of the first term can only be explained with contributions of the acoustic branches with linear (LA and TW branches) and quadratic (TA branches) dispersion.

The specific heat of MWNTs is expected to have similar regimes as SWNT bundles. To verify this, calculations of $C(T)$ were carried out for MWNTs with 1 to 5 layers of the type $(5m, 5m)$, $m = 1, 2, \ldots, 5$. It can be seen in Fig. 19 that starting from a single layer (5, 5) and adding more layers, the $T^{1/2}$ part diminishes and disappears and is replaced by a linear T dependence. The part of the $C(T)$ curve with predominant contribution of TA phonons depend again on the tube radius R since the slope of the TA branch is proportional to R. In the limit of infinite tube radius, the specific heat should behave as that of a 3D system with $C \sim T^3$. The theoretical curves disagree with the experimental data possibly because of the reduced interlayer coupling [45] or presence of MWNTs with a large number of layers [46].

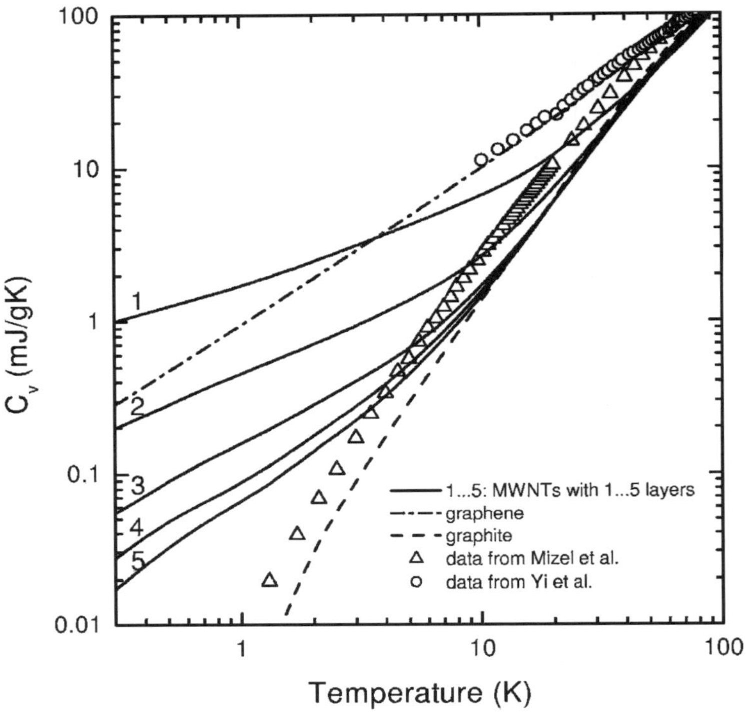

Fig. 19. Calculated specific heat of MWNTs consisting of tubes $(5m, 5m)$, $m = 1, 2, \ldots, 5$ [49] in comparison with experimental data [45, 46].

9. Conclusions

The presented study encompasses theoretical models and simulations of the vibrational, elastic and thermal properties of various nanotube systems. Special attention is paid to the low-frequency phonons and, in particular, to the breathing-like phonons that give rise to high light-scattering peaks and can serve as markers for sample characterization. The elastic and low-temperature thermal properties of nanotubes are dependent on the acoustic phonon dispersion near the zone center. It is argued here that, while the LA and TW phonons have a linear dispersion, the TA phonons have a quadratic dispersion. This behavior is shown to have crucial consequences for the elastic and thermal properties of nanotubes. The results of the calculations are discussed in comparison with available experimental data.

Acknowledgments

V.N. Popov, was partly supported by a scholarship from the Belgian Federal Science Policy Office for promoting the S&T co-operation with Central and Eastern Europe and by a Marie-Curie Intra-European Fellowship.

References

[1] Iijima, S., *Nature (London)* **354** (1991) 56–58.
[2] Iijima, S. and Ichihashi, T., *Nature (London)* **363** (1993) 603–605.
[3] Bethune *et al.*, *Nature (London)* **363** (1993) 605–607.
[4] Thess, A. *et al.*, *Science* **273** (1996) 483–487.
[5] Fischer, J. *et al.*, *Phys. Rev. B* **55** (1997) R4921–4924.
[6] Rao, A.M. *et al.*, *Science* **275** (1997) 187–191.
[7] Wildöer, J.W.G. *et al.*, *Nature (London)* **391** (1998) 59–62.
[8] Odom, T.W. *et al.*, *Nature (London)* **391** (1998) 62–64.
[9] Jishi, R.A., Venkataraman, L., Dresselhaus, M.S. and Dresselhaus, G., *Chem. Phys. Lett.* **209** (1993) 77–82.
[10] Saito, R., Takeya, T., Kimura, T., Dresselhaus, G. and Dresselhaus, M.S., *Phys. Rev. B* **57** (1998) 4145–4153.
[11] Yu, J., Kalia, R.K. and Vashishta, P., *J. Chem. Phys.* **103** (1995) 6697–6705.
[12] Sánchez-Portal, D., Artacho, E., Soler, J.M., Rubio, A. and Ordejón, *Phys. Rev. B* **59** (1999) 12678–12688.
[13] Popov, V.N., Van Doren, V.E. and Balkanski, M., *Phys. Rev. B* **59** (1999) 8355–8358.
[14] Robertson, D.H., Brenner, D.W. and Mintmire, J.W., *Phys. Rev. B* **45** (1992) 12592–12595.
[15] Hamada, N., Sawada, S.–I. and Oshiama, A., *Phys. Rev. Lett* **68** (1992) 1579–1581.
[16] Saito, R., Fujita, M., Dresselhaus, G. and Dresselhaus, M.S., *Phys. Rev. B* **46** (1992) 1804–1811.
[17] White, C.T., Robertson, D.H. and Mintmire, J.W., *Phys. Rev. B* **47** (1993) R5485–5488.
[18] Kürti, J., Kresse, G. and Kuzmany, H., *Phys. Rev. B* **58** (1998) R8869–8872.
[19] Henrard, Hernandez, E., Bernier, P. and Rubio, A., *Phys. Rev. B* **60** (1999) R8521–8524.
[20] Yi, J.-Y. and Bernholc, *Phys. Rev. B* **47** (1993) R1708–1711.
[21] Aizawa, T., Souda, R., Otani, S., Ishizawa, Y. and Oshima, C., *Phys. Rev. B* **42** (1990) 11469–11478.
[22] Popov, V.N., Van Doren, V.E. and Balkanski, M., *Phys. Rev. B* **61** (2000) 3078–3084.
[23] Venkateswaran, U.D. *et al.*, *Phys. Rev. B* **59** (1999) 10928–10934.
[24] Kahn, D. and Lu, J.P., *Phys. Rev. B* **60** (1999) 6535–6540.
[25] Popov, V.N. and Henrard, L., *Phys. Rev. B* **63** (2001) 233407-1/4.
[26] Henrard, L., Popov, V.N. and Rubio, A., *Phys. Rev. B* **64** (2001) 205403-1/10.

[27] Dresselhaus, M.S. and Eklund, P.C., *Adv. Phys.* **49** (2000) 705–814.
[28] Popov, V.N. and Henrard, L., *Phys. Rev. B* **65** (2002) 235415-1/6.
[29] Bandow, S., Takizawa, M., Hirahara, K., Yudasaka, M. and Iijima, S., *Chem. Phys. Lett.* **337** (2001) 48–54.
[30] Treacy, M.M.J., Ebbese, T.W. and Gilson, J.M., *Nature (London)* **381** (1996) 678–680.
[31] Krishnan, A., Dujardin, E., Ebbesen, T.W., Yianilos, P.N. and Treacy, M.M.J., *Phys. Rev. B* **58** (1998) 14013–14019.
[32] Wong, E.W., Sheehan, P.E. and Lieber, C.M., *Science* **277** (1997) 1971–1975.
[33] Salvetat, J.-P. *et al.*, *Phys. Rev. Lett.* **82** (1999) 944–947.
[34] Lu, J.P., *Phys. Rev. Lett.* **79** (1997) 1297–1300.
[35] Yakobson, B.I., Brabec, C.J. and Bernholc, J., *Phys. Rev. Lett.* **76** (1996) 2511–2514.
[36] Cornwell, C.F. and Wille, L.T., *Solid State Commun.* **101** (1997) 555–558.
[37] Hernández, E., Goze, C., Bernier, P. and Rubio, A., *Phys. Rev. Lett.* **80** (1998) 4502–4505.
[38] Born, M. and Huang, K., *Dynamical Theory of Crystal Lattices* (Oxford University Press, Oxford, 1954).
[39] Landau, L.D. and Lifshitz, E.M., *Theory of Elasticity* (Pergamon, Oxford, 1995).
[40] Blakslee, O.L., Proctor, D.G., Seldin, E.J., Spence, G.B. and Weng, T., *J. Appl. Phys.* **41** (1970) 3373–3382.
[41] Tersoff, J. and Ruoff, R.S., *Phys. Rev. Lett.* **73** (1994) 676–679.
[42] Popov, V.N., Van Doren, V.E. and Balkanski, M., *Solid State Commun.* **114** (2000) 395–399.
[43] Lu, J.P. and Yang, W., *Phys. Rev. B* **49** (1994) 11421–11424.
[44] Chesnokov, S.A., Nalimova, V.A., Rinzler, A.G., Smalley, R.A. and Fischer, J.E., *Phys. Rev. Lett.* **82** (1999) 343–346.
[45] Yi, W., Lu, L., Zhang, D.-L., Pan, Z.W. and Xie, S.S., *Phys. Rev. B* **59** (1999) R9015–9018.
[46] Mizel, A. *et al.*, *Phys. Rev. B* **60** (1999) 3264–3270.
[47] Hone, J., Batlogg, B., Benes, Z., Johnson, A.T. and Fischer, J.E., *Science* **289** (2000) 1730–1733.
[48] Lasjaunias, J.C., Biljaković, K., Benes, Z., Fischer, J.E. and Monceau, P., *Phys. Rev. B* **65** (2002) 113409–1/4.
[49] Popov, V.N., *Phys. Rev. B* **66** (2002) 153408-1/4.
[50] Benedict, L.X., Louie, S.G. and Cohen, M.L., *Solid State Commun.* **100** (1996) 177–180.
[51] Maultzsch, J. *et al.*, *Solid State Commun.* **121** (2002) 471–474.
[52] Dresselhaus, M.S., Dresselhaus, G. and Avouris, Ph., Carbon Nanotubes: Synthesys, Structure, Properties and Applications, in *Topics in Applied Physics* Vol. 80 (Springer–Verlag, Berlin, 2001).

CHAPTER 8

GLASSY BEHAVIOR DUE TO KINETIC CONSTRAINTS: FROM TOPOLOGICAL FOAM TO BACKGAMMON

David Sherrington[a]

Department of Physics, Theoretical Physics, University of Oxford, 1 Keble Road, Oxford, OX1 3NP, UK

A study is reported of a series of simple model systems with only non-interacting Hamiltonians, and hence simple equilibrium thermodynamics, but with constrained kinetics of a type initially suggested by topological considerations of foams and two-dimensional covalent glasses. It is demonstrated that macroscopic dynamical features characteristic of real glasses, such as two-time decays in energy and auto-correlation functions, arise and may be understood in terms of annihilation-diffusion concepts and theory. This recognition leads to a sequence of further models which (i) encapsulate the essense but are more readily simulated and open to easier analytic study, and (ii) allow generalization and extension to higher dimension. Fluctuation–dissipation relations are also considered and show novel aspects. The comparison is with strong glasses.

1. Introduction

Glasses are amorphous solid-like systems, produced when liquids are supercooled at a rate which is too fast to permit equilibration or crystallization on normal time scales. They are characterized by a combination of fast and slow temporal evolution; for example, by correlation functions having a fast initial (β) decay, followed by a plateau and then a slow (α) decay whose

[a]This chapter reports a collection of studies done in collaboration with Tomas Aste, Arnaud Buhot, Lexie Davison and Juanpe Garrahan.

effective onset becomes later and slower either with lower temperature in an equilibrated system or as a function of the time since quench (from the liquid state) in a non-equilibrated scenario. Conventional real glasses have interactions between their atoms or molecules, but in this paper it is shown that such behavior occurs due to purely kinetic constraints in some simple many-body systems with non-interacting Hamiltonians.

The idea that glassiness can occur due to kinetic constraints is not new, and a recent review of such studies is given in [1], but the formulation of such models has been mainly heuristic rather than devised by a formal transformation of variables. Here are considered a sequence of minimalist models [2–5] inspired by an idealization of a covalent network glass and a topological foam. They exhibit the classic features of a glass mentioned above and can be understood in terms of annihilation-diffusion processes. Within the nomenclature of Angell [6] they are strong glasses, having Arrhenius long-time-relaxation behavior, but also, unusually, in the final distillation can be studied within mean-field theory while still exhibiting this behavior.

2. Topological Network Models

The first model we consider [2, 3] is based on the topology of a foam; i.e., is characterized by a fully connected network of three-armed vertices, as shown in Fig. 1(a). By Euler's theorem, the average number of edges per cell is six. The crystalline form is hexagonal. To ensure the latter as ground state in a minimalist manner, one may choose an energy function

$$E = \sum_{i=1}^{N} (6 - n_i)^2, \tag{1}$$

where $i = 1, \ldots, N$ label the cells, and n_i refers to the number of sides of cell i. Clearly such a model has trivial thermodynamics and no finite temperature phase transition. The interest comes from the dynamics. Again we choose as simply as possible, permitting only T1 moves as illustrated in Fig. 1(b); these conserve the total number of cells, edges and vertices. Thermal effects are introduced by making the T1 moves stochastic with acceptance probabilities determined by $\mathrm{Min}[1, \exp(-\Delta E/T)]$, where ΔE is the energy change which would ensue. To avoid unphysical features, moves which would produce two-sided or self-neighboring cells are forbidden.

This model can be related to an idealization of a two-dimensional covalently-bonded glass, for which the vertices of Fig. 1(a) are sp^2 hybrids

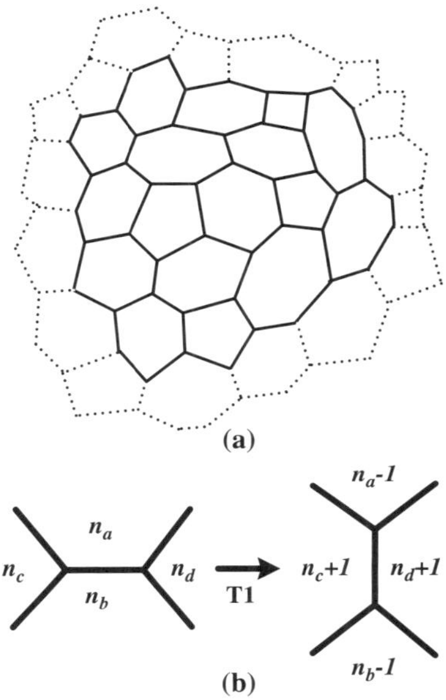

Fig. 1. (a) A topologically stable cellular partition (froth). (b) A T1 move. (From [2].)

and the edges are the covalent bonds. The preferred angle between the sp^2 lobes is $2\pi/3$ and, in a harmonic approximation which ignores correlations between vertex angles, a perturbation to an angle θ costs an energy $\sim (\theta - 2\pi/3)^2$. If further each θ is approximated by the average angle within the cell containing it, namely $(4\pi/n)$ in a cell of n sides, there results the total energy

$$E \sim \sum_{i=1}^{N} (6 - n_i)^2/(6n_i)^2. \tag{2}$$

At low temperatures n remains close to six so that the variation in the denominator is secondary to that in the numerator and its ignorance leads to Eq. (1). The T1 process corresponds to locally changing the interatomic bond connections.

Simulations [2, 3] show that the system equilibrates easily at high temperatures (the "liquid state") but below a temperature of order unity (in the above units) it exhibits glassy behavior of the type discussed in the introduction, namely rapid equilibration at high temperatures but with severe slowing-down before equilibration at lower temperatures and correspondingly in the low-temperature regime the cell side-number temporal auto-correlation dropping rapidly to a plateau with only much slower eventual decay to its asymptotic limit. This behavior can be understood in terms of a picture of annihilating and diffusing defects. These defects are cells which deviate from the ground state values $n_i = 6$, i.e., of non-zero topological charge $q_i = (6 - n_i)$. A T1 process increases the topological charges on each of the initially adjacent cells and decreases the charges on the other two cells (which become adjacent). By so doing it provides mechanisms of defect annihilation, creation and diffusion. At low temperatures only $q = \pm 1, 0$ are present in significant numbers and hence we restrict explicit consideration to these (but generalization is easy). A brief consideration then convinces one that (i) energy reduction is only possible through the annihilation of a pair of adjacent oppositely charged topological charges, i.e., a dimer, in a process in which the other two cells either comprise an oppositely oriented dimer or one has an appropriate non-zero and the other a zero charge, (ii) a pair of adjacent opposite charges in a zero background can move without energy cost, and (iii) an isolated non-zero charge can move only by increasing the overall energy, metamorphosing into three non-zero charges on neighboring sites, two of the same sign and one opposite to that originally present. Since no energy cost is involved the first two of these processes can occur even at zero temperature with the microscopic time scale, while the third is activated with a time scale of Arrhenius form, which becomes large as the temperature is lowered.

3. Lattice Models

3.1. *Hexagonal lattice*

The changing topologies of the "foam" model complicate both simulation and theoretical analysis. Hence, to simplify, we pass to a lattice analogue [4] which consists of a set of 3-state "spins" $s_i = 0, \pm 1$ on the cells of a hexagonal lattice, with energy function

$$E = D \sum_{i=1}^{N} s_i^2 \tag{3}$$

and Euler's theorem emulated by $\sum_i s_i = 0$. The analog of a T1 process consists of picking an edge on the hexagonal lattice and randomly increasing (decreasing) by one the spins on the adjacent cells and decreasing (increasing) by one those on the cells at its ends; see Fig. 2. As before, these moves

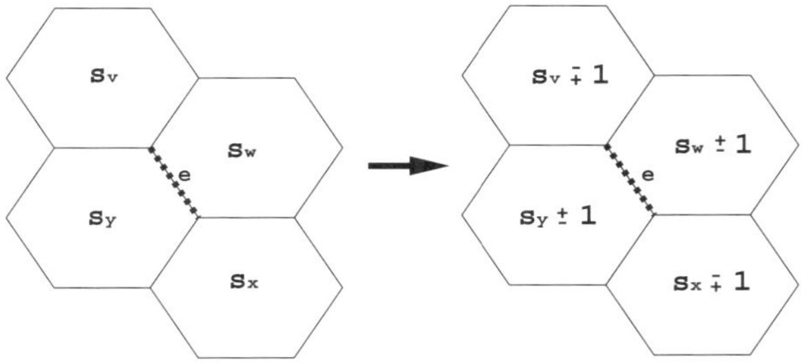

Fig. 2. The analog of a T1 move for the hexagonal lattice model. The upper or lower signs are chosen randomly at each attempted move, consistently for all four cells. (From [4].)

are executed stochastically with an acceptance probability determined by $\mathrm{Min}[1, \exp(-\Delta E/T)]$, with moves which would place a spin outside the range $0, \pm 1$ forbidden.

3.1.1. $D > 0$

The original topological foam model is emulated by $D > 0$, for which the ground state is unique. Figure 3 shows results of simulations together with theoretically-inspired fits, to be explained below, respectively for (i) the energy as a function of time for a system started from a random high-temperature state but evolving with Metropolis–Kawasaki dynamics corresponding to a range of low temperatures, and (ii) the autocorrelation

$$C(t_w, t_w + t) = \frac{\sum_{i=1}^{N} s_i(t_w) s_i(t_{w+t})}{\sum_{i=1}^{N} s_i^2(t_w)} \tag{4}$$

for systems at various temperatures and in equilibrium (so that there is no dependence on t_w). In both cases one has an initial rapid decay, followed by a plateau and then a slow temperature-dependent decay. As Fig. 3(a) shows clearly, the time scale of the initial decay is not temperature-dependent. The slower decay from the plateau has an Arrhenius characteristic time scale.

These results can be understood in terms of annihilation-diffusion modelling. Let us start by associating two particle types A, B and a null-state $\emptyset$, with the "spin" states $s = 0, \pm 1$ as $\emptyset, A, B$. In this language, annihilation involves moves expressible as

$$2A + 2B \to \emptyset, \tag{5}$$

$$2A + B \to A, \quad A + 2B \to B \tag{6}$$

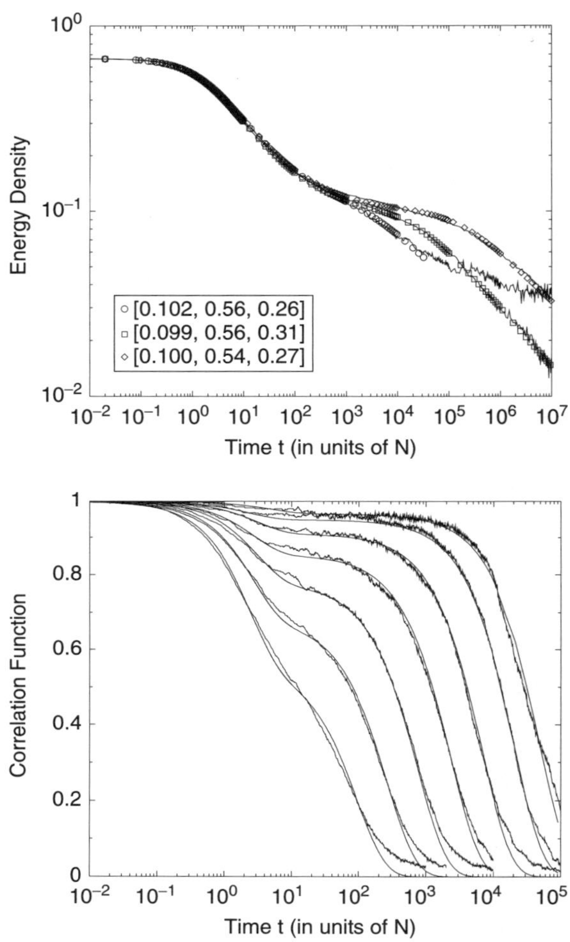

Fig. 3. (a) The energy density E/N of the hexagonal system with $D > 0$ as a function of time after the quench for inverse temperatures $\beta = 4, 5, 6$ (from left to right), fitted with Eq. (10). (b) The equilibrium auto-correlation function as a function of time for, from left to right, $\beta = 3, 3.5, 4, 4.5, 5, 5.5, 6$. The solid lines superimposed are fits of the form of Eq. (11). (From [4].)

over appropriately inter-configured sets of four sites and choices of ± 1 in the moves of Fig. 2.[b] Let us consider the fast processes first. Here the characteristic time scale is the microscopic one (2 in attempt time units). Fast

[b]Recall that in the lattice analog of T1 moves one chooses randomly the ± 1 combinations, but in any particular instance of the initial state only one of these two choices can proceed.

diffusion occurs for AB or BA neighbor dimers moving into a neighboring $2\emptyset$. The processes which drive the initial decays of the energy in Fig. 3(a) are the combination of dimer diffusion and dimer annihilation. The plateau corresponds to where essentially all the dimers have been eliminated by these processes; more precisely, at finite temperature there remain some dimers due to thermal excitation but they are negligible on the scale and at the temperatures shown. In the case of the equilibrium $C(t)$ (Fig. 3(b)) the initial fast decay is due to dimer diffusion alone, movement of a spin altering its autocorrelation and reducing C even though the total dimer (and singleton) density remains unchanged.

The slower decays from the plateaux require the motion of isolated defects (± 1 spins). This occurs by the inverse process to that of Eq. (6) with a time scale $\exp(-2\beta)$ times that of the fast process; the dimers created in this process diffuse quickly and annihilate so that they can be effectively ignored. For $C(t)$ the motion of isolated defects is itself sufficient to affect the value, whereas for $E(t)$ to decay the initially separated defects must also pair up as $(+\ -)$ dimers which then diffuse quickly and annihilate; these dimers can again be effectively ignored. We are therefore left with a picture of the slow process as one involving effective $\tilde{A}, \tilde{B}$ particles diffusing with the slow Arrhenius time scale and annihilating via

$$\tilde{A} + \tilde{B} \to \emptyset \,. \tag{7}$$

We are now in a position to utilize results from the field theory of annihilation-diffusion processes [7, 8] to provide fits to the simulations. Considering $E(t)$ first, this can be related to the densities of particles in the usual annihilation-diffusion theory. For the effective slow particles $\tilde{A}$ and $\tilde{B}$ standard annihilation-diffusion field theory can be imported directly with $E(t)$ decaying asymptically as $(t/\tau_2)^{-1/2}$, where τ_2 goes as $\exp(2\beta)$.

For the fast processes, the analogy is somewhat more complex but nevertheless useful. In usual annihilation-diffusion studies the characteristic diffusion is that of the fundamental particles. Here in the fast region the effective diffusing particles are the dimers, which are of two types corresponding to $(AB), (BA)$, while the annihilation processes are

$$2A + 2B \to 4\emptyset, \tag{8}$$

$$2A + B + \emptyset \to 3\emptyset + A, \quad 2B + A + \emptyset \to 3\emptyset + B. \tag{9}$$

There are clearly several ways to view the "dimers" in these processes. However, it is natural to anticipate the aymptotic field-theory behavior

$(t/\tau_1)^{-\alpha}$, where τ_1 is the microscopic time. Further considerations of the identifications suggest $\alpha = 1/2$.

Hence, one is we left with the "predicted" form for $E(t)$:

$$\frac{E(t)}{N} = \left(\frac{2}{3} - a\right)\left(1 + \frac{t}{2}\right)^{-b} + (a - e_{\mathrm{eq}})\left(1 + \frac{t}{e^{2\beta}}\right)^{-c} + e_{\mathrm{eq}}, \qquad (10)$$

where a is the plateau value, e_{eq} is the energy per spin in equilibrium and we expect both b and c to be close to 0.5. The fit values are shown in the key to Fig. 3(a) and are seen to be in very good accord with expectations. Correspondingly for the correlation function the "prediction" is

$$C(t) = ae^{-t/\tau_1} + (1 - a)e^{-t/\tau_2}. \qquad (11)$$

3.1.2. $D < 0$

In the lattice analog one can also consider $D < 0$, although this no longer emulates the original foam model. In this case the ground state is highly degenerate, any site having its S arbitrarily ± 1. Nevertheless, $E(t)$ and $C(t)$ again have the same form of fast decay to a plateau followed by slow decay characterized by an Arrhenius time scale. There are however important differences of details: one is that there is now only a single defect type, 0; another is that the 00 dimers cannot move so easily through the ± 1 background as was the case of the $+, -$ dimers in the $\emptyset$ non-degenerate background of the case $D > 0$. Because of the single type of defect, the slower processes are now of the type

$$\tilde{A} + \tilde{A} \to \emptyset\,. \qquad (12)$$

In a free background this would be expected to yield a decay as $(t/\tau_2)^{-d/2}$, but in fact a fit to $(t/\tau_2)^{-\kappa}$ yields $\kappa \sim 0.6$, possibly due to the hindering of dimer motion by the $\pm$ ground state background. Similarly, in $C(t)$ the slow decay is better fit with a stretched exponential $\sim \exp(-(t/\tau_2)^{\gamma})$ (with $\gamma \sim 0.8$).

3.2. *Square lattice*

Although the hexagonal lattice "matches" the original topological foam model in its general structure and vertex character, the annihilation-diffusion picture does not require it. Hence, it is interesting to simplify the model further, while hopefully retaining the fundamental essentials.

One natural simplification is to a square lattice [4], again with $s_i = 0, \pm 1$ associated with the cells. We replace the "Feynman-diagram" T1 process

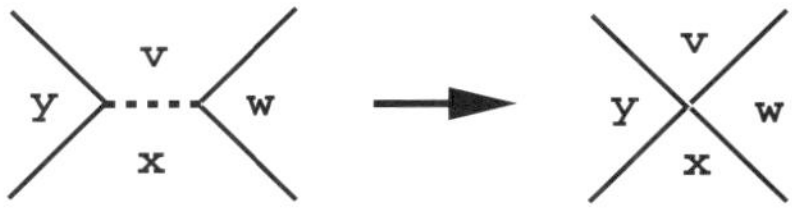

Fig. 4. The square lattice (right) can be considered equivalent to a hexagonal lattice (left) in which the central bond, denoted by a dashed line, has been shrunk to a point [9].

by one involving four cells around a four-vertex, as illustrated in Fig. 4. $E(t)$ and $C(t)$ still show the same characteristic behavior (Fig. 5) and the annihilation-diffusion explanations continue to apply.

4. Summary so Far and Encapsulation

Clearly all the processes discussed so far capture the same essence, at least qualitatively.[c] Let us therefore recall the common features of the models so far and look to simplify and extend modelling and analysis further through appropriate encapsulation.

Schematically the key processes we have been considering are

(i) annihilation of dimers via

$$2A + B + \emptyset \to 3\emptyset + A, \tag{13}$$

$$2B + A + \emptyset \to 3\emptyset + B, \tag{14}$$

$$2A + 2B \to 4\emptyset; \tag{15}$$

(ii) diffusion of dimers via

$$A + B + 2\emptyset \to 2\emptyset + A + B; \tag{16}$$

(iii) movement of isolated defects through dimer creation

$$A + 3\emptyset \to 2A + B + \emptyset, \tag{17}$$

where A and B denote the defects (non-ground state sites), allowing for one or two types by $B = A$ and $B \neq A$, and with the movement in location in the "equation" indicating a corresponding motion in real space. Note that in each case four neighbors are affected.

[c]and, for example, the fact that the original foam model is non-abelian while the lattice analogs are abelian is not of obvious consequence in the simulations.

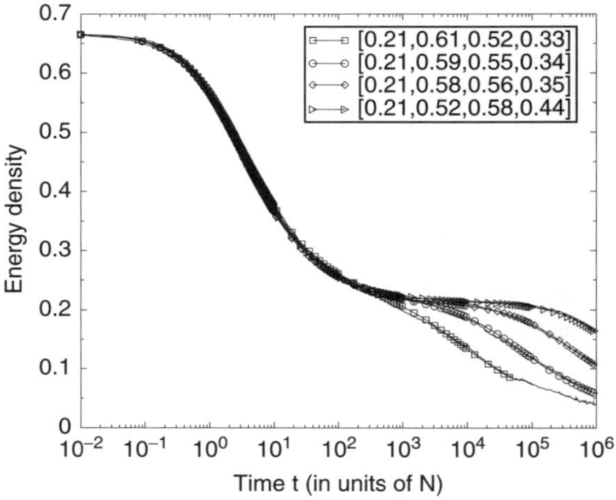

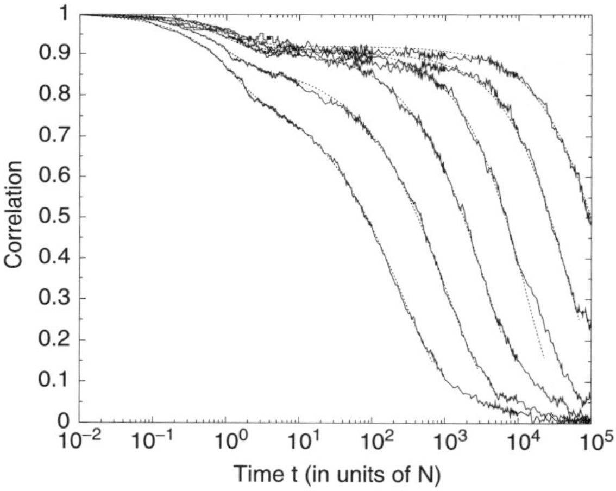

Fig. 5. (a) The energy function of the square system fitted with $E/N = (\frac{2}{3} - a)\,(1 + mt)^{\kappa_1} + a(1 + te^{-2\beta})^{\kappa_2}$ for, from left to right, $\beta = 4, 5, 6, 7$. Values in the inset are for $[a, m, \kappa_i, \kappa_2]$. (b) The equilibrium correlation functions for, from left to right, $\beta = 3, 3.5, 4, 4.5, 5, 5.5$. The dotted lines are fits of the form $C(t) = \alpha\,\mathrm{e}^{-t/\tau_1} + (1 - \alpha)\,\mathrm{e}^{-(t/\tau_2)^{\gamma}}$. (From [9].)

For the next stage of simplification and generalization one would like

(i) fast annihilation of dimers in appropriate environments,
(ii) fast diffusion of dimers, with the possibility of being different from the fast annihilation time,
(iii) slow diffusion of isolated defects,

(iv) fast annihilation of appropriate defect pairs after they have been brought together,
(v) all processes involving 4 units,
(vi) non-degenerate absorbing ground states,
(vii) either one (A) or two (A, B) defect types,
(viii) extension to arbitrary spatial dimensionality.

These ideals can be achieved in the following encapsulation.

5. Generalized Backgammon Models

The minimalist models we study here are based on a coarse-grained simplification of the ideas described above. They correspond to defects (or particles in our language below), which can be either of a single kind A, or different kinds, A and B, and live in a d-dimensional lattice. They can also be considered as a generalization of backgammon [10] or urn models [11] to a non-constant number of particles, with energetic barriers rather than entropic ones, thus allowing for the existence of activated processes. The analogy with the models discussed above is obtained through dynamical rules which mimic the processes of Eqs. (13)–(17). Two versions are considered below, relating to single and double defect types.

5.1. *Single type of particles*

The model system consists of a hypercubic lattice[d] on each site of which can exist a non-negative integer number n_i of particles less than or equal to some maximum $n_{\max}$. Associated with any configuration is the Hamiltonian

$$H = \sum_{i=1}^{N} n_i; \quad 0 \le n_i \le n_{\max}. \tag{18}$$

Due to its non-interacting nature, the equilibrium properties of the model are trivial. There are no thermodynamic phase transitions and the ground state is fully absorbing with $n_i = 0$.

The dynamical rules are inspired by the T1 moves discussed above. Three different kinds of moves are considered, based on an analogy in which all the defect sites of a 4-set in an original process are combined on one site of the new model and all the $\emptyset$ sites of the original process are combined

[d]This could easily be extended to other lattices but here the objective is simplification.

on a second site of the new model:

(i) Annihilation of two particles: in analogy with the processes of Eqs. (13) and (14), three particles are removed from a site i and one is created on a neighboring site j:

$$(n_i, n_j) \to (n_i - 3, n_j + 1). \tag{19}$$

This process is to have a rate 1.

(ii) "Dimer diffusion" analogous to the process of Eq. (16): two particles move from site i to a neighboring site j:

$$(n_i, n_j) \to (n_i - 2, n_j + 2). \tag{20}$$

This process is taken to have a rate D.

(iii) Creation of two particles analogous to the processes of Eq. (17): a particle disappears from site i to create three particles on a neighboring site j:

$$(n_i, n_j) \to (n_i - 1, n_j + 3). \tag{21}$$

This process is taken to have rate $e^{-2\beta}$, corresponding to detailed balance with move-type (i), ensuring that the normal Boltzmann equilibrium properties will be reached asymptotically by the dynamics. Since the interesting region is low temperature, and particularly longer times, analogs of Eq. (8c) and its complement are excluded. Hence $n_{\max} = 3$ suffices and will be taken henceforth. Processes which would take any n_i outside the range $0 \le n_i \le 3$ are forbidden. All these processes are of the form $(n_i, n_j) \to (n_i - x, n_j + y)$ with $x + y = 4$, which reflects the four cell character of the original model transitions. The dynamics considered explicitly is random sequential, in which a site i is chosen randomly, a j is chosen randomly among the neighbors of i and then one of the above processes is chosen according to the probabilities indicated.

In a direct analogy with the processes considered in the earlier sections, the dimer diffusion rate D would be the same as the dimer annihilation rate. However, in the extension this imposition is unnecessary and $D \ll 1$ may also be considered to allow explicit separation of the time scales for annihilation and diffusion of dimers. Taking the temperature $T \ll 2(\ln D)^{-1}$ ensures that the diffusion rate for isolated defects is slower than that for dimers. Furthermore, whereas the original models were based in two-dimensional space, the new model can be in any dimension, including infinite dimension or arbitrary neighboring where mean-field analysis should hold.

Since in this case there is only one type of particle, it represents an analog of the lattice model of Sec. 3 with $D < 0$ but with a simpler non-degenerate ground state.

5.2. *Two different types of particles*

A modification to provide analogy with the original foam model and the $D > 0$ lattice model with their two types of defect consists of considering two different types of particles (A and B) on each site. In the simplest form it suffices to restrict the total number of particles per site to $n_{\max} = 3$, irrespective of their nature, with in addition the difference between the numbers of A and B particles on a site limited to -1, 0 or 1. Again the thermodynamic equilibrium properties are straightforwardly calculable and there is no thermodynamic phase transition.

The dynamical rules are again a straightforward generalization of those above:

(i) annihilation of an AB dimer: an AB dimer and another particle (A or B) are removed from site i but only the extra particle appears on a neighboring site j with a rate 1:

$$[(AAB)_i, (X)_j] \to [(\emptyset)_i, (AX)_j], \tag{22}$$

$$[(ABB)_i, (X)_j] \to [(\emptyset)_i, (BX)_j]; \tag{23}$$

(ii) AB dimer diffusion: a dimer moves from site i to a neighboring site j with a diffusive rate D:

$$[(ABX)_i, (Y)_j] \to [(X)_i, (ABY)_j]; \tag{24}$$

(iii) creation of an AB dimer: a single particle from site i moves to a neighboring site j and additionally creates a dimer there with a rate $e^{-2\beta}$

$$[(AX)_i, (\emptyset)_j] \to [(X)_i, (AAB)_j], \tag{25}$$

$$[(BX)_i, (\emptyset)_j] \to [(X)_i, (ABB)_j]. \tag{26}$$

In all of these processes, symbols X and Y stand for possible A, B or $\emptyset$ particles respecting the restrictions in the number of particles on each site. The rates again satisfy detailed balance conditions, ensuring equilibration.

5.3. *General features*

We now show that these two models share common behavior in all dimensions and with the models considered previously and are in accord with the expectations of the asymptotic field theory of annihilation-diffusion. In what follows we discuss equilibrium dynamical properties, in particular the existence of two different time scales, as well as out-of-equilibrium features like the multi-stage decay of the energy density after a quench.

5.4. *Dynamics in equilibrium*

Let us first consider the auto-correlation function:

$$C(t,t') = \langle n_i(t) n_i(t') \rangle \tag{27}$$

with the brackets denoting ensemble average. In equilibrium this two-time function reduces to a single time equilibrium correlation $C_{\text{eq}}(t-t')$ due to the time translational invariance. From it we can define a relaxation time τ from $C^c_{\text{eq}}(\tau) = C^c_{\text{eq}}(0)/e$, where the connected correlation $C^c_{\text{eq}}(t) = C_{\text{eq}}(t) - c^2_{\text{eq}}$. At low temperatures the temperature dependent and slower process is the creation of particles which has energy barrier $\Delta E = 2$. As a consequence we expect the Arrhenius law for the relaxation time:

$$\tau(\beta) \propto e^{2\beta}. \tag{28}$$

This is confirmed by numerical simulations for all dimensions and all diffusive constants (see Fig. 6 for one type of particle; similar results are found for the model with two types of particles).

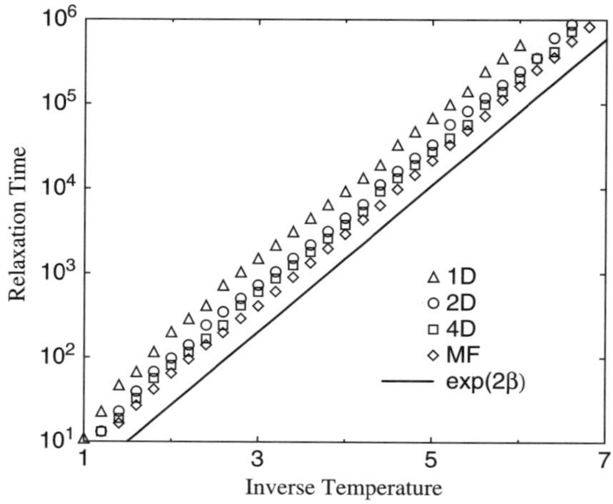

Fig. 6. Relaxation time as a function of the inverse temperature for different dimensions ($d = 1, 2, 4$ and ∞) and a diffusive constant $D = 10^{-4}$. The line corresponds to the expected $\tau \propto e^{2\beta}$ behavior. (From [5].)

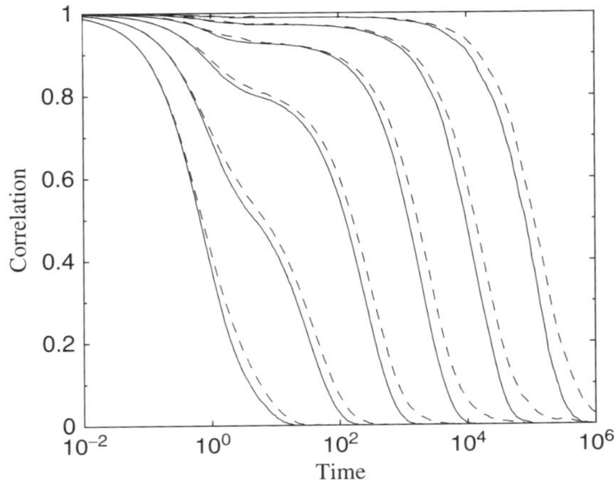

Fig. 7. Normalized equilibrium autocorrelation $C^c_{\rm eq}(t)/C^c_{\rm eq}(0)$ for the model with a single type of particles and a diffusive constant $D = 1$. Different temperatures are considered (from left to right, $\beta = 1, 2, 3, 4, 5$ and 6) as well as dimensions: $d = 2$ (dashed curves) and mean-field $d = \infty$ (full curves). (From [5].)

The equilibrium correlation for $D = 1$ explicitly shows the expected two time scale behavior (see Fig. 7): (i) an initial fast temperature-dependent decay, and (ii) a second decay, on the relaxation time scale. At low temperatures when the two time scales are well separated a plateau appears between the two relaxing regions. This general structure applies for all dimensions, although there are differences of detail especially in the final asymptotic decay to zero which is exponential for the mean-field case ($d = \infty$) and algebraic for finite dimensions (shown for $d = 2$ in Fig. 7 for one type of particle). This difference is explainable in terms of the probability for a particle to come back to the same place (which depends on the dimension of the system).

5.5. *Out-of-equilibrium dynamics*

We now consider the out-of-equilibrium behavior of the models, in particular the decay of the concentration of particles (or equivalently the energy density) $c(t) \equiv N^{-1}\langle H(t)\rangle$, after a quench from an infinite temperature to a low temperature T at time $t = 0$. Taking $e^{-2\beta} \ll D \ll 1$ yields an interesting structure with two intermediate plateaux. The first regime is dominated by the "dimer annihilation" process which eliminates sites with three particles and leads to a configuration with less than three particles on the same site. This first regime occurs on a time scale of order 1. Then, the "dimer diffusion" process comes into play on a time scale of order D^{-1} and remaining dimers (pairs of particles on a single site) diffuse until they reach a singly-occupied site and annihilate.[e] At this stage the system reaches a configuration with mainly isolated particles. Finally, in order to reach the equilibrium concentration of particles, the activated regime involving the effective motion of isolated particles through the creation or annihilation of dimers is necessary and occurs on a time scale of order $e^{2\beta}$.

The last regime in the concentration decay (before the equilibrium concentration is reached) may also be seen as either $A + A \to \emptyset$ or $A + B \to \emptyset$ reaction–diffusion processes, depending on the models, one particle or two-particle respectively, since the particles have to pair up in order to be annihilated. Although these two processes, $A+A \to \emptyset$ and $A+B \to \emptyset$, are both predicted by field theory [7, 8] to behave asymptotically as $(t/\tau)^{\alpha}$, where τ is the characteristic effective microscopic time (here expected to behave as $\exp(2\beta)$) they are predicted to have different critical α and different critical dimensions: $\alpha = d/2$ and $d_c = 2$ for the former and $\alpha = d/4$ and $d_c = 4$ for the latter. As a consequence, we expect a power law decay:

$$c(t) \sim \left(e^{2\beta}/t\right)^{\alpha} \tag{29}$$

with $\alpha = 1$ above the critical dimension d_c and as the above predictions below the critical dimension $d < d_c$.

Figure 8 presents numerical simulations of the concentration decay for the model with a single type of particles for different dimensions and a diffusive constant $D = 10^{-4}$. The temperature after the quench is $T = 1/10$, so the different time scales are well separated ($1 \ll D^{-1} \ll e^{2\beta}$), and the decay presents a two plateau structure. The first plateau is roughly independent of the dimension whereas the second plateau decreases with increasing dimensions to reach the mean-field value. Notice that the qualitative behavior is maintained even in the mean-field limit. The dynamics during the last stage of the decay corresponds to a power law decay with the expected critical exponents α. Figure 8 (right) shows similar results for the model with two types of particles, but with a diffusive constant $D = 1$ (so there is only one plateau). The diffusive time scale is now equivalent to the annihilation one, and the decay presents a single plateau structure.

[e] Taking $D = 1$ combines these effects and consequently eliminates the first plateau.

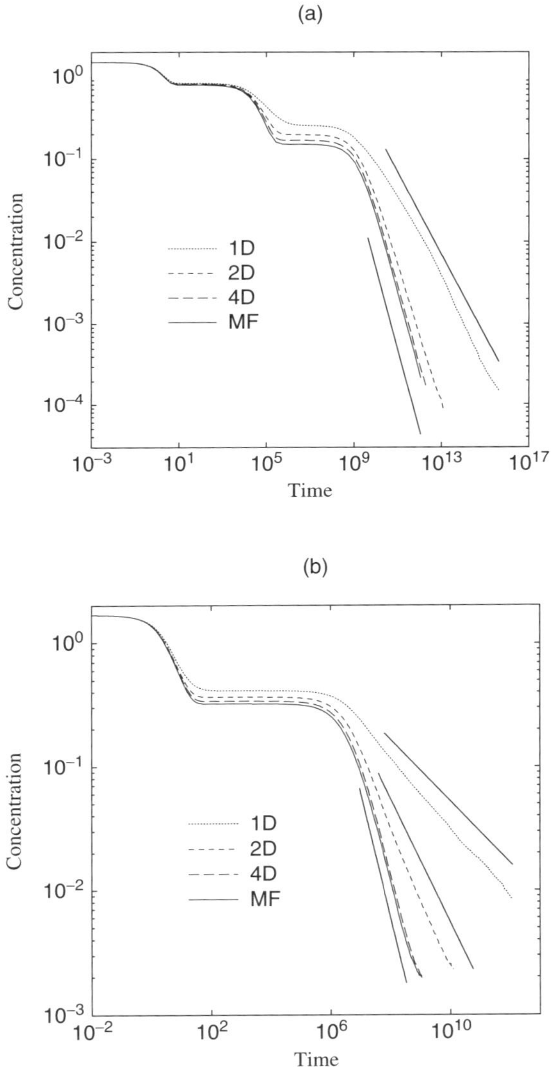

Fig. 8. (a) Concentration of particles for the model with a single type of particles after a quench from $T = \infty$ to $T = 1/10$ with a diffusive constant $D = 10^{-4}$ and for different dimensions ($d = 1, 2$ and 4 and mean-field). The $A + A \to \emptyset$ reaction–diffusion process during the second decay is illustrated by the change in power law for $d < d_c = 2$. The two straight lines are guides with slopes corresponding to the expected exponents $\alpha = 1$ for $d = 2$ and $\alpha = 1/2$ for $d = 1$. (b) Model with two different types of particles, and diffusive constant $D = 1$. The straight lines are guides with the expected power law decays corresponding to $A + B \to \emptyset$ reaction–diffusion processes in different dimensions: for $d < d_c = 4$, $\alpha = d/4$, while $\alpha = 1$ for $d \geq d_c$. (From [5].)

Again, we see the critical behavior during the last stage of the decay, with the critical exponents α expected from the theory.

Below we concentrate mainly on the model with one kind of particle.

6. Mean-Field Solution

For $d = \infty$, where all sites are mutual neighbors, mean-field theory applies and one may easily write exact evolution equations for the probability $p_n(t)$ of a site to be occupied by n particles at time t.[f]

In the mean-field limit there is no i-dependence of the probabilities and conservation of probability, $\sum_n p_n^i = 1$, reduces the number of independent variables to only three, for example p_0, p_1 and p_2. The resultant (three) coupled equations can be solved numerically to arbitrary accuracy. They can also usefully be considered analytically regime by regime to a very good approximation for the case of well separated time scales, to better illustrate the underlying physics, but again for details the reader is referred to [5]. Figure 9 shows a comparison of simulation and analysis using

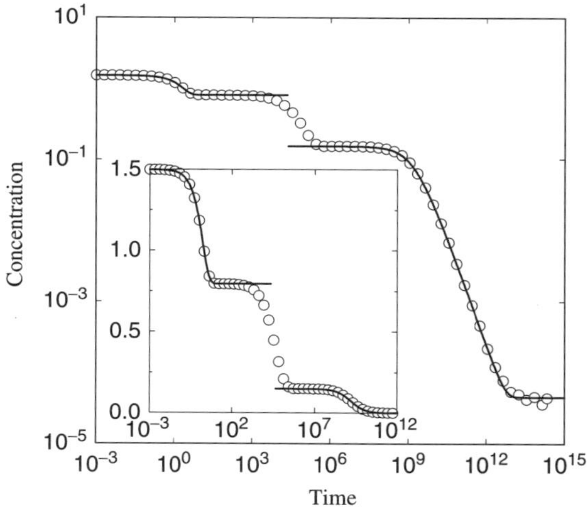

Fig. 9. Concentration decay as a function of time after a quench to the temperature $T = 1/10$ and for a diffusive constant $D = 10^{-4}$. Symbols correspond to numerical simulations, and lines to the analytical results for the first and third regimes. Inset: $T = 1/6$. (From [5].)

[f]For explicit details, the reader is referred to [5].

approximations appropriate to each regime; complete numerical analysis of the above equations would fit the simulations perfectly.

6.1. *Out-of-equilibrium correlation and response*

We now turn to the behavior of two-time correlation and response functions in the out-of-equilibrium regime.

6.1.1. *Correlation functions*

From the two-time out-of-equilibrium auto-correlation functions

$$C_{n,n'}(t,t_w) = \langle \delta_{n_i(t),n}\delta_{n_i(t_w),n'} \rangle \tag{30}$$

with initial conditions

$$C_{n,n'}(t_w,t_w) = p_n(t_w)\delta_{n,n'}, \tag{31}$$

it is possible to construct all relevant two point autocorrelations, and in particular

$$C(t,t_w) \equiv \langle n_i(t)n_i(t_w)\rangle = \sum_{n,n'} nn'C_{n,n'}(t,t_w). \tag{32}$$

The correlation functions $C_{n,n'}$ correspond to the probabilities of having n particles at time t on a given site when there were n' particles at time $t_w \leq t$ on this particular site. They satisfy a finite set of explicitly time-dependent coupled linear equations.

Concentrating on the regime in which $t_w \gg D^{-1}$ and after the second plateau in the concentration decay, Fig. 10 shows the autocorrelations $C_{1,1}(t,t)$ and $C_{1,1}(t,t_w)$ for two different waiting times $t_w = 10^4$ and 10^6. There is good agreement between simulations and theory.

6.1.2. *Response functions*

It is also of interest to consider out-of-equilibrium response. To this end one introduces a perturbation at time t_w after the quench. To get a auto-response one may consider the application of a small randomly-signed field on each site coupled to the corresponding observable and sign-multiply the measurement similarly [12]. The simplest possibility is to couple the random field to the single occupancy operator $\delta_{n_i,1}$, leading to the perturbation

$$\delta H = -h\sum_i \epsilon_i \delta_{n_i,1}. \tag{33}$$

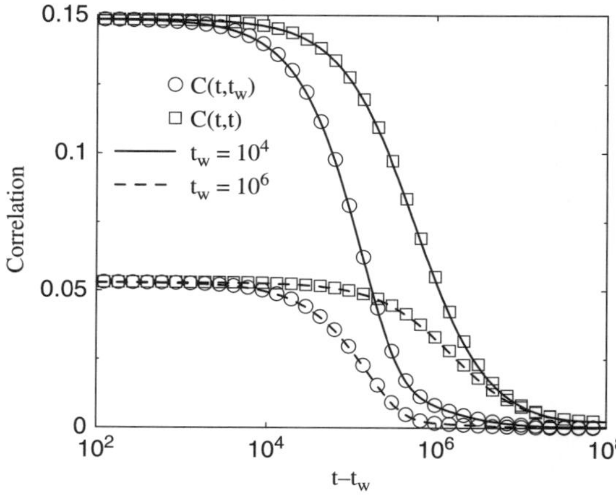

Fig. 10. Out-of-equilibrium $C_{1,1}(t, t_w)$ (circles) and $C_{1,1}(t, t)$ (squares) after a quench to temperature $T = 1/6$. Symbols correspond to simulations and lines to the analytical result. Waiting times are $t_w = 10^4$ (full lines) and 10^6 (dashed lines), and the diffusion constant $D = 10^{-2}$. (From [5].)

h is the strength of the field and is taken small to stay in the linear regime. The ϵ_i are taken as ± 1 randomly. The corresponding (integrated) response function is the change in the expectation value of $\delta_{n(t),1}$ due to the perturbation,

$$\chi_1(t, t_w) = \overline{h^{-1} N^{-1} \sum_i \epsilon_i \langle \delta_{n_i(t),1} \rangle_h}, \tag{34}$$

where the overline stands for the average over the random field variables. This response is conjugate to the autocorrelation $C_{1,1}(t, t')$, which is the relevant one for long times and low temperatures,

$$\overline{N^{-1} \sum_{i,j} \epsilon_i \epsilon_j \langle \delta_{n_i(t),1} \delta_{n_j(t'),1} \rangle} = C_{1,1}(t, t'). \tag{35}$$

We also have to define how this perturbation affects the dynamical rules, maintaining the detailed balance conditions in order to ensure equilibrium asymptotically. Different definitions are possible, of which two are considered. The natural definition is to use for the rates a Metropolis rule with the perturbed Hamiltonian $H + \delta H$,

$$\min\left(1, e^{-\beta \Delta(H+\delta H)}\right) \quad \text{(M)}, \tag{36}$$

where $\Delta(H + \delta H)$ corresponds to the change in the perturbed Hamiltonian under the corresponding transition. A disadvantage is that this definition only extracts a response from unoccupied sites. A second possibility is to modify the dynamical rules by multiplying the unperturbed rates by another Metropolis factor:

$$\min\left(1, e^{-\beta\Delta(\delta H)}\right) \times \min\left(1, e^{-\beta\Delta(H)}\right) \quad \text{(MM)}. \tag{37}$$

This modification of the dynamical rules preserves detailed balance with respect to $H + \delta H$ and has the advantage that it allows extraction of a response from occupied and unoccupied sites. For simple spin facilitated models the two dynamics yield equivalent responses, the second one being more efficient from the numerical point of view, but this equivalence turns out not to hold for the present models.

Both dynamics are soluble in the mean-field ($d = \infty$ case) and calculations and simulations agree. Results for M dynamics are shown in are shown in Fig. 11. The non-monotonic behavior is given by the fact that the response is the product of a decreasing function, $p_1(t)$, corresponding to the number of defects able to respond, and an increasing one, $1 - e^{-(t-t_w)/\tau_c}$, corresponding to the monotonic rescaled equilibrium response function. The reduction with increasing t_w reflects the fact that the total number of particles decays after the quench.

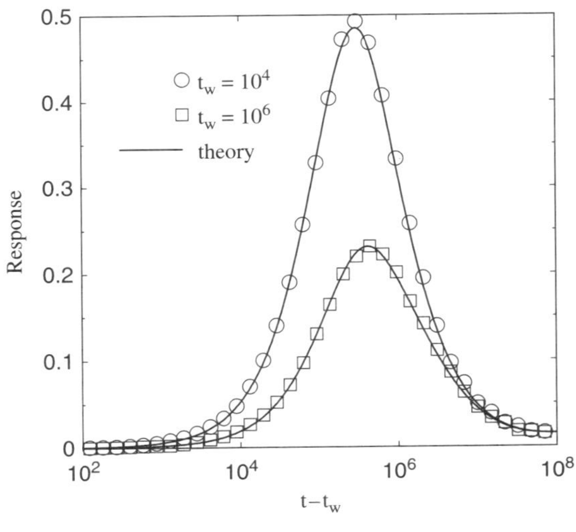

Fig. 11. Out-of-equilibrium response $\chi_1(t, t_w)$ for MM dynamics as a function of $t - t_w$, at temperature $T = 1/6$, for waiting times $t_w = 10^4$ (circles) and 10^6 (squares), and a diffusive constant $D = 10^{-2}$. The lines correspond to the analytical result. (From [5].)

6.1.3. *Fluctuation–dissipation relations*

Having obtained correlation and response functions, we can now study out-of-equilibrium fluctuation–dissipation (FD) relations, which have been a topic of much interest in recent glass studies. Since we are considering the case of long but finite times, and therefore one-time quantities are still changing with time, FD relations have to be considered between the integrated response, $\chi_1(t, t_w)$, and the difference of the conjugate connected correlation functions, $C_{1,1}^{(c)}(t,t) - C_{1,1}^{(c)}(t,t_w)$, where $C_{1,1}^{(c)}(t,t') \equiv C_{1,1}^{(c)}(t,t') - p_1(t)p_1(t')$. In Figs. 12(a) and 12(b) we show for mean-field ($d = \infty$) the FD plots for the case of MM and M dynamics, respectively, for temperature $T = 1/6$ and waiting times $t_w = 10^4$ and 10^6 (inset).

Several things are worthy of note. First, despite the fact that both response functions and the differences of connected correlations are non-monotonic in t, to a very good approximation $\chi_1(t,t_w) = \chi_1\big[C_{1,1}^{(c)}(t,t) - C_{1,1}^{(c)}(t,t_w)\big]$, similarly to what has been found for other simple strong glass formers. Second, the FD curves approach the fluctuation–dissipation theorem (FDT) value as the waiting time is increased, as expected. Third, the FD relations look almost linear (although this may be just a consequence of the fact that the departure from FDT is relatively small). In this case the FDT violation ratio $X(t, t_w)$ [13, 14] is just a function of the waiting time, $X = X(t_w)$. $X > 1$ for the case of MM dynamics, while $X < 1$ for the case of M dynamics.

Finally, in Fig. 12(c) we compare the behavior in the mean-field model with that at finite dimensions. For $d = 1$, FDT is obeyed, similar to what happens in the Fredrickson–Andersen model [15]. For $d \geq d_c = 2$ however, the FD plots coincide with the mean-field ones. This indicates that the aging behavior is controlled by the out-of-equilibrium critical point of the underlying diffusion–annihilation process, and that mean-field serves as a good approximation for the physically relevant dimensions $d = 2, 3$.

7. Conclusions

Through a sequence of simple models it has been shown that glassy behavior can arise due to purely kinetic constraints which generate dynamical frustration even in the absence of interactions in the Hamiltonian (and consequently trivial thermodynamics without equilibrium phase transitions). The models discussed have been inspired by considerations of an idealized foam and covalently-bonded networks, originally in two spatial dimensions, but have been further simplified and extended to allow easier simulation,

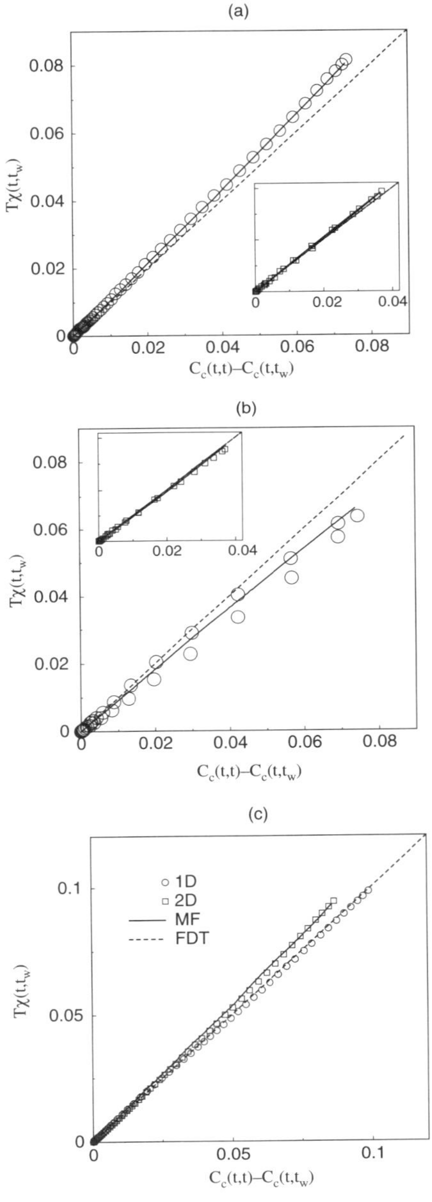

Fig. 12. (a) FD plot for MM dynamics at $T = 1/6$ and waiting time $t_w = 10^4$ (inset: $t_w = 10^6$). The symbols correspond to simulations, the full lines to the analytical result, and the dashed line to FDT. (b) Similar plot for M dynamics. (c) FD plot in various dimensions for MM dynamics compared with MF (full line); $T = 1/6$, $t_w = 10^4$.

analysis and comprehension, application in arbitrary dimensions and, in the limit $d \to \infty$, to mean-field analysis. They exhibit strong glass behavior, the connection has been explored with underlying diffusion–annihilation processes, and it has been shown that the aging dynamics of these models is dominated by the critical out-of-equilibrium fixed point of the associated diffusion–annihilation theory.

Acknowledgments

The author would like to thank his collaborators in the research reported here, T. Aste, A. Buhot, L. Davison, J. and P. Garrahan. Original reports are to be found in the reference list below.

References

[1] Ritort, F. and Sollich, P., *Adv. Phys.* **52** (2003) 219.
[2] Aste, T. and Sherrington, D., *J. Phys. A* **32** (1999) 7049.
[3] Davison, L. and Sherrington, D., *J. Phys. A* **33** (2000) 8615.
[4] Davison, L., Sherrington, D., Garrahan, J.P. and Buhot, A., *J. Phys. A* **34** (2001) 5147.
[5] Buhot, A., Garrahan, J.P. and Sherrington, D., *J. Phys. A* **34** (2003) 307.
[6] Angell, C.A., *Science* **267** (1995) 1924.
[7] Cardy, J., *Field Theory and Non-equilibrium Statistical Mechanics*, Lectures presented at the Troisieme Cycle de la Suisse Romande (1999).
[8] Hinrichson, H., *Adv. Phys.* **49** (2000) 815.
[9] Sherrington, D., Davison, L., Buhot, A. and Garrahan, J.P., *J. Phys. Cond. Matt.* **14** (2002) 1673.
[10] Ritort, F., *Phys. Rev. Lett.* **75** (1995) 1190.
[11] Godreche, C. and Luck, J.M., *Eur. Phys. J.* **23** (2001) 473.
[12] Barrat, A., *Phys. Rev. E* **57** (1998) 3629.
[13] Cugliandolo, L.F., Kurchan, J. and Peliti, L., *Phys. Rev. E* **55** (1997) 3898.
[14] Bouchaud, J.-P., Cugliandolo, L.F., Kurchan, J. and Mezar, M., in *Spin-Glasses and Random Fields*, ed. Young, A.P. (World Scientific, Singapore, 1997).
[15] Buhot, A. and Garrahan, J.P., *Phys. Rev. Lett.* **88** (2002) 225702.

CHAPTER 9

ON GLASS TRANSITION WITH RAPID COOLING EFFECTS

Richard Kerner

Laboratoire de Physique Théorique des Liquides, Université Pierre et Marie Curie — CNRS URA 7600, Tour 22, 4-ème étage, Boite 142, 4, Place Jussieu, 75005 Paris, France

Ondřej Mareš

Department of Mathematics, FJFI — ČVUT, Trojanova 13, 120 00 Prague 2, Czech Republic

The glass transition process can be interpreted in terms of stochastic growth and an agglomeration process described by means of algebraic or differential equations which rule the evolution and behavior of the probabilities of local configurations. This method has been successfully used for the description of several covalent glasses. We discuss in more detail a version of the model which takes into account the effect of rapid cooling.

1. Introduction

Analytic treatment of glass transition has remained one of the challenges in theoretical physics for a long time, and the results obtained until now with quite different approaches and methodology can be considered as only partially satisfactory. Glass is a very strange thing, extremely homogeneous on the macroscopic scale, and apparently displaying a large variety of local structures, all different, but similar when it comes to essential physical parameters. But apparently there is no unique and clear answer to the question as to which physical parameters in glass should be considered as "essential." However, without much doubt, the geometry of local configurations, involving dozens of atoms including not only the closest neighbors, must play a very important role. These configurations define what many authors refer to as "the energy landscape" in which single atoms or certain stable clusters seem to evolve. The geometry of such a "landscape" influences the geometry of local configurations, and *vice versa.* This mutual influence seems to be the key to understanding both the formation process and the resulting structure of glass.

During the past decade, in a series of published papers [1–4], original models of growth by agglomeration of smaller units have been elaborated, and applied to many important physical systems, such as quasicrystals [5], fullerene molecules [6, 7], and oxide and chalcogenide glasses [8–11]. In all these applications, it was tacitly supposed that temperature variations are slow enough to be neglected in the model. However, many glasses, chalcogenide-based in particular, are obtained with rapid quenching, so that the temperature variations during the glass transition are very important. It is also well known that rapid quenching modifies the value of glass transition temperature T_{g}, the main tendency observed being its increase with increasing rapidity of quenching. In contrast, a very slow decrease in temperature tends to lower the transition temperature, as long as spontaneous crystallization can be avoided.

We shall make our presentation as concise as possible. To this end, the physical example we shall choose to illustrate this approach will be one of the simplest covalent amorphous networks known to physicists, the binary chalcogenide glass $\mathrm{As}_x\mathrm{Se}_{(1-x)}$, where x is the concentration of arsenic atoms in the basic glass-former, which in this case is pure selenium. The generalization to other covalent networks, e.g., $\mathrm{Ge}_x\mathrm{Se}_{(1-x)}$, is then quite straightforward. These glasses (in the form of thin and elastic foils) are used in photocopying devices.

Let us stress the fact that our model's predictions depend very strongly on the connectivity of the considered network, which is a function of the valence of atoms which agglomerate to produce the network. In the case of the $\mathrm{As}_x\mathrm{Se}_{(1-x)}$ glass the situation is much more complex than in the case of the $\mathrm{Ge}_x\mathrm{Se}_{(1-x)}$ glass, because a certain amount of As atoms appears with higher valence (5 instead of 3), which we do not take into account here for the sake of simplicity. The discussion of connectivity and valence variations in the $\mathrm{As}_x\mathrm{Se}_{(1-x)}$ glass can be found in [19].

When the formation of a solid network of atoms or molecules occurs in a liquid which is cooled more or less rapidly, the most important feature of the process is the progressive agglomeration of small and mobile units (which may be just single atoms, or stable molecules, or even small clusters already present in the liquid state) into an infinite (from the nanoscale point of view) stable network, whose topology cannot be modified anymore unless the temperature is raised again, leading to the inverse (melting or evaporation) process.

To describe such an agglomeration with all geometrical and physical parameters, such as bond angles and lengths, and the corresponding chemical and mechanical energies stored in each bond, is beyond the possibilities of any reasonable model. This is why stochastic theory represents an ideal tool for the description of random agglomeration and growth processes. Instead of reconstructing all local configurations, it takes into account only the *probabilities* of them being found in the network, and then the probabilities of higher order, corresponding to local correlations. This is achieved by using the *stochastic matrix* technique. A *stochastic matrix* $\mathbf{M}$ represents an operator transforming given finite distribution of probabilities, $[p_1, p_2, \ldots, p_N]$, with $p_1 + p_2 + \cdots + p_N = 1$, into another distribution of probabilities, $[p'_1, p'_2, \ldots, p'_N]$. It follows immediately that such a matrix must have all its entries real, positive or null, and the elements of each single column must sum up to 1.

The algebraic properties of such matrices are very well known. The main feature that we shall use here is the fact that any stochastic matrix has at least one eigenvalue equal to 1. The remaining eigenvalues have their absolute value always less than 1. This means that if we continue to apply a stochastic matrix to any initial probability distribution, after some time only the distribution corresponding to the unit eigenvalue will remain, all other contributions shrinking exponentially. This enables us to find the asymptotic probability distribution.

In what follows, we shall identify these probability distributions with stable or meta-stable states of the system, describing the statistics of characteristic sites in the network. Taking into account Boltzmann factors (with chemical potentials responsible for the formation of bonds), we are able to draw conclusions concerning the glass transition temperature in various compounds. In particular, one is able to predict the initial slope of the curve $T_g(c)$, i.e., the value of $dT_g/dc|_{c=0}$ [12, 13].

The stochastic matrix depends on the temperature T via Boltzmann factors entering the transition probabilities. In all models of agglomeration discussed in our previous articles, the creation of consecutive layers on the surfaces of growing clusters were symbolized by the consecutive action of *the same* stochastic matrix, i.e., with constant temperature T, which meant that the cooling was very slow compared with the speed of spontaneous agglomeration. This statement can be given a clear mathematical formulation. Let us denote the cooling rate (dT/dt) as q; the logarithmic cooling rate is then q/T and has the dimension of $\sec^{-1}$. Let τ denote the characteristic time that is needed to form a new layer on the surface of an average cluster. We can define *slow cooling rate* when the dimensionless number $(\tau q/T)$ is very small compared with the probability factors entering the stochastic matrix, and *rapid cooling* when these numbers are comparable.

2. Cluster Agglomeration Described with the Stochastic Matrix

Consider a binary selenium–arsenic glass, in which selenium is the basis glass former, and arsenic is added as modifier (although its concentration can be as high as 30%). The chemical formula denoting this compound is $As_cSe_{(1-c)}$, where c is the As concentration. As mentioned in the introduction, the real behavior of the $As_xSe_{(1-x)}$ glass is more complicated due to the appearance of a certain proportion of five-valenced As atoms; we shall not take this phenomenon into account here for the sake of simplicity. What is important here is the fact that the glass is formed out of two distinct chemical components, with the coordination numbers (valences) clearly defined and constant during the glass formation process.

In a hot liquid, prior to solidification, the basic building blocks that agglomerate, are just selenium and arsenic atoms, indicated respectively by (—○—) and (-<). When the temperature goes down, clusters of atoms start to appear everywhere, growing by agglomeration of new atoms on their rim. Consider a growing cluster: one can distinguish three types of situations (we shall call them "*sites*") on the cluster's rim. The concentration of free As atoms in the liquid will be called c and that of Se, $(1-c)$. In principle, three kinds of situations can be found on the surface, as shown in the left columns of Fig. 1 below: a selenium atom linked by one of its valences to the bulk, with one valence still available for further agglomeration; an arsenic atom with one valence engaged in the bulk and two free valences available for further agglomeration; finally, an arsenic atom with two valences already saturated and the remaining one still avaliable to the agglomeration.

The probabilities constituting the entries of the stochastic matrix will depend on the following factors:

(i) the coordination number (i.e., the valence) of the incoming particle;
(ii) the number of free valences available at a given rim site;
(iii) the concentration of the corresponding species;
(iv) the binding energies involved in each case, entering the corresponding Boltzmann factors as contributions to the chemical potential.

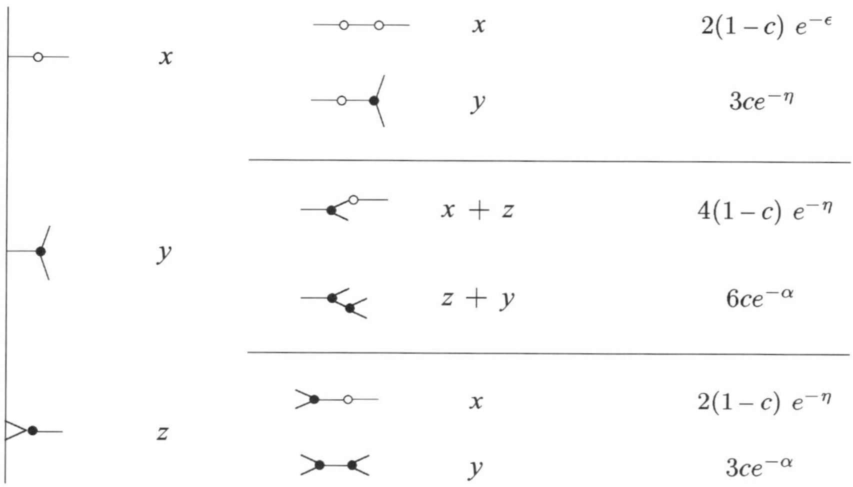

Fig. 1. States, steps and matrix entries in the one-bond saturation case.

We shall suppose that there are Boltzmann factors of three kinds, corresponding to three types of chemical bonds to be formed:

$$
\begin{aligned}
\circ\!\!-\!\!\circ \Leftrightarrow e^{E_{\mathrm{SeSe}}/kT} &= e^{-\epsilon},\\
\circ\!\!-\!\!\bullet \Leftrightarrow e^{E_{\mathrm{SeAs}}/kT} &= e^{-\eta},\\
\bullet\!\!-\!\!\bullet \Leftrightarrow e^{E_{\mathrm{AsAs}}/kT} &= e^{-\alpha}.
\end{aligned}
$$

For the one-atom-at-a-time approach, the unnormalized probability factors are shown in the last column.

The probability of obtaining state y after one step, starting from state x, is proportional to $3ce^{-\eta}$; that to get y from z, to $3ce^{-\alpha}$; and so on.

Now we build up a provisional transition matrix, with not-yet normalized columns,

$$
\begin{pmatrix}
2(1-c)e^{-\epsilon} & 4(1-c)e^{-\eta} & 2(1-c)e^{-\eta}\\
3ce^{-\eta} & 6ce^{-\alpha} & 3ce^{-\alpha}\\
0 & 4(1-c)e^{-\eta}+6ce^{-\alpha} & 0
\end{pmatrix}, \tag{1}
$$

and then proceed to normalize the entries in each column to have their sum equal to 1. To get the stochastic matrix in a simple final form, it will be convenient to introduce the combinations

$$
A = \frac{2(1-c)}{2(1-c)+3ce^{\epsilon-\eta}} \quad \text{and} \quad B = \frac{2(1-c)}{2(1-c)+3ce^{\eta-\alpha}}. \tag{2}
$$

The transition matrix governing the process takes on the following form:

$$M = \begin{pmatrix} A & \frac{B}{2} & B \\ 1-A & \frac{1-B}{2} & 1-B \\ 0 & \frac{1}{2} & 0 \end{pmatrix}. \tag{3}$$

A general state on which M acts is described by a probability distribution

$$P = (p_x, p_y, 1 - p_x - p_y).$$

The matrix M has the eigenvalues

$$\lambda_1 = 1, \quad \lambda_{2,3} = \frac{2A - B - 1 \mp \sqrt{(2A-B-1)^2 - 8(B-A)}}{4}.$$

We can check that, for $A \leq 1$ and $B \leq 1$, the absolute values of λ_2 and λ_3 are smaller then 1. The eigenvector corresponding to the unit eigenvalue, which is

$$v = \frac{1}{3 - 3A + 2B} \begin{pmatrix} 2B \\ 2(1-A) \\ (1-A) \end{pmatrix}, \tag{4}$$

will be preserved by higher powers of M, while the other will be progressively damped. Thus, the system evolution will move preferentially along that stable eigenvector, which represents the asymptotic state. Because one can start with any vector, and apply the matrix M indefinitely in order to approach the final state, we can determine the limit of the matrix itself, M^∞, and then find its eigenvector. After some calculus (see, e.g., [14]) we get:

$$M^\infty = \frac{2}{3(1-A) + 2B} \begin{pmatrix} B & B & B \\ 1-A & 1-A & 1-A \\ \frac{1-A}{2} & \frac{1-A}{2} & \frac{1-A}{2} \end{pmatrix}. \tag{5}$$

Note that each column is exactly the stationary eigenvector (4), and the matrix, being a projector, is singular of rank 1.

Some extreme cases are illustrative. The pure selenium case $c = 0$ gives $A = B = 1$, eigenvalues $\lambda_1 = 1$, $\lambda_2 = 0$ and $\lambda_3 = 0$. The stable eigenvector corresponding to the eigenvalue $\lambda_1 = 1$ is the pure selenium state $(1, 0, 0)$, consistent with what is seen in Fig. 1. It is immediate clear that M^∞, applied to an arbitrary distribution $P = (p_x, p_y, 1 - p_x - p_y)$, gives precisely that eigenvector.

Though the realistic cases involve the values of $c < 0.4$, it is instructive to consider also the pure arsenic case $c = 1$. This leads to $A = B = 0$, with eigenvalues $\lambda_{1,2,3} = 1, -1/2, 0$. The stable eigenvector, which is then $(0, 2/3, 1/3)$, is also the result of applying M^∞ to an arbitrary initial distribution.

Another extreme case is that of high temperatures. When kT is much larger than the energy differences appearing in (2), the Boltzmann factors tend to 1, and we have $A = B = 2(1-c)/(2+c)$, $\lambda_1 = 1$, $\lambda_2 = (a-1)/2$, $\lambda_3 = 0$ and

$$M^\infty = \frac{1}{4+5c}\begin{pmatrix} 4(1-c) & 4(1-c) & 4(1-c) \\ 6c & 6c & 6c \\ 3c & 3c & 3c \end{pmatrix}.$$

The stable eigenvector, corresponding to $\lambda_1 = 1$, is

$$v = \frac{1}{4+5c}\begin{pmatrix} 4(1-c) \\ 6c \\ 3c \end{pmatrix}.$$

In all the cases above, we find that the asymptotic transition matrix is an idempotent, that is, it satisfies $(M^\infty)^2 = M^\infty$.

Notice from Fig. 1 that, on the agglomerate's surface, p_x is the Se concentration and $p_y + p_z$ is the As concentration. Now, the high homogeneity exhibited by known glass structures suggests that, even in relatively small clusters, deviations from the average modifier concentration c must be negligible. Thus, *in the bulk*, the As concentration should be equal to c. To compute the bulk concentration, however, we should count only the y-sites created on the surface during the agglomeration process. Indeed, counting the z-sites would amount to counting the same atoms twice, because all the z-sites are created out of previously counted y-sites. Therefore, the condition of minimal fluctuations in the bulk concentration would read

$$\frac{p_y}{p_x} = \frac{2(1-A)}{2B} = \frac{c}{1-c},$$

which leads to the explicit dependence $c(T_\mathrm{g})$:

$$c = \frac{2(3\xi - 2)}{12\xi - 9\mu - 4}, \tag{6}$$

where we have introduced the abbreviated notation $\xi = e^{\epsilon - \eta}$ and $\mu = e^{\epsilon - \alpha}$.

This equation can be checked against experiment. For example, we can evaluate the derivative $\partial T/\partial c = (\partial c/\partial T)^{-1}$ for a given value of c. In particular, as $c \to 0$, when we can neglect the As–As bond creation (equivalent to putting $\mu = 0$ in (6)), we get

$$\left[\frac{\partial T}{\partial c}\right]_{c=0} = \frac{T_{g_0}}{\ln(3/2)},$$

(where T_{g_0} is the glass transition temperature of pure Se). This is the present-case expression of the general formula given by the stochastic approach,

$$\left[\frac{\partial T}{\partial c}\right]_{c=0} = \frac{T_{g_0}}{\ln(m'/m)} \tag{7}$$

(where m and m' are the valences of the basic glass former and of the modifier), which is in very good agreement with experimental data (see [15, 16]).

Another choice is possible, in what concerns the states and the transition matrix. In the first case considered above, shown in the Fig. 1, the attachment of *one single basic unit*, or the saturation of one single bond, is a step in the evolution. In the second choice, illustrated by Fig. 2 below, each step is obtained by the complete saturation of all the bonds at the rim, so that only two types of sites (denoted by x and y) are seen on the cluster's rim, assuming that the growth is of dendritic type (no small rings

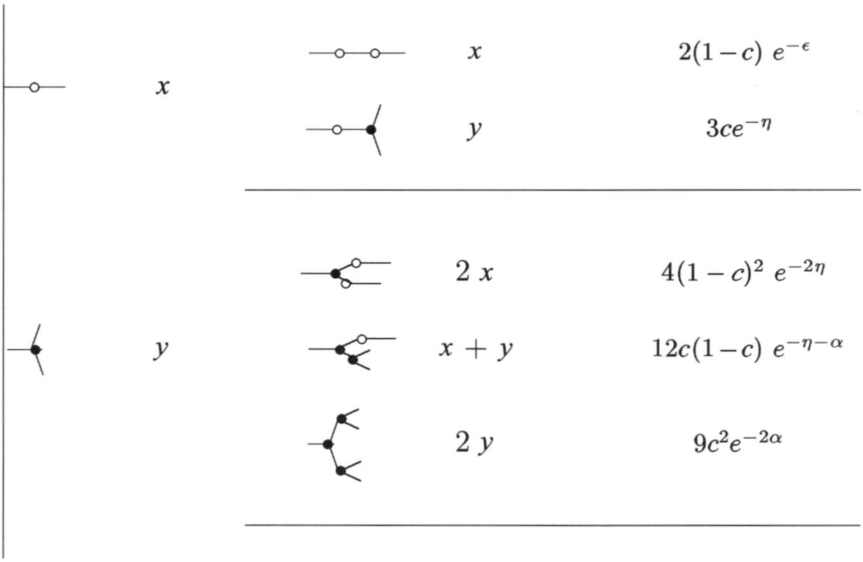

Fig. 2. States, steps and un-normalized probability factors.

present). It can be shown [14] that the two approaches lead to the same results, which may be considered as proof of the ergodicity of the proposed model. In what follows, we shall choose the second version of the model which leads to a two-by-two transition matrix.

Observing that from the site z only the sites of x and y type can be produced, we can forget it and consider the dendritic growth with only two types of sites appearing all the time. Given an arbitrary initial state (p_x, p_y), the new state results from taking into account all possible ways of saturating the bonds of the previous state's sites by the available external atoms. The un-normalized probability factors are displayed in the figure; some of them are quadratic in the concentrations and Boltzmann factors, but they become simpler after normalization. The non-normalized probability factors can be arranged to give a matrix:

$$\begin{pmatrix} 2(1-c)e^{-\epsilon} & 4(1-c)^2 e^{-2\eta} \\ 8(1-c)^2 e^{-2\eta} + 12c(1-c)e^{-\eta-\alpha} & 12c(1-c)e^{-\eta-\alpha} + 18c^2 e^{-2\alpha} \end{pmatrix}. \tag{8}$$

The normalized transition matrix is written as

$$M = \begin{pmatrix} M_{xx} & M_{xy} \\ M_{yx} & M_{yy} \end{pmatrix} = \begin{pmatrix} M_{xx} & 1 - M_{yy} \\ 1 - M_{xx} & M_{yy} \end{pmatrix}, \tag{9}$$

where M_{xx} and M_{yy} are obtained by normalizing to 1 the columns of the matrix (8):

$$M_{xx} = \frac{2(1-c)\xi}{2(1-c)\xi + 3c} \quad \text{and} \quad M_{yy} = \frac{3c\mu}{2(1-c) + 3c\mu}, \tag{10}$$

where we have introduced the abbreviated notation $\xi = e^{\eta-\epsilon}$ and $\mu = e^{\eta-\alpha}$.

The eigenvalues of this matrix are 1 and $M_{xx} - M_{yy} = M_{xy} - M_{yx}$. The asymptotic form is

$$M^{\infty} = \frac{1}{M_{xy} + M_{yx}} \begin{pmatrix} M_{xy} & M_{xy} \\ M_{yx} & M_{yx} \end{pmatrix}, \tag{11}$$

and the stationary eigenvector is

$$v = \frac{1}{M_{xy} + M_{yx}} \begin{pmatrix} M_{xy} \\ M_{yx} \end{pmatrix}, \tag{12}$$

which appears as the columns in M^{∞}. We also find here that $(M^{\infty})^2 = M^{\infty}$. We again impose that the asymptotic state is fixed by the external concentration, which means that the above eigenvector must equal the medium distribution vector $(1-c, c)$. The solutions are $c = 0$, $c = 1$ and the non-trivial one:

$$c = \frac{M_{yx}}{M_{xy} + M_{yx}}, \tag{13}$$

which reduces to (6) when M_{xy} and M_{yx} are replaced by their values.

Note that it is not actually necessary to produce the asymptotic eigenvector to arrive at this condition. It is enough to apply M directly to any average state of cluster's rims. This coincidence can be interpreted as an example of ergodicity in the system under consideration: randomness in time is equivalent to randomness in the space of states.

It can be seen from the figure that on the surface of an average cluster, p_x is the Se concentration and p_y is the As concentration. Now, the high homogeneity exhibited by known glass structures suggests that even in relatively small clusters, deviations from the average modifier concentration c must be negligible. Thus, *in the bulk*, the As concentration should be equal to c. Therefore, the condition of minimal fluctuations in the bulk concentration would read

$$\frac{p_y^\infty}{p_x^\infty} = \frac{(M_{yx})}{M_{xy}} = \frac{c}{1-c},$$

which leads to the explicit dependence $c(T_{\mathrm{g}})$:

$$c = \frac{6 - 4\xi}{12 - 4\xi - 9\mu}. \tag{14}$$

This equation can be checked against experiment. For example, we can evaluate the derivative $\partial T/\partial c = (\partial c/\partial T)^{-1}$ for a given value of c. In particular, as $c \to 0$, when we can neglect the As–As bond creation (equivalent to putting $\mu = 0$ in (14)), we get

$$\left[\frac{\partial T}{\partial c}\right]_{c=0} = \frac{T_{\mathrm{go}}}{\ln(3/2)},$$

(where T_{go} is the glass transition temperature of pure Se). This is the present-case expression of the general formula given by the stochastic approach,

$$\left[\frac{\partial T}{\partial c}\right]_{c=0} = \frac{T_{\mathrm{go}}}{\ln(m'/m)} \tag{15}$$

(where m and m' are the valences of the basic glass former and of the modifier), which is in a very good agreement with the experimental data (see [17–19]).

3. Generalization and the Low Concentration Limit

The above scheme can be easily generalized to the case of arbitrary valence, say m_A and m_B. In that case, the stochastic 2×2 matrix has the form

$$\begin{pmatrix} M_{xx} & M_{xy} \\ M_{yx} & M_{yy} \end{pmatrix} = \begin{pmatrix} M_{xx} & 1 - M_{yy} \\ 1 - M_{xx} & M_{yy} \end{pmatrix} \tag{16}$$

with

$$M_{yx} = \frac{m_B c}{m_A(1-c)\xi + m_B c}, \quad M_{xy} = \frac{m_A(1-c)}{m_A(1-c) + m_B c\mu}.$$

The asymptotic probability has the same form as before, as well as the zero fluctuation condition relating c with T (interpreted as the glass transition temperature). The derivation of c with respect to the temperature T gives the "magic formula":

$$\frac{dc}{dT} = \frac{1}{T} \frac{\left(\frac{m_A}{m_B} - \mu\right) \xi \ln \xi - \left(\frac{m_B}{m_A} - \xi\right) \mu \ln \mu}{\left[\left(1 - \frac{m_A}{m_B}\xi\right) + \left(1 - \frac{m_B}{m_A}\mu\right)\right]^2}, \tag{17}$$

where we used the fact that $d\xi/dT = (-1/T)\xi \ln \xi$, and $d\mu/dT = (-1/T)\mu \ln \mu$. This defines the slope of the function $T_g(c)$, which is an important measurable quantity:

$$\frac{dT_g}{dc} = T_g \frac{\left[\left(1 - \frac{m_A}{m_B}\xi\right) + \left(1 - \frac{m_B}{m_A}\mu\right)\right]^2}{\left(\frac{m_A}{m_B} - \mu\right)\xi + \ln \xi - \left(\frac{m_B}{m_A} - \xi\right)\mu \ln \mu}. \tag{18}$$

The initial slope, at $c = 0$, is of particular interest. Its expression is very simple, taking into account that in the limit when $c = 0$, we have also $\xi = m_B/m_A$, which leads to

$$\left[\frac{dT_g}{dc}\right]_{c=0} = \frac{T_{g0}\left(1 - \frac{m_B}{m_A}\mu\right)}{\ln\left(\frac{m_B}{m_A}\right)}. \tag{19}$$

Its value has been checked against experiment very successfully, in more than 30 different compounds. In some cases the formula does not seem to work well; usually it comes from the change of valence of certain atoms provoked by the influence of the surrounding substrate.

Now, one could be worried about the apparent singularity in the above formula when $m_A = m_B$, i.e., when one deals with a mixture of two different glass formers with the same coordination number. It is not difficult to show that also in such a case the reasonable limit can be defined, as has been recently suggested by M. Micoulaut [21].

As a matter of fact, suppose that the glass transition temperature of the pure glass-former A is T_{g0}, and that of the pure glass-former B is T_{g1}.

We can re-write our minimal fluctuation condition in a very symmetric manner, manifestly invariant with respect to the simultaneous substitution $m_A \leftrightarrow m_B$, $c \leftrightarrow (1-c)$ and $\xi \leftrightarrow \mu$:

$$c(1-c)\left[(1-c)\left(1-\frac{m_A}{m_B}\,\xi\right)-c\left(1-\frac{m_B}{m_A}\,\mu\right)\right]=0. \tag{20}$$

Obviously, the "pure states" $c=0$ or $c=1$ represent stationary solutions of (20) and can be factorized out. The non-trivial condition for the glass forming is thus

$$(1-c)\left[1-\frac{m_A}{m_B}\xi\right]-c\left[1-\frac{m_B}{m_A}\mu\right]=0. \tag{21}$$

Now, using the limit conditions at $c\to 0, T_{\rm g}=T_{\rm g0}$ and $c\to 1, T_{\rm g}=T_{\rm g1}$, and introducing the generalized Boltzmann factors with the energy barriers for corresponding bond creations as E_{AA}, E_{AB} and E_{BB}, we can write

$$E_{AB}-E_{AA}=kT_{\rm g0}\ln\left(\frac{m_B}{m_A}\right),\quad E_{AB}-E_{BB}=kT_{\rm g1}\ln\left(\frac{m_A}{m_B}\right), \tag{22}$$

so that the expressions ξ and μ at the arbitrary temperature T can be written as

$$\begin{aligned}\xi(T)&=e^{\frac{E_{AB}-E_{AA}}{T_{\rm g0}}\cdot\frac{T_{\rm g0}}{T}}=\left(\frac{m_B}{m_A}\right)^{\frac{T_{\rm g0}}{T}};\\ \mu(T)&=e^{\frac{E_{AB}-E_{BB}}{T_{\rm g1}}\cdot\frac{T_{\rm g1}}{T}}=\left(\frac{m_A}{m_B}\right)^{\frac{T_{\rm g1}}{T}}.\end{aligned} \tag{23}$$

Substituting these expressions into (18) and taking the limit $c\to 0$, we get

$$\left.\frac{dT_{\rm g}}{dc}\right|_{c=0}=\frac{T_{\rm g0}\left[1-\left(\frac{m_B}{m_A}\right)^{\frac{T_{\rm g0}-T_{\rm g1}}{T_{\rm g0}}}\right]}{\ln\left(\frac{m_B}{m_A}\right)}. \tag{24}$$

It is easy to see now that even when $m_A=m_B$, this formula has a well defined limit. Indeed, if we set at first $m_B/m_A=1+\epsilon$, and then develop the numerator and the denominator of the above equation in powers of ϵ, then in the limit when $\epsilon\to 0$, we arrive at a very simple linear dependence which is in agreement with common sense and with experiment as well, namely

$$\left.\frac{dT_{\rm g}}{dc}\right|_{c=0}=T_{\rm g1}-T_{\rm g0}. \tag{25}$$

The utility and value of this simple model can be illustrated by the results it gives in a situation when one should take into account rapid cooling, often called *quenching*, which influences the measured glass transition temperature. We shall introduce the basic idea on the model of rim saturation, because it is simpler to work with 2×2 stochastic matrices.

Consider the agglomeration process defined by the above stochastic matrix, $\vec{p}' = M\vec{p}$, with $\vec{p}$ representing a normalized column (a "vector") with two entries, p_x and $p_y = 1 - p_x$. After one agglomeration step, representing on the average one new layer formed on the rim of a cluster, we can write

$$\Delta\vec{p} = \vec{p}' - \vec{p} = (M - \mathbf{1})\vec{p}. \tag{26}$$

Let us introduce a symbolic variable s defining the progress in the agglomeration process; obviously, $s(t)$ should be a monotonously increasing function during the glass transition. If the temperature variation is so slow that the derivative $dT/dt = (dT/ds)(ds/dt)$ can be neglected (which is often called the *annealing* of glass), the master equation of our model can be written as

$$\Delta\vec{p} = \frac{\partial\vec{p}}{\partial s}\Delta s = (M - \mathbf{1})\vec{p}\Delta s,$$

where the variation Δs represents one complete agglomeration step.

Now, if we want to describe the process using real time t as an independent parameter, we should write

$$\frac{d\vec{p}}{dt} = \frac{\Delta\vec{p}}{\Delta s}\frac{ds}{dt} = \tau^{-1}\frac{\Delta\vec{p}}{\Delta s} = \frac{1}{\tau}(M - \mathbf{1})\vec{p}. \tag{27}$$

We have introduced here the new entity $\tau = (ds/dt)^{-1}$ which can be interpreted as the average time needed to complete a new layer in any cluster, or alternatively, the time needed for an average bond creation. Now, if the temperature varies rapidly enough, the matrix M cannot be considered as constant anymore. Equation (27) must be modified according to the well known principle of "moving target." That is, the *total* derivative of $\vec{p}$ with respect to t should read:

$$\frac{d\vec{p}}{dt} = (M - \mathbf{1})\vec{p}\frac{ds}{dt} + \frac{dM}{dT}\frac{dT}{dt}\vec{p}. \tag{28}$$

We shall suppose a linear dependence of the temperature on time, so that the derivative dT/dt can be denoted by constant cooling rate q, and write (28) as

$$\frac{d\vec{p}}{dt} = \left[\frac{1}{\tau}(M - \mathbf{1}) + q\frac{\partial M}{\partial T}\right]\vec{p}. \tag{29}$$

In the two-dimensional case only one component of $\vec{p}$ is independent, because $p_x + p_y = 1$. Let us choose p_y (whose asymptotic value should be equal to c) as an independent variable. Then (29) will reduce to the single equation:

$$\frac{dp_y}{dt} = \frac{1}{\tau}\Big[(M_{yy}-1)p_y + M_{yx}(1-p_y)\Big] + q\left[\frac{\partial M_{yy}}{\partial T}p_y + \frac{\partial M_{yx}}{\partial T}(1-p_y)\right], \tag{30}$$

where we used the fact that $p_x = 1-p_y$, $M_{xx} = 1-M_{yx}$ and $M_{yy} = 1-M_{xy}$.

What remains is just simple algebra. After a few operations we find the asymptotic value of p_y, denoted p_y^∞, obtained when we set $dp_y/dt = 0$:

$$p_y^\infty = \frac{M_{yx} + \tau q\left(\frac{\partial M_{yx}}{\partial T}\right)}{(M_{xy}+M_{yx}) + \tau q\left(\frac{\partial(M_{xy}+M_{yx})}{\partial T}\right)}. \tag{31}$$

As in the former case, we define the glass transition temperature by solving the zero-fluctuation condition $p_y^\infty = c$. The quasi-equilibrium condition thus obtained can be written in a form displaying an apparent symmetry between the two ingredients of binary glass:

$$\begin{aligned}&\frac{m_B c(1-c)}{m_A(1-c)\xi + m_B c} - \frac{m_A c(1-c)}{m_A(1-c)+m_B c\mu}\\ &\quad = \frac{\tau q}{T} m_A m_B c(1-c)\left[\frac{c\mu\ln\mu}{[m_A(1-c)+m_B c\mu]^2} - \frac{(1-c)\xi\ln\xi}{[m_A(1-c)\xi + m_B c]^2}\right].\end{aligned} \tag{32}$$

As in the previous case (when $q = 0$), the extreme values $c = 0$ and $c = 1$ represent stationary solutions, which is obvious (no local fluctuations of concentration c are possible when there is no other ingredient than A or B atoms alone). After factorizing out $c(1-c)$, we get

$$\begin{aligned}&\frac{m_B}{m_A(1-c)\xi + m_B c} - \frac{m_A}{m_A(1-c)+m_B c\mu}\\ &\quad = \frac{\tau q}{T} m_A m_B \left[\frac{c\mu\ln\mu}{[m_A(1-c)+m_B c\mu]^2} - \frac{(1-c)\xi\ln\xi}{[m_A(1-c)\xi + m_B c]^2}\right],\end{aligned} \tag{33}$$

where we have used the fact that

$$\frac{\partial(\ln\xi)}{\partial T} = -\frac{\ln\xi}{T}, \quad \frac{\partial(\ln\mu)}{\partial T} = -\frac{\ln\mu}{T}.$$

The above formula seems quite cumbersome, but it becomes much simpler in the low concentration limit, $c \to 0$. Close to $c = 0$ we get

$$\frac{m_B}{m_A} - \xi + \frac{\tau q}{T}\frac{m_B}{m_A}\ln\xi = 0. \tag{34}$$

(Quite obviously, in the limit $c \to 1$ one gets the same formula switching m_A with m_B and replacing ξ by μ.) Replacing ξ by expression (23), we arrive at

$$\left[1 - \left(\frac{m_B}{m_A}\right)^{\frac{T_{g0}-T}{T}}\right] + \left(\frac{\tau q}{T}\right)\frac{T_{g0}}{T}\ln\left(\frac{m_B}{m_A}\right) = 0. \tag{35}$$

It is easy to see that independently of the ratio m_B/m_A, for temperatures T *above* T_{g0} we must have $q < 0$, and *vice versa*: during rapid cooling, the glass transition occurs at the temperature $T > T_{g0}$.

This can be seen as follows. If $T > T_{g0}$, then the exponent of (m_B/m_A) is negative. Suppose that (m_B/m_A) is greater then one. Then the first term is positive, and so is $\ln(m_B/m_A)$, therefore, q must be negative in order to satisfy Eq. (35). In contrast, if (m_B/m_A) is less than one, then the first term becomes negative, but now so is $\ln(m_B/m_A)$, and again q must be negative.

By this we also prove that the glass transition temperature with slow annealing (i.e., very slow cooling, $q \sim 0$), is the lowest T_g one can obtain.

The dimensionless combination $(\tau q)/T$ defines the quenching rate as the product of $(1/T)(dT/dt) = d(\ln T)/dt$ by the time constant τ, characterizing the kinetics of the agglomeration process, i.e., the average time it takes to create a new bond. It may depend weakly on the temperature, but for the sake of simplicity suppose it is constant. It can be determined by comparing formula (35) with the experimental data.

To take an example, let us again consider the selenium–arsenic glass at $c \to 0$ (almost pure Se with a small addition of As). We know that in this case $T_g \to T_{g0} = 318°\text{K}$. The formula (35) then gives the quasi-linear dependence of $\Delta T = T - T_{g0}$ on the quenching rate q.

Next, if we want to establish the formula for a pure glass-former, without any modifier, we should take the limit $(m_A/m_B) \to 1$ and $\mu \to \xi$; we then get

$$\frac{T - T_{g0}}{T} + \left(\frac{\tau q}{T}\right)\frac{T_{g0}}{T} = 0 \quad \text{or} \quad T - T_0 = \Delta T_g = -(\tau q)\frac{T_{g0}}{T}. \tag{36}$$

The deviations from this simple dependence may indicate that the characteristic time τ depends on T. This can shed more light on the agglomeration kinetics in various glass-forming liquids. The experiment shows that the dependence of T_g on q is of the logarithmic and not linear type. Also the characteristic time's dependence is exponential:

$$\tau(T) = \tau_0\, e^{\frac{E}{kT}}.$$

Therefore, the relation above can be considered as a linear approximation of a more realistic one,

$$T - T_0 = \Delta T_{\mathrm{g}} = \frac{T_{\mathrm{g}0}}{C - \ln\left(\frac{\tau_0|q|}{T_{\mathrm{g}0}}\right)} \simeq \frac{T_{\mathrm{g}0}}{C}\left(1 + \frac{1}{C}\ln\left(\frac{\tau_0|\,q\,|}{T_{\mathrm{g}0}}\right)\right). \qquad (37)$$

Here C is a constant that can be determined by experiment, and the linearized version of the formula gives a good approximation under the condition that the quantity $\frac{1}{C}\ln\left(\frac{\tau_0|q|}{T_{\mathrm{g}0}}\right)$ remains small compared to 1. Also the bulk viscosity of the liquid melt about to undergo glass transition displays the well-known Arrhenius behavior, i.e., exponential growth towards infinity when $T \to T_{\mathrm{g}}$:

$$\eta(T) \simeq \eta_0 e^{\frac{B}{kT}},$$

which is in direct relation with the characteristic agglomeration time τ. More details can be found in [20, 23].

Finally, let us investigate the influence of rapid cooling on the initial slope of the curve $T_{\mathrm{g}}(c)$. This can be done as follows. Rewriting the glass-forming condition (33) equivalently as vanishing of a certain function of the two variables, T and c:

$$\Phi(T,c) = \frac{m_B}{m_A(1-c)\xi + m_B c} - \frac{m_A}{m_A(1-c) + m_B c\mu},$$

$$-\frac{\tau q}{T} m_A m_B \left[\frac{c\mu \ln \mu}{[m_A(1-c) + m_B c\mu]^2} - \frac{(1-c)\xi \ln \xi}{[m_A(1-c)\xi + m_B c]^2}\right] = 0. \qquad (38)$$

The derivative (dT/dc) can be found using the implicit function formula

$$\frac{dT}{dc} = -\left[\left(\frac{\partial \Phi}{\partial c}\right)\Big/\left(\frac{\partial \Phi}{\partial T}\right)\right]_{c=0}. \qquad (39)$$

The calculus is quite tedious, and even in the limit $c = 0$, when we can use the reduced version of the identity (38) which amounts to

$$\xi = \frac{m_B}{m_A}\left(1 - \left(\frac{\tau q}{T_{\mathrm{g}0}}\right)\frac{\ln \xi}{\xi}\right),$$

it remains quite cumbersome:

$$\left(\frac{dT}{dc}\right)_{c=0} = \frac{T\left[\left(1 - \frac{m_B}{m_A}\mu\right) + \frac{m_B}{m_A}\left(\frac{\tau q}{T}\right)\left[\mu \ln \mu - \frac{m_B}{\xi m_A}\frac{\tau q}{T}\frac{(\ln \xi)^2}{\xi^2}\right]\right]}{\ln \xi\left[1 - \left(\frac{\tau q}{T}\right)\left(1 - \frac{m_B}{m_A}\frac{\tau q}{T}\frac{\ln \xi}{\xi}\right)(2 - \ln \xi)\right]}. \qquad (40)$$

This formula becomes more transparent if we suppose a relatively low quenching rate, $\frac{\tau q}{T} \ll 1$, and consequently linearize the above formula in the vicinity of $c = 0, T = T_{g0}$. Keeping only linear terms we get the simplified version:

$$\left(\frac{dT}{dc}\right)_{c=0} = \frac{T\left[\left(1-\frac{m_B}{m_A}\mu\right)+\left(\frac{\tau q}{T}\right)\left[\frac{m_B}{m_A}\mu\ln\mu+\left(1-\frac{m_B}{m_A}\mu\right)\left(2+\frac{m_A}{m_B}+\ln\frac{m_A}{m_B}\right)\right]\right]}{\ln\left(\frac{m_B}{m_A}\right)}$$

from which we recover the formula (19) when $q \to 0$.

Although we know that in order to satisfy the glass transition condition, the temperature must go down, and the cooling rate is indeed negative as it should be, $q < 0$, it is not easy to decide what is the actual slope when $q \neq 0$. The influence of the cooling on the initial slope of the curve $T_g(c)$ becomes obvious when we examine the case of two different glass formers having the same valence (e.g., the mixture of sulfur with selenium, both being of valence 2). In such a case, if the glass transition temperature of the first glass former is T_{g0} (i.e., T_g at $c = 0$) and that of the second glass former is T_{g1} (i.e., T_g at $c = 1$), and if $m_B = m_A$, then using the same limit procedure as before we obtain the following simple relation:

$$\left(\frac{dT}{dc}\right)_{c=0} = (T_{g1} - T_{g0})\left[1 + \frac{2\tau q}{T_{g0}}\right] \tag{41}$$

and as q must be always negative, this means that rapid cooling tends to diminish the initial slope of the curve $T_g(c)$, letting at the same time the glass transition temperature become higher than in the case of the very slow cooling rate.

More realistic models should take into account the dependence of τ and E on the temperature, which will modify the master equation (38).

Acknowledgments

Enlightening discussions with R. Aldrovandi, R.A. Barrio and M. Micoulaut are gratefully acknowledged.

References

[1] Kerner, R., *Physica B* **215** (1995) 267.
[2] Kerner, R. and Micoulaut, M., *J. Phys. Cond. Matt.* **9** (1997) 2551–2562.
[3] Kerner, R., A theory of glass formation, in *Atomic Diffusion in Amorphous Solids*, eds. Balkanski, M. and Elliott, R.J. (World Scientific, Singapore, 1998), pp. 25–80.
[4] Kerner, R., The principle of self-similarity and its applications to the description of non-crystalline matter, in *Proc. Workshop in Cocoyoc*, 1997, ed. Moran-Lopez, J.L. (Plenum Press, 1998), pp. 323–337.

[5] dos Santos-Loff, D.M. and Kerner, R., *J. Physique* **I**(4) (1994) 1491–1511.

[6] Kerner, R., Bennemann, K.H. and Penson, K., *Europhys. Lett.* **19**(5) (1992) 363–368.

[7] Kerner, R., *Comput. Mat. Sci.* **2** (1994) 500–508.

[8] dos-Santos-Loff, D.M., Kerner, R. and Micoulaut, M., *J. Phys. C* **7** (1995) 8035–8052.

[9] Barrio, R.A., Duruisseau, J.P. and Kerner, R., *Phil. Mag. B* **72**(5) (1995) 535–550.

[10] Barrio, R.A., Kerner, R., Micoulaut, M. and Naumis, G.G., *J. Phys.: Cond. Matt.* **9** (1997) 9219–9234.

[11] Kerner, R. and Naumis, G.G., *J. Physics: Cond. Matt.* **12**(8) (2000) 1641–1648.

[12] Kerner, R., *J. Non-Cryst. Solids* **182** (1995) 9–21.

[13] Kerner, R. and Micoulaut, M., *J. Non-Cryst. Solids* **210** (1997) 298–305.

[14] Aldrovandi, R. and Kerner, R., in "New Symmetries and Integrable Models", in *Proc. XIV Max Born Symposium*, eds. Frydryszak, A., Lukierski, J. and Popowicz, Z. (World Scientific, Singapore, 2000), pp. 153–169.

[15] Kerner, R. and Micoulaut, M., *J. Non-Cryst. Solids* **210** (1997) 298.

[16] Gerardo G. Naumis and Kerner, R., *J. Non-Cryst. Solids* **231** (1998) 111.

[17] Boolchand, P. and Bresser, W.J., *Phil. Mag. B* **80** (2000) 1757–1772.

[18] Boolchand, P., Feng, X. and Bresser, W.J., Rigidity transition in binary Ge–Se glasses (2000), preprint, University of Cincinnati, to appear in *J. Non-Cryst. Solids.*

[19] Georgiev, D.G., Boolchand, P. and Micoulaut, M. *Phys. Rev. B* **62** (2000) R9228.

[20] Kerner, R., Mathematical models of glass formation, in *Proc. Conf.* "Glasses and Solid Electrolytes", St. Petersburg, May, 1999, *Glass Phys. Chem.* **26**(4) (2000) 313–324.

[21] Micoulaut, M., private communication (2001).

[22] Kerner, R. and Micoulaut, M., *Acad. C. R. Sci. Paris* **315**(II) (1992) 1307–1313.

[23] Kerner, R. and Phillips, J.C., *Solid State Commun.* **117** (2000) 47–56, Kerner, R., in preparation.

CHAPTER 10

THE DIELECTRIC LOSS FUNCTION AND THE SEARCH FOR SIMPLE MODELS FOR RELAXATION IN GLASS FORMERS

A.P. Vieira

Instituto de Física, Universidade de São Paulo, C.P. 66318, 05315-970, São Paulo, Brazil

M. López de Haro and J. Tagüeña-Martínez

Centro de Investigación en Energía, UNAM, Temixco, Morelos, 62580, Mexico

L.L. Gonçalves

Departamento de Física, Universidade Federal do Ceará, C.P. 6030, 60451-970, Fortaleza, CE, Brazil

We present a brief review of recent advances in experimental investigations of dielectric relaxation in glass-forming systems, especially with regard to the feature of the excess wing and the applicability of Nagel scaling. Based on the new findings described, we analyze our contributions to the search for microscopic models yielding glass-like relaxation behavior, and suggest future investigations.

In recent years, much effort has been put into attempts to describe the behavior of glasses and glass-forming systems. Despite all the work, there is an ongoing debate even on such fundamental aspects as the existence of an ideal glass transition (for a recent review see [1]). In particular, relaxation properties of these systems have been the subject of intensive experimental work, often probing dielectric relaxation, for which a very broad range of frequencies can be studied.

Dielectric spectra of glass formers exhibit a structural, or α relaxation, characterized by a time scale τ_α, which is related to viscous properties and in many systems seems to diverge below the glass temperature $T_{\rm g}$, according to the Vogel–Fulcher–Tammann expression

$$\tau_\alpha \propto \exp\left(\frac{E_\alpha}{T - T_{\rm K}}\right), \tag{1}$$

where $T_{\rm K}$, the Kauzmann temperature, would correspond to the ocurrence of an ideal glass transition. In some systems, sometimes classified as type-B [2], a secondary, faster process is also present, called β or Johari–Goldstein relaxation. In contrast to the α relaxation, the thermal dependence of the time scale τ_β associated with the β relaxation follows an Arrhenius form,

$$\tau_\beta \propto \exp\left(\frac{E_\beta}{T}\right). \tag{2}$$

The nature of the microscopic processes giving rise to the β relaxation is still a matter of controversy (see [3] and references therein).

In general, owing to its non-Debye character, the frequency-dependent dielectric response corresponding to the α relaxation is fitted by empirical expressions such as the Cole–Davidson function

$$\varepsilon_{\rm CD}(\omega) \propto \frac{1}{(1 + i\omega\tau_\alpha)^\beta}, \tag{3}$$

or the Kohlrausch–Williams–Watts (streched-exponential) function

$$\varepsilon_{\rm KWW}(\omega) \propto \text{the Fourier transform of } \left\{-\frac{d}{dt}\exp\left[-\left(\frac{t}{\tau_\alpha}\right)^\beta\right]\right\}. \tag{4}$$

In type-A systems [2], for which a β relaxation seems to be absent (however, see below), the dielectric loss function $\varepsilon''(\omega) = -\text{Im}\,\varepsilon(\omega)$, exhibits an excess contribution to the high-frequency power law $\varepsilon''(\omega) \propto \omega^{-\beta}$ expected from the fits. This "excess wing," which appears a few decades above the peak-frequency ω_p, is well described by a second power law $\varepsilon''(\omega) \propto \omega^{-\gamma}$, with $\gamma < \beta$. For many small-molecule glass formers, $\varepsilon''(\omega)$ curves for different temperatures and materials can be collapsed onto a single master curve by the so-called Nagel scaling [4–6], which plots the ordinate $w^{-1}\log(\omega_p\varepsilon''/\omega\Delta\varepsilon)$ versus the abscissa $w^{-1}(1 + w^{-1})\log(\omega/\omega_p)$. Here, w denotes the half-maximum width of the loss curve normalized to that of a Debye peak and $\Delta\varepsilon$ is the relaxation strength. More recently, it has been shown that the scaling is violated in various situations [7, 8], thereby challenging the universal character of Nagel's master function.

Another line of research concentrates on understanding the very nature of the excess wing. Recent experiments in glycerol and propylene carbonate [9] provide evidence that the excess wing is a nonequilibrium process, since dielectric spectra obtained along five-day measurements show the excess wing progressively turning into a secondary relaxation process, as indicated by the appearance of a high-frequency shoulder. This gives support to the hypothesis that the excess wing is in fact the high-frequency flank of a β relaxation, hidden under the dominant α peak. In type-A systems, the characteristic relaxation times τ_α and τ_β would be too close for the α and β relaxations to be clearly resolved, except at temperatures well below T_{g}, where τ_α becomes of the order of days, and equilibrium is reached only very slowly (i.e., ageing is present). In the materials studied in [9], as well as in glassy or supercooled ethanol [7], a fitting curve composed of a sum of Cole–Davidson and Cole–Cole functions satisfactorily describes experimental data. Since β relaxations are usually well parametrized by a Cole–Cole function,

$$\varepsilon''_{\mathrm{CC}}(\omega) \propto \frac{1}{1+(i\omega\tau_\beta)^\gamma}, \tag{5}$$

this constitutes further evidence in favor of interpreting the excess wing as a hidden β process. It is also checked that a Cole–Cole function can be obtained from a sum of heterogeneous Debye terms with relaxation times corresponding to a Gaussian distribution of activation energies [2, 10–12].

Experiments in polychlorinated biphenyls with different chemical structures have shown that the α relaxation remains unaltered, while the excess wing is strongly sensitive to variations of molecular weight, leading to violations of the Nagel scaling [8]. Although the $\varepsilon''(\omega)$ curves do not exhibit secondary peaks, the region corresponding to the excess wing can once more be adjusted by a Cole–Cole function with a parameter τ_β varying with temperature in an Arrhenius form. Similarly, measurements taken in dimethyl ethers with variable chain lengths, but comparable fragilities, revealed the transformation of an excess wing into a β relaxation with increasing molecular weight [13]. Finally, very recent experiments in polypropylene glycol terpolymer [14], whose dielectric spectra exhibit both an excess wing and a clear secondary relaxation peak, have shown that, under high pressure, the excess wing changes into a second β relaxation, although its properties seem to be highly correlated with those of the α relaxation.

With such a fast-changing experimental picture, it is not surprising that no comprehensive microscopic description of glassy behavior is available. Instead, most theoretical approaches account for a limited set of features, sometimes providing good qualitative descriptions (for an up-to-date overview see [15]). In the last few years, looking for a bona fide microscopic model capable of yielding the Nagel scaling and the excess wing, for which competing relaxation mechanisms are presumably essential, we have been investigating the frequency-dependent susceptibility of inhomogeneous, but periodic, one-dimensional kinetic Ising systems, employing both the usual Glauber dynamics [16, 17] and a restricted dynamics [18] introduced in the context of lynear polymer chains [19, 20]. The competing relaxation processes are generated by the presence of spins with two or more bare relaxation times (i.e., relaxation times of the isolated spins), which become coupled and temperature-dependent due to interactions between particles. Unfortunately, a mistake was made in the way we computed the half-maximum width of the imaginary part of the susceptibility in the Nagel plots, so that the conclusions regarding the ability of our models to yield the Nagel scaling had to be put in the proper perspective [21–23]. In any event, the analytical results we obtained for the susceptibility and also for the dynamical exponent are not affected by such an error. On the one hand, we could check that introducing n different values for the coupling constants leads to non-universal behavior of the dynamical exponent of the Glauber–Ising chain [23], as in the alternating chain [24]; we could also verify this non-universal behavior in more complex one-dimensional structures, such as chains with rings [25], when the spins are subject to the restricted dynamics. On the other hand, the presence of more than one relaxational process still allowed us to investigate whether those systems were able to produce an excess wing.

Motivated by encouraging results obtained using both dynamics in periodic structures [25], we proceeded by considering the effects of randomness or quasiperiodicity on the dynamical response. However, we could not significantly improve over our previous results: the presence of widely different bare relaxation times yields a dynamical response exhibiting two relaxation processes characterized by two time scales, mimicking τ_α and τ_β, but their thermal dependences are too distinct for a consistent excess wing to appear. Figure 1 shows a typical example of the behavior of the imaginary part of the dynamical susceptibility $\chi(\omega)$ in a Glauber–Ising chain with bare relaxation times chosen either as $\tau_0 = 1/\omega_0$ (with probability $p = 9/10$) or $\tau_i = 1/\omega_i$ (with probability $1 - p$), where ω_i is a Gaussian-distributed random variable with mean $\bar{\omega}$ and width σ, with $\bar{\omega} = \sigma = 100 \gg \omega_0 = 1$; for details of the analytical calculation see [23]. In the light of these findings, and the recent experimental evidence on the possible non-equilibrium nature of the excess wing, we are led to believe that more sophisticated models are needed in order to faithfully reproduce the relaxation processes in glass formers. For instance, it would be interesting to check whether

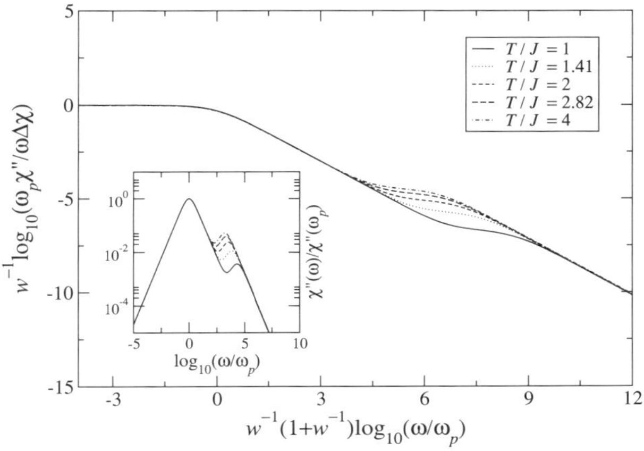

Fig. 1. Nagel plots of the imaginary part of the frequency-dependent susceptibilty $\chi(\omega)$ of a Glauber–Ising chain with uniform couplings J and various temperatures T (in units of the Boltzmann constant). The inset shows a Debye plot. The bare relaxation times are chosen either as $\tau_0 = 1/\omega_0$ (with probability $p = 9/10$) or $\tau_i = 1/\omega_i$ (with probability $1 - p$), where ω_i is a random Gaussian variable with mean $\bar{\omega}$ and width σ, with $\bar{\omega} = \sigma = 100 \gg \omega_0 = 1$.

models with random p-spin interactions, which are known to exhibit phenomena analogous to glassy systems (see the article by L. Cugliandolo in [15]), especially ageing, also present a time-dependent excess wing in their dynamical magnetic response.

Acknowledgments

We acknowledge financial support from DGAPA-UNAM under Projects Nos. IN101100 and IN103100, and from the Brazilian agencies CNPq, Finep and Fapesp.

References

[1] Debenedetti, P.G. and Stillinger, F.H., *Nature* **410** (2001) 259.
[2] Kudlik, A. *et al.*, *J. Mol. Struct.* **479** (1999) 201.
[3] Hensel-Bielowka, S. and Paluch, M., *Phys. Rev. Lett.* **89** (2002) 025704.
[4] Dixon, P.K., Wu, L., Nagel, S.R., Williams, B.D. and Carini, J.P., *Phys. Rev. Lett.* **65** (1990) 1108.
[5] Dixon, P.K., *Phys. Rev. B* **42** (1990) 8179.
[6] Leslie-Pelecky, D.L. and Birge, N.O., *Phys. Rev. Lett.* **72** (1994) 1232.
[7] Brand, R., Lunkenheimer, P., Schneider, U. and Loidl, A., *Phys. Rev. B* **62** (2000) 8878.

[8] Casalini, R. and Roland, C.M., *Phys. Rev. B* **66** (2002) 180201 (R).
[9] Schneider, U., Brand, R., Lunkenheimer, P. and Loidl, A., *Phys. Rev. Lett.* **84** (2000) 5560.
[10] Wu, L., *Phys. Rev. B* **43** (1991) 9906.
[11] Wu, L. and Nagel, S.R., *Phys. Rev. B* **46** (1992) 11198.
[12] Deegan, R.D. and Nagel, S.R., *Phys. Rev. B* **52** (1995) 5653.
[13] Mattsson, J., Bergman, R., Jacobsson, P. and Börjesson, L., *Phys. Rev. Lett.* **90** (2003) 075702.
[14] Casalini, R. and Roland, C.M., *Phys. Rev. Lett.* **91** (2003) 015702.
[15] Barrat, J., Feigelman, M.V., Kurchan, J. and Dalibard, J. (eds.), *Slow Relaxations and Nonequilibrium Dynamics in Condensed Matter*, Vol. 77 of *Les Houches — Ecole d'Ete de Physique Theorique* (Springer, New York, 2003).
[16] Gonçalves, L.L., López de Haro, M., Tagüeña-Martínez, J. and Stinchcombe, R.B., *Phys. Rev. Lett.* **84** (2000) 1507.
[17] Gonçalves, L.L., López de Haro, M. and Tagüeña-Martínez, J., *Braz. J. Phys.* **30** (2000) 731.
[18] Gonçalves, L.L., López de Haro, M. and Tagüeña-Martínez, J., *Phys. Rev. E* **63** (2001) 026114.
[19] López de Haro, M., Tagüeña-Martínez, J., Espinosa, B. and Gonçalves, L.L., *J. Phys. A: Math. Gen.* **26** (1993) 6697.
[20] López de Haro, M., Tagüeña-Martínez, J. and Gonçalves, L.L., *Mod. Phys. Lett. B* **10** (1996) 1441.
[21] Gonçalves, L.L., López de Haro, M., Tagüeña-Martínez, J. and Stinchcombe, R.B., *Phys. Rev. Lett.* **88** (2002) 089901 (E).
[22] Gonçalves, L.L., López de Haro, M. and Tagüeña-Martínez, J., *Phys. Rev. E* **68** (2003) 049903 (E).
[23] Gonçalves, L.L., López de Haro, M., Tagüeña-Martínez, J. and Vieira, A.P., in *Recent Developments in Mathematical and Experimental Physics*, eds. Macías, A., Uribe, F. and Díaz, E., Vol. B: Statistical Physics and beyond (Kluwer Academic/Plenum, New York, 2003), p. 215.
[24] Droz, M., Kamphorst Leal da Silva, J. and Malaspinas, A., *Phys. Lett. A* **115** (1986) 448.
[25] López de Haro, M., Tagüeña-Martínez, J., Gonçalves, L.L. and Vieira, A.P., *J. Non-Cryst. Solids* **329** (2003) 82.

CHAPTER 11

THE THEORY OF TURING PATTERN FORMATION

Teemu Leppänen
Helsinki University of Technology, Laboratory of Computational Engineering, P.O. Box 9203, FIN-02015 Hut, Finland
teemu.leppanen@hut.fi

The theory behind Turing instability in reaction-diffusion systems is reviewed. The use of linear analysis and nonlinear bifurcation analysis with center manifold reduction for studying the behavior of Turing systems is presented somewhat meticulously at an introductory level. The symmetries that are considered here in the context of a generic Turing model are the two-dimensional hexagonal lattice and three-dimensional SC- and BCC-lattices.

1. Introduction

The self-organization of dissipative structures is a phenomenon typical to non-equilibrium systems. These structures exist far from equilibrium and differ from typical equilibrium structures (e.g., crystals) in that they are kept in the steady-state by competing ongoing dynamic processes, which feed energy into the system. The structures persist by dissipating the input energy (and thus generating entropy), which makes the process irreversible. Dissipative structures are typically macroscopic and the characteristic length scale of the structure is independent of the size of the individual constituents (e.g., molecules) of the system. The systems representing self-organized dissipative structures vary from growing bacterial colonies to fluids with convective instabilities (e.g., Rayleigh-Bénard convection) [1, 2].

The formal theory of self-organization is based on non-equilibrium thermodynamics [3] and was pioneered by chemist Ilya Prigogine. Most of the research was made in Brussels from the 1940s to 1960s by Prigogine and co-workers. They extended the treatment of non-equilibrium thermodynamic systems to nonlinear regime far from equilibrium and applied bifurcation theory to analyze the state selection [4]. Already in 1945, Prigogine had suggested that a system in non-equilibrium tries to minimize the rate of entropy production and chooses the state accordingly [5]. This condition was proved inadequate by Rolf Landauer in 1975, who argued that minimum entropy production is not in general a necessary condition for the steady-state and that one cannot determine the most favorable state of the system based on the behavior in the vicinity of the steady state, but one must consider the global non-equilibrium dynamics [6]. Nevertheless, in 1977 Prigogine was awarded the Nobel prize in chemistry for his contribution to the theory of dissipative structures.

This article does not deal with the so far incomplete theory of non-equilibrium thermodynamics, but with chemical systems, which can exhibit instabilities resulting in either oscillatory or stationary patterns [7, 8]. The difficulties of non-equilibrium thermodynamics are also present in the theory of chemical pattern formation. We will specifically concentrate on systems showing so-called Turing instability, which arises due to different diffusion rates of reacting chemical substances. Turing instability can be thought of as a competition between activation by a slow diffusing chemical (activator) and inhibition by a faster chemical (inhibitor). The idea of diffusion-driven instability was first discussed by Nicolas Rashevsky in 1938 [9], but the renowned British mathematician and computer scientist Alan Turing has earned most of his fame for giving the first mathematical treatment and analysis of such a model in 1952 [10].

Turing's motivation for studying the chemical system was biological, and consequently, the seminal paper was titled *The Chemical Basis of Morphogenesis*, where he called the reacting and diffusing chemicals *morphogens*. Turing emphasized that his model is very theoretical and a severe simplification of any real biological system, but he was still confident that his model could explain some of the features related to spontaneous symmetry-breaking and morphogenesis, i.e., the growth of form in nature. Turing neglected the mechanical and electrical aspects and considered the diffusion and reaction of morphogens in the tissue to be more important [10]. Nowadays, there is some qualitative evidence of the capability of Turing models to imitate biological self-organization [11–13], but the conclusive proof that morphogenesis is indeed a Turing-like process is still elusive.

The problem with Turing's theory was that the existence of chemical spatial patterns as predicted by his mathematical formulations could not be confirmed experimentally. The existence of Turing patterns in any chemical system could be questioned, not to mention the biological systems. Actually, in the early 1950s a Russian biochemist, Boris Belousov, observed an oscillation in a chemical reaction, but he could not get his results published in any journal, because he could not explain the results, which were claimed to contradict with the Second Law of Thermodynamics. Prior to Prigogine's work it was held that entropy always has to increase in a process and thus a chemical oscillation was deemed to be impossible with entropy increasing and decreasing by turns. It was not before the late 1960s, when it was noticed that the reaction first observed by Belousov exhibits a pattern formation mechanism with similarities to the mechanism Turing had proposed. However, it is important to note that the Belousov–Zhabotinsky reaction forms traveling waves, whereas Turing patterns are time-independent.

The first experimental observation of stationary Turing patterns was preceded by theoretical studies [14, 15] and the practical development of a new kind of continuously stirred tank reactors [16]. It was not until 1990, when Patrick De Kepper's group observed a stationary spotty pattern in a chemical system involving the reactions of chlorite ions, iodide ions and malonic acid (CIMA reaction) [17]. In particular the required difference in the diffusion rates of the chemical substances delayed the first experimental observation of Turing patterns. The condition was achieved by carrying out the experiment in a slab of polymer gel and using a starch indicator, which decreases the diffusion rate of the activator species (iodide ions). In 1991 stripes were also observed in the CIMA reaction and it was shown that experimental Turing patterns can be grown also over large domains [18].

The observation of Turing patterns initiated a renewed interest in Turing pattern formation. What had previously been only a mathematical prediction of an ingenious mathematician was 40 years later confirmed to be a real chemical phenomenon. This article introduces the reader to the theoretical background and mathematical formalism related to the study of Turing pattern formation. The reader will be introduced to the idea of Turing instability and to the mathematical models that generate Turing patterns in Sec. 2. In Sec. 3 the same topics will be discussed from a more mathematical point of view. In Sec. 4 the methods of the bifurcation theory will be applied to the study of the pattern (2D) or structure (3D) selection in a generic Turing model.

2. Turing Instability

The traveling wave patterns generated by the Belousov–Zhabotinsky reaction are not due to Turing instability since the diffusion rates of the chemicals involved in that reaction are usually more or less the same. The difference in the diffusion rates of the chemical substances is a necessary, but not a sufficient condition for the diffusion-driven or Turing instability. Turing's idea that diffusion could make a stable chemical state unstable was innovative since usually diffusion has a stabilizing effect (e.g., a droplet of ink dispersing into water). Intuitively Turing instability can be understood by considering the long-range effects of the chemicals, which are not equal due to a difference in the pace of diffusion and thus an instability arises. Some exhilarating common sense explanations of the mechanism of the instability can be found from the literature: Murray has discussed sweating grasshoppers on a dry grass field set alight [11] and Leppänen *et al.* have tried to illustrate the mechanism by using a metaphor of biking missionaries and hungry cannibals on an island [19].

Turing begun his reasoning by considering the problem of a spherically symmetrical fertilized egg becoming a complex and highly structured organism. His purpose was to propose a mechanism by which the genes could determine the anatomical structure of the developing organism. He assumed that genes (or proteins and enzymes) act only as catalysts for spontaneous chemical reactions, which regulate the production of other catalysts or morphogens. There was not any new physics involved in Turing's theory. He merely suggested that the fundamental physical laws can account for complex physico-chemical processes. If one has a spherically symmetrical egg, it will remain spherically symmetrical forever notwithstanding the chemical diffusion and reactions. Something must make the stable spherical state unstable and thus cause spontaneous symmetry-breaking. Turing hypotetized that a chemical state, which is stable against perturbations, i.e., returns to the stable state in the absence of diffusion may become unstable against perturbations in the presence of diffusion. The diffusion-driven instability initiated by arbitrary random deviations results in spatial variations in the chemical concentration, i.e., chemical patterns [10].

No egg in the blastula stage is exactly spherically symmetrical and the random deviations from the spherical symmetry are different in two eggs of the same species. Thus one could argue that those deviations are not of importance since all the organisms of a certain species will have the same anatomical structure irrespective of the initial random deviations. However, Turing emphazised and showed that "it is important that there are some deviations for the system may reach a state of instability in which these irregularities tend to grow" [10]. In other words, if there are no random deviations, the egg will stay in the spherical state forever. In biological systems the random deviations arise spontaneously due to natural noise and distortions. A further unique characteristic of the Turing instability is that the resulting chemical pattern (or biological structure) is not dependent on the direction of the random deviations, which are necessary for the pattern to arise. The morphological characteristics of the pattern, e.g., whether it is stripes or spots, are determined by the rates of the chemical reactions and diffusion. That is, the morphology selection is regulated intrinsically and not by external length scales (as in the case of many convective instabilities [2]).

The random initial conditions naturally have an effect on the resulting pattern, but only with respect to the phase of the pattern. The intrinsic parameters determine that the system will evolve to stripes of a fixed width, but the random initial conditions determine the exact positions and the alignment of the stripes (the phase). The relevance of Turing instability is not confined to chemical systems, but also many other physical systems exhibiting dissipative structures can be understood in terms of diffusion-driven instability. Turing instability has been connected to gas discharge systems [20], catalytic surface reactions [21], semiconductor nanostructures [22], nonlinear optics [23], irradiated materials [24] and surface waves on liquids [25]. In this article we will concentrate solely on reaction-diffusion Turing systems.

3. Linear Theory

In this section the mathematical formulation of Turing models is introduced, the Turing instability is discussed using mathematical terms and analyzed by employing linear stability analysis. We will witness the strength of linear analysis in predicting the instability and the insufficiency of it in explaining pattern selection. The nonlinear analysis required for explaining pattern selection will be presented in the next section.

Let us denote two space- and time-dependent chemical concentrations by $U(\vec{x},t)$ and $V(\vec{x},t)$, where $\vec{x} \in \Omega \subset \mathbb{R}^n$ denotes the position in an n-dimensional space, $t \in [0,\infty)$ denotes time and Ω is a simply connected bounded domain. Using these notations one can derive a system of reaction-diffusion equations from first principles [11] and they are given as follows

$$\begin{aligned} U_t &= D_U \nabla^2 U + f(U,V), \\ V_t &= D_V \nabla^2 V + g(U,V), \end{aligned} \tag{1}$$

where $U \equiv U(\vec{x}, t)$ and $V \equiv V(\vec{x}, t)$ are the morphogen concentrations, and D_U and D_V the corresponding diffusion coefficients setting the time scales for diffusion. The diffusion coefficients must be unequal for the Turing instability to occur in two or more dimensions [26]. One should note that Eq. (1) is actually a system of two diffusion equations, which are coupled via the kinetic terms f and g describing the chemical reactions. This reaction-diffusion scheme is generalizable to any number of chemical species, i.e., equations.

The form of the reaction kinetics f and g in Eq. (1) determines the behavior of the system. These terms can be derived from the chemical formulae describing the reaction by using the law of mass action [11] or devised based on phenomenological considerations. There are numerous possibilities for the exact form of the reaction kinetics including the Gray–Scott model [27–30], the Gierer–Meinhardt model [31], the Selkov model [32, 33], the Schnackenberg model [34, 35], the Brysselator model [4, 36] and the Lengyel–Epstein model [14, 37, 38]. All these models consist of two coupled reaction-diffusion equations and exhibit the Turing instability within a certain parameter range. The Lengyel–Epstein model is the only one which corresponds to a real chemical reaction (the experimentally observed CIMA reaction). The presence of some nonlinear term in f and g is common to all these models. A nonlinearity is required since it bounds the growth of the exponentially growing unstable modes.

The stationary state (U_c, V_c) of a model is defined by the zeros of the reaction kinetics, i.e., $f(U_c, V_c) = g(U_c, V_c) = 0$. Typically the models are devised in such a manner that they have only one stationary state, but some models have more. With certain parameters the stationary state is stable against perturbations in the absence of diffusion, but in the presence of diffusion the state becomes unstable against perturbations. If we initialize the system to the stationary state, it will remain there forever. If we perturb the system around the stationary state in the absence of diffusion, the system will return to its original state, i.e., the state is stable. However, when we perturb the system in the presence of diffusion arbitrarily around the stationary state, the perturbations will grow due to the diffusion-driven instability, i.e., the state is unstable. Mathematical definitions of stability can be found from the literature [40].

All the analysis in this article will be carried out using a generic Turing model introduced by Barrio *et al.* [39] This model is a phenomenological model, where the reaction kinetics are obtained by Taylor expanding the nonlinear functions around a stationary solution (U_c, V_c). If terms of the fourth and higher order are neglected, the reaction-diffusion equations can be written as

$$\begin{aligned} u_t &= D\delta\nabla^2 u + \alpha u(1 - r_1 v^2) + v(1 - r_2 u), \\ v_t &= \delta\nabla^2 v + v(\beta + \alpha r_1 uv) + u(\gamma + r_2 v), \end{aligned} \tag{2}$$

where the concentrations have been normalized so that $u = U - U_c$ and $v = V - V_c$, which makes $(u_c, v_c) = (0, 0)$ a stationary state. The parameters r_1, r_2, α, β and γ have numerical values and they define the reaction kinetics. The diffusion coefficients are written in terms of a scaling factor δ and the ratio D. We must always have $D \neq 1$ for the instability.

To reduce the number of parameters and simplify the analysis we carry out non-dimensionalization [11] of Eq. (2) by rescaling the parameters, concentrations and the time and length scales. This yields the system

$$\begin{aligned} \bar{u}_t &= D\nabla^2\bar{u} + \nu(\bar{u} + a\bar{v} - \overline{uv}^2 - C\overline{uv}), \\ \bar{v}_t &= \nabla^2\bar{v} + \nu(b\bar{v} + h\bar{u} + \overline{uv}^2 + C\overline{uv}), \end{aligned} \tag{3}$$

where the concentrations are scaled such that $(u, v) = \frac{1}{\sqrt{r_1}}(\bar{u}, \bar{v})$ and the time–space relation is given by $T = L^2/\delta$ ($t = T\tau$ and $x = L\bar{x}$). In terms of the original parameters, the new parameters are $C = r_2/(\alpha\sqrt{r_1})$, $a = 1/\alpha$, $b = \beta/\alpha$, $h = \gamma/\alpha$ and $\nu = \alpha T$. The term C adjusts the relative strength of the quadratic and cubic nonlinearities favoring either stripe or spot selection [39]. From now on we omit the overlines for simplicity.

One can easily see that the system of Eq. (3) has a stationary state at $(u_c, v_c) = (0, 0)$. For $h \neq -1$ the system has also two other stationary states defined by $f(u_c, v_c) = g(u_c, v_c) = 0$, and they are given by

$$u_c^i = -v_c^i/K \tag{4}$$

and

$$v_c^i = \frac{-C + (-1)^i \pm \sqrt{C^2 - 4(h - bK)}}{2}, \tag{5}$$

where $K = \frac{1+h}{a+b}$ and $i = 1, 2$.

Linear analysis is a general method used for evaluating the behavior of perturbations in a nonlinear system in the vicinity of a stationary state [11, 40, 41]. In the linear analysis one takes into account only the linear terms and thus the results are insufficient. However, within its limitations the method is typically quite efficient in predicting the existence of an instability and the characteristic wavelength of it. The linearized system in the absence of diffusion can be written in the form $\vec{w}_t = \mathbf{A}\vec{w}$ and reads as

$$\begin{pmatrix} u_t \\ v_t \end{pmatrix} = \begin{pmatrix} f_u & f_v \\ g_u & g_v \end{pmatrix}_{u_c, v_c} \begin{pmatrix} u \\ v \end{pmatrix}, \tag{6}$$

where f_u, f_v, g_u and g_v in the matrix $\mathbf{A}$ denote the partial derivatives of the reaction kinetics evaluated at the stationary state (u_c, v_c). In the case of Eq. (3) the linearized matrix is given as

$$\mathbf{A} = \nu \begin{pmatrix} 1 - v_c^2 - Cv_c & -2u_c v_c + a - Cu_c \\ v_c^2 + h + Cv_c & b + 2u_c v_c + Cu_c \end{pmatrix}, \tag{7}$$

where u_c and v_c are given by Eqs. (4) and (5). In the linear analysis the spatial and time variance are taken into account by substituting a trial solution of the form $w(\vec{r}, t) = \sum_k c_k e^{\lambda t} w_k(\vec{r}, t)$ into the linearized system in the presence of diffusion. The eigenvalues of this linearized system are obtained from the equation

$$|\mathbf{A} - \mathbf{D}k^2 - \lambda \mathbf{I}| = 0, \tag{8}$$

where $\mathbf{A}$ is given by Eq. (6), $D_{11} = D_u, D_{22} = D_v, D_{12} = D_{21} = 0$ and $\mathbf{I}$ is the identity matrix in the general case. In the case of Eq. (3) $\mathbf{A}$ is given by Eq. (7), and $D_{11} = D$ and $D_{22} = 1$. The determinant in Eq. (8) can be solved, which yields the equation

$$\begin{aligned} \lambda^2 + \left[(D_u + D_v)k^2 - f_u - g_v\right] \lambda + D_u D_v k^4 \\ - k^2(D_v f_u + D_u g_v) + f_u g_v - f_v g_u = 0, \end{aligned} \tag{9}$$

where $k^2 = \vec{k} \cdot \vec{k}$. The dispersion relation $\lambda(k)$ predicting the unstable wave numbers can be solved from Eq. (9). One can obtain an estimate for the most unstable wave number and the critical value of the bifurcation parameter by considering the fact that at the onset of the instability $\lambda(k_c) = 0$. Thus the term independent of λ in Eq. (9) must be zero at k_c. In the case of the generic Turing model this condition reads as

$$Dk_c^4 - k_c^2 \nu (Db + 1) + \nu^2 (b - ah) = 0. \tag{10}$$

At the onset this equation has only one solution given by $k_c^2 = \nu(Db + 1)/(2D)$, which takes place for the bifurcation parameter value

$a = a_c = -(Db-1)^2/(4Dh)$. An instability exists for $a < a_c$. The conditions for the Turing instability are widely known [8, 11] to be the following

$$\begin{aligned} f_u + g_v &< 0, \\ f_u g_v - f_v g_u &> 0, \\ D_v f_u + D_u g_v &> 0. \end{aligned} \tag{11}$$

Based on linear analysis and numerical observations it can be deduced that for $h < -1$ stationary state $(0, 0)$ is no more unstable and one of the other stationary states shows a Hopf-like bifurcation to a damped time-dependent oscillatory state without a characteristic length scale. On the other hand, for $h > -1$ there is a coupling of Turing instability with a characteristic length (bifurcation of the state $(0, 0)$) and a time-dependent instability with $k = 0$ unstable (bifurcation of the state defined by Eq. (5) with $i = 1$), which results in a transient large amplitude competition between oscillatory and fixed length scale instability. To simplify the study of the generic model of Eq. (3), we will fix $h = -1$ from now on. This makes (0, 0) the only stationary state and simplifies our analysis.

In the case of the generic model with $h = -1$ the above conditions (Eq. (11)) and the previous reasoning yield the constraints $-b < a < a_c = (Db-1)^2/(4D)$ and $-1/D < b < -1$ for the Turing instability. Based on this one can sketch the stability diagram for the generic Turing model. This is presented in Fig. 1. From the stability diagram one notices that the number of the parameter sets exhibiting Turing instability, i.e., the size of the Turing space (shaded region), is relatively small. For explanations of the other regions of the diagram see the caption.

Based on the stability diagram one can choose parameters that result in the Turing instability. Here we use two sets of parameters that have been used earlier in the case of the generic Turing model [39, 42]. For the non-dimensionalized generic model (with $L = 1$) the parameters are given as $D = 0.516, a = 1.112, b = -1.01$ and $\nu = 0.450$ for $k_c = 0.46$ ($a_c = 1.121$) and $D = 0.122, a = 2.513, b = -1.005$ and $\nu = 0.199$ for $k_c = 0.85$ ($a_c = 2.583$). The dispersion relation is obtained by solving Eq. (9) with respect to λ and plotting the real part of the solution. The dispersion relations corresponding to the onset of the instability (at $a = a_c$) and the above parameter sets resulting in the instability are shown in Fig. 2. The growing modes are of the form $Ae^{i\vec{k}\cdot\vec{r}}e^{\lambda(k)}$. Thus the wave numbers k with $\mathrm{Re}\{\lambda(k)\} < 0$ will be damped, whereas the wave numbers with $\mathrm{Re}\{\lambda(k)\} > 0$ will grow exponentially until the nonlinearities bound the growth. The dispersion relations in Fig. 2 tell us the growing modes

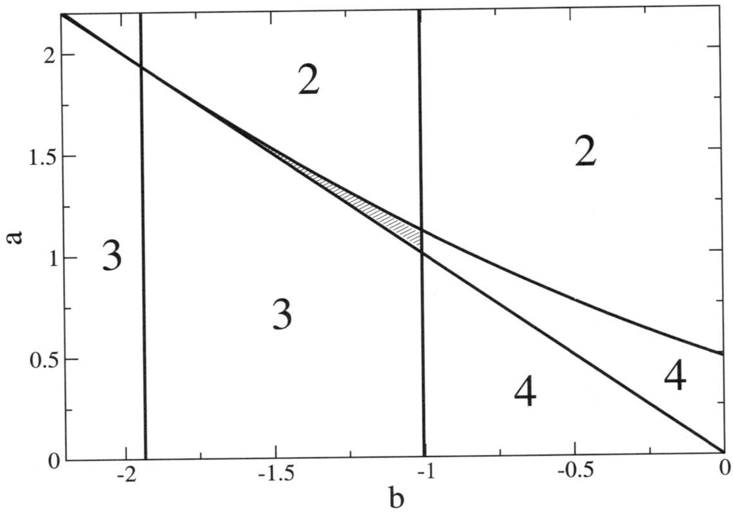

Fig. 1. The stability diagram of the generic Turing model (Eq. (3)). The Turing space (shaded region) is bounded by lines $b = -1/D, b = -1, b = -a$ and the curve $a = (Db - 1)^2/(4D)$. For the plot, parameters were fixed to $D = 0.516$. The other regions: stable state (2), other instabilities (3) and Hopf instability (4).

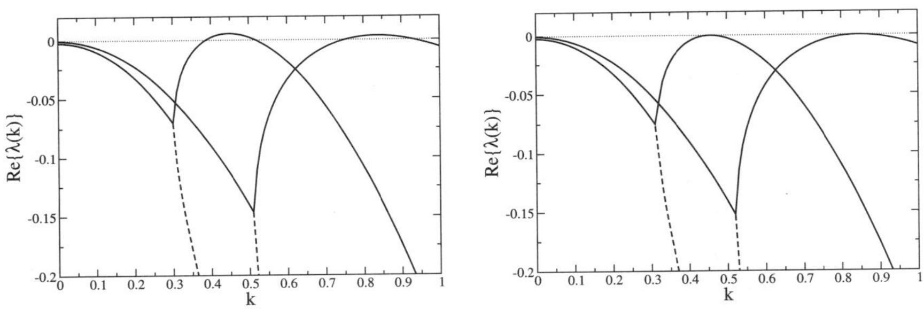

Fig. 2. The dispersion relations $\lambda(k)$ corresponding to two different parameter sets with $k_c = 0.46$ and $k_c = 0.85$, respectively. Left: At the onset of instability $a = a_c$ and there are no unstable modes. Right: For $a < a_c$ there is a finite wavelength instability corresponding to wave numbers k for which $\text{Re}\{\lambda(k)\} > 0$.

and predict the characteristic length of the pattern. The wave number and the wavelength are related by $\lambda = 2\pi/k$.

Due to the width of the unstable wave window (Fig. 2), there is more than one unstable mode. The unstable modes that do not correspond to k_c, i.e., the highest point of the dispersion relation, are called the sideband.

The sideband widens with the distance to onset $a_c - a$ as the bifurcation parameter a is varied. The linear analysis does not tell which of these unstable modes will be chosen. In addition, there is degeneracy due to isotropy, i.e., there are many wave vector $\vec{k}$ having the same wave number $k = |\vec{k}|$. In a discrete three-dimensional system the wave number is given by

$$|\vec{k}| = 2\pi\sqrt{\left(\frac{n_x}{L_x}\right)^2 + \left(\frac{n_y}{L_y}\right)^2 + \left(\frac{n_z}{L_z}\right)^2}, \tag{12}$$

where $L_{x/y/z}$ denote the system size in respective directions and $n_{x/y/z}$ the respective wave number indices. For a one-dimensional system $n_y = n_z = 0$ and for a two-dimensional system $n_z = 0$. Based on Eq. (12) one notices that e.g., two-dimensional vectors $(k_c, 0)$, $(0, k_c)$ and $(k_c/\sqrt{2}, k_c/\sqrt{2})$ all correspond to the wave number k_c and thus they are all simultaneously unstable. Actually, in a continuous system there would be an infinite number of unstable wave vectors pointing from the origin to the perimeter of a circle with a radius k_c. To tackle the problem of pattern selection, i.e., which of the degenarate modes will contribute to the final pattern, a nonlinear analysis is required.

From the results of the linear analysis we can identify the parameter domain, which results in the Turing instability and approximate the characteristic wavelength of the patterns. Figure 3 shows a stripe pattern and a spotty pattern with different characteristic lengths obtained from a numerical simulation with different parameter values. These two morphologies are typical for reaction-diffusion systems in two dimensions [43]. The generic

Fig. 3. Chemical concentration patterns obtained from numerical simulations of Eq. (3) in a system sized 128×128. Left: Stripe pattern ($k_c = 0.85, C = 0$). Right: Hexagonal spotty pattern ($k_c = 0.46, C = 1.57$). See the text for the other parameters.

Turing model has been devised in such a way that by adjusting parameter C one can favor either stripe or spotty patterns [39, 42]. The rigorous proof of this requires nonlinear bifurcation analysis and is the subject of the next section.

4. Nonlinear Bifurcation Theory

Bifurcation theory is a mathematical tool generally used for studying the dynamics of nonlinear systems [44–46]. The result of bifurcation analysis or weakly nonlinear analysis is a qualitative approximation for the changes in the dynamics of the system under study. In the case of Turing systems the bifurcation analysis answers the question concerning the changes in the stability of different simple morphologies as a parameter is varied. The bifurcation analysis has previously been applied in the cases of the Brysselator model [47] and the Lengyel–Epstein model [48]. The problem of these published analyses is that they omit the mathematical details required for understanding at an elementary level. In this section we will try to illustrate the idea of bifurcation analysis and the related mathematical techniques in a meticulous manner.

The idea of the bifurcation analysis is to find a presentation for the concentration field $\vec{w} = (u, v)^T$ in terms of the active Fourier modes, i.e.,

$$\vec{w} = \vec{w_0} \sum_{\vec{k}_j} \left(W_j e^{i\vec{k}_j \cdot \vec{r}} + W_j^* e^{-i\vec{k}_j \cdot \vec{r}} \right), \tag{13}$$

where $\vec{w_0}$ defines the direction of the active modes. W_j and W_j^* are the amplitudes of the corresponding modes $\vec{k}_j$ and $-\vec{k}_j$. Notice that the sum of complex conjugates is real. The unstable modes have slow dynamics, whereas the stable modes relax quickly and are said to be slaved to the unstable modes. Typically the bifurcation analysis is carried out by observing changes as a function of the bifurcation parameter, i.e., the distance to onset. In the case of the generic Turing system (Eq. (3)) there is an additional quadratic nonlinearity, which is adjusted by parameter C. This parameter governs the morphology selection between linear (stripe) and radial (spot) structures instead of the bifurcation parameter and forces some additional algebraic manipulations at the end of the bifurcation analysis.

The bifurcation analysis can be divided into three parts: derivation of the normal form for the amplitude equations in a particular symmetry, determining the parameters of the amplitude equations (there are various techniques for this) and finally analyzing the stability of different morphologies by applying the linear analysis (Sec. 3) on the system of amplitude equations. These three phases will be the respective topics of the following subsections.

4.1. *Derivation of the amplitude equations*

A system of n amplitude equations describes the time variation of the amplitudes W_j of the unstable modes $\vec{k}_j (j = 1, \ldots, n)$. The amplitude equations contain a linear part corresponding to the linear growth predicted by the positive eigenvalue of the linearized system defined by Eq. (9) and a nonlinear part due to nonlinear coupling of the unstable modes. Thus the most general form of an amplitude equation is given by

$$\frac{dW_j}{dt} = \lambda_c W_j + f_j(W_1, \ldots, W_n). \tag{14}$$

The eigenvalue may be approximated in the vicinity of the onset by a linear approximation defined by

$$\lambda_c = \left.\frac{d\lambda}{da}\right|_{a=a_c} (a - a_c) = \frac{\nu^2(\nu - 2R)}{(\nu(1+b) - 2R)(\nu - R)}, \tag{15}$$

where $R = \nu(Db+1)/2$ with notations of the generic Turing model (Eq. (3)). The exact form of the term $f(W_1, \ldots, W_n)$ in Eq. (14) depends on the symmetries under study and may be constructed by geometrical arguments. In two dimensions reaction-diffusion systems typically exhibit either stripes or a hexagonally arranged spots (Fig. 3). Thus the natural selection for the study of 2D patterns is a hexagonal lattice. In three dimensions there are various possibilities: One can study the simple cubic lattice (SC), base-centered cubic lattice (BCC) or face-centered cubic lattice (FCC) [49]. Callahan and Knobloch have been among the first to address the problem of bifurcations in three-dimensional Turing systems [50–52]. In the following we will derive the amplitude equations for the two-dimensional hexagonal lattice and three-dimensional SC- and BCC-lattices.

2D hexagonal lattice

The base vectors of a two-dimensional hexagonal lattice can be chosen to be $\vec{k}_1 = k_c(1, 0), \vec{k}_2 = k_c(-1/2, \sqrt{3}/2)$ and $\vec{k}_3 = k_c(-1/2, -\sqrt{3}/2)$ with $|\vec{k}_{1,2,3}| = k_c$. Since $-\vec{k}_1 - \vec{k}_2 = \vec{k}_3$ we can say that a hexagonal lattice exhibits resonant modes which one must take into account while deriving the form of the amplitude equations. In a simple square lattice there would not be any resonant modes, since any subset of the base vectors does not

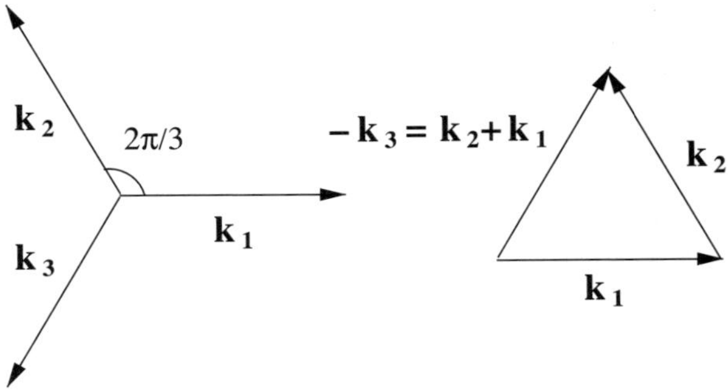

Fig. 4. The base vectors of a two-dimensional hexagonal lattice are not linearly independent and thus there are resonant modes.

sum into another base vector. The base vectors of a hexagonal lattice and the idea of resonant modes is illustrated in Fig. 4.

Based on the above reasoning one would suggest that there must be a quadratic coupling term in the amplitude equation; since the negative sum of two other modes may contribute to any one mode, there must be a term of the form $(-W^*_{j+1})$ $(-W^*_{j+2})$ in $f_j(W_1, W_2, W_3)$ $(j = 1, 2, 3 \pmod 3)$. The other combinations of wave vectors that sum up to $\vec{k}_j$ are $\vec{k}_j - \vec{k}_j + \vec{k}_j$, $\vec{k}_{j+1} - \vec{k}_{j+1} + \vec{k}_j$, and $\vec{k}_{j+2} - \vec{k}_{j+2} + \vec{k}_j$. The respective contributions to $f_j(W_1, W_2, W_3)$ are $-|W_j|^2 W_j, -|W_{j+1}|^2 W_j$ and $-|W_{j+2}|^2 W_j$. We assume that the saturation occurs at the third order and thus take into account only the sums of maximum three vectors. Now the full amplitude equation for the two-dimensional hexagonal lattice may be written as

$$\frac{dW_j}{dt} = \lambda_c W_j + \Gamma W^*_{j+1} W^*_{j+2} - g\big[|W_j|^2 + \kappa\big(|W_{j+1}|^2 + |W_{j+2}|^2\big)\big] W_j, \quad (16)$$

where the coefficients Γ, g and κ can be presented in terms of the parameters of the original reaction-diffusion system (Eq. (3)). The coefficients are obtained via complicated mathematical techniques, which will be discussed in the next subsection.

3D SC-lattice

The base vectors of a three-dimensional simple cubic lattice can be chosen to be $\vec{k}_1 = k_c(1, 0, 0), \vec{k}_2 = k_c(0, 1, 0)$ and $\vec{k}_3 = k_c(0, 0, 1)$ with $|\vec{k}_{1,2,3}| = k_c$. Now the base vectors are linearly independent and thus there are no resonant modes. By following the same reasoning as in the case of 2D hexagonal lattice, one can deduce the form of the amplitude equations to be the following

$$\frac{dW_j}{dt} = \lambda_c W_j - g\big[|W_j|^2 + \kappa\big(|W_{j+1}|^2 + |W_{j+2}|^2\big)\big]W_j. \tag{17}$$

3D BCC-lattice

The three-dimensional base-centered cubic lattice is a little bit more tricky. The set of base vectors is $\vec{k}_1 = k_c(1,1,1)/\sqrt{3}, \vec{k}_2 = k_c(1,1,-1)/\sqrt{3}, \vec{k}_3 = k_c(1,-1,1)/\sqrt{3}$ and $\vec{k}_4 = k_c(1,-1,-1)/\sqrt{3}$ with $|\vec{k}_{1,2,3,4}| = k_c$. These are not linearly independent since e.g., $\vec{k}_2 + \vec{k}_3 - \vec{k}_4 = \vec{k}_1$. Thus there is a cubic resonant coupling term and in addition there are the other nonlinear terms. The amplitude equations are given by

$$\begin{aligned}\frac{dW_j}{dt} &= \lambda_c W_j + \Gamma W_{j+1}^* W_{j+2}^* W_{j+3}^* - g\big[|W_j|^2 \\ &\quad + \kappa\big(|W_{j+1}|^2 + |W_{j+2}|^2 + |W_{j+3}|^2\big)\big]W_j,\end{aligned} \tag{18}$$

where ($j = 1,2,3,4 \pmod 4$)).

4.2. *Center manifold reduction*

There are various methods for determining the parameters for the amplitude equation. In the most used method, the multiscale expansion [44, 47], the bifurcation parameter and the chemical concentrations are expanded in a small parameter (e.g., $a - a_c = \epsilon a_1 + \epsilon^2 a_2 + \cdots$) and the coefficients are obtained based on the solvability conditions of the resulting linear differential equations at different degrees of ϵ. In this article we will not use multiscale expansion, but a related method called the center manifold reduction [45, 52]. We will take the method of center manifold reduction as given and for the mathematical justification of it we refer the reader elsewhere [45, 51, 53].

The purpose of the center manifold reduction is to devise a mapping from the concentration space (Eq. (3)) to a high-dimensional equivariant amplitude space (Eq. (14)). The center manifold is a surface separating the unstable and stable manifolds in the wave vector space. The center manifold reduction confines the nonlinear effects in the system to the center manifold and thus one can obtain good approximations for the stability of different structures. In the following, we will sketch the general procedure to obtain the amplitude equations following Callahan and Knobloch [52] to whom we refer the reader for details.

In general, we can write the component h of a Turing system with n chemical species ($h \in \{1, \ldots, n\}$) as

$$\frac{dX^h}{dt} = D^h \nabla^2 X^h + \sum_{i=1}^{n} A^{h,i} X^i + \sum_{i=1}^{n} \sum_{j=1}^{n} A^{h,ij} X^i X^j + \sum_{i=1}^{n} \sum_{j=1}^{n} \sum_{k=1}^{n} A^{h,ijk} X^i X^j X^k + \cdots, \tag{19}$$

where $X^h = X^h(\vec{x}, t)$ is the spatially varying concentration of one chemical species and $X^h = 0$ in the uniform stationary state. The tensors $A^{h,i}$, $A^{h,ij}$ and $A^{h,ijk}$ define the parameters for the component h and are symmetric with respect to permutations of the indices. In a discrete system we can write the concentration in a certain position of the lattice as

$$X^h(\vec{x}, t) = \sum_{l \in L} \tilde{X}_l^h(t) e^{i\vec{k}_l \cdot l}, \tag{20}$$

where L is the set of all lattice points. From now on we write species indices (e.g., h) as superscripts and lattice point indices (e.g., l) as subscripts. Substituting Eq. (20) into Eq. (19) yields

$$\frac{dX_l^h}{dt} = -D^h |\vec{k}_l|^2 X_l^h + A^{h,i} X_l^i + A^{h,ij} \sum_{l_1 + l_2 = l} X_{l_1}^i X_{l_2}^j + A^{h,ijk} \sum_{l_1 + l_2 + l_3 = l} X_{l_1}^i X_{l_2}^j X_{l_3}^k, \tag{21}$$

where we have used the Einstein summation convention for the indices i, j and k, and included only the terms up to cubic order. The linear part the previous equation defines the unstable modes (see Sec. 3) and the linear matrix can be written in the form

$$J_l^{h,i} = -D^h k_l^2 \delta^{h,i} + A^{h,i}, \tag{22}$$

where it is assumed that there is no cross-diffusion ($\delta^{h,i} = 1$ only when $h = i$). For each lattice point we may now choose a matrix

$$S_l = \begin{pmatrix} \alpha_l^{11} & \ldots & \alpha_l^{1n} \\ \vdots & \ddots & \vdots \\ \alpha_l^{n1} & \ldots & \alpha_l^{nn} \end{pmatrix}, \tag{23}$$

with $\det(\mathbf{S}_l) = 1$. In addition, we require that it has an inverse matrix $S_l^{-1} = \{\beta_l^{ij}\}$ such that

$$S_l^{-1} J_l S_l = \begin{pmatrix} \lambda_l^1 & & \\ & \ddots & \\ & & \lambda_l^n \end{pmatrix}. \tag{24}$$

The conditions for this similarity transformation are widely known [52, 54]. Now we can map the original concentrations (Eq. (20)) to a new basis defined by $S_l^{-1} X_l = W_l$. In this new basis Eq. (21) reads as

$$\begin{aligned} \frac{dW_l^g}{dt} &= \lambda_l^g W_l^g + \beta_l^{gh} A^{h,ij} \sum_{l_1+l_2=l} \alpha_{l_1}^{ii'} W_{l_1}^{i'} \alpha_{l_2}^{jj'} W_{l_2}^{j'} \\ &\quad + \beta_l^{gh} A^{h,ijk} \sum_{l_1+l_2+l_3=l} \alpha_{l_1}^{ii'} W_{l_1}^{i'} \alpha_{l_2}^{jj'} W_{l_2}^{j'} \alpha_{l_3}^{kk'} W_{l_3}^{k'}. \end{aligned} \tag{25}$$

The coefficient of the linear term is defined by Eq. (15). The coefficients of the nonlinear terms can be calculated at the onset by fixing $a = a_c$ and using the information about stable and unstable modes. Using the fact that the only contribution to the growth of the stable modes comes via a nonlinear coupling, one can derive relations for the parameters [52]. Further simplification of Eq. (25) for a critical wave vector m at the onset ($\lambda_l^g = 0$) yields

$$\begin{aligned} \frac{dW_m^1}{dt} &= \beta^{1h} A^{h,ij} \alpha^{i1} \alpha^{j1} \sum_{m_1+m_2=m} W_{m_1}^1 W_{m_2}^1 \\ &\quad + \sum_{m_1+m_2+m_3=m} F(m_2+m_3) W_{m_1}^1 W_{m_2}^1 W_{m_3}^1, \end{aligned} \tag{26}$$

where

$$F(r) \equiv -2\beta^{1h} A^{h,ij} \alpha^{i1} \left(J_r^{-1}\right)^{ja} A^{a,bc} \alpha^{b1} \alpha^{c1} + \beta^{1h} A^{h,ijk} \alpha^{i1} \alpha^{j1} \alpha^{k1}. \tag{27}$$

One should note that the coefficient $F(r)$ depends on the argument r only through the square of its length. Thus the previous treatment has been general and not specific to any particular symmetry. In the following, we will derive the form of the function $F(r)$ for three different lattices in two or three dimensions. In order to survive the horrendous linear algebra involved in the calculation, we follow a computation procedure that has been used earlier [52]. The derivation is based on finding the number and type of resonant modes that contribute to the amplitude of a particular mode as shown in Eq. (26). Due to symmetry, the coefficient of all the amplitude equations in a particular amplitude system (Eq. (16) or (17) or (18)) are the same.

2D hexagonal lattice

In the hexagonal lattice two wave vectors sum up to another wave vector. The base vectors for the hexagonal lattice are given by $\vec{k}_1 = (1,0)$, $\vec{k}_2 = (-1/2, \sqrt{3}/2)$ and $\vec{k}_3 = (-1/2, -\sqrt{3}/2)$. The strength of the quadratic coupling term is determined by the first term in Eq. (26). Since there are two possible selections (permutations) of m_1 and m_2, i.e., $-\vec{k}_3 - \vec{k}_2 = -\vec{k}_2 - \vec{k}_3 = \vec{k}_1$ one has to take into account both of them. Thus the quadratic coupling parameter in Eq. (16) is given by $\Gamma = 2\beta^{1h} A^{h,ij} \alpha^{i1} \alpha^{j1}$. The strength of the cubic coupling terms can be found by similar arguments. However, there are two cases that have to be treated separately, case 1: $m_1 = m_2 \neq m_3$ and case 2: $m_1 = m_2 = m_3$.

In the first case the coupling is of the type $k_1 + k_2 - k_2 = k_1$. There are three different combinations of m_2 and m_3 with two corresponding permutations. The combinations are

(i) $m_2 = k_2$ and $m_3 = -k_2$ with $|m_2 + m_3|^2 = 0$,
(ii) $m_2 = k_1$ and $m_3 = k_2$ with $|m_2 + m_3|^2 = 1$,
(iii) $m_2 = k_1$ and $m_3 = -k_2$ with $|m_2 + m_3|^2 = 3$,

which defines the coefficient $g\kappa$ in Eq. (16) to have the value $g\kappa = -2F(0) - 2F(1) - 2F(3)$.

In the second case the coupling is of the type $k_1 + k_1 - k_1 = k_1$. There are three possible permutations with

(i) $m_2 = k_1$ and $m_3 = -k_1$ with $|m_2 + m_3|^2 = 0$,
(ii) $m_2 = -k_1$ and $m_3 = k_1$ with $|m_2 + m_3|^2 = 0$,
(iii) $m_2 = k_1$ and $m_3 = k_1$ with $|m_2 + m_3|^2 = 4$,

which results in $g = -2F(0) - F(4)$ for Eq. (16).

Based on the above reasoning and Eq. (27) one may calculate the exact form of the coefficient in Eq. (16) with respect to the parameters of the generic Turing model of Eq. (3). The parameters of the amplitude equations are given by

$$\Gamma = \frac{-2bC\nu R\sqrt{\nu(\nu - 2R)}}{(\nu + b\nu - 2R)\sqrt{(\nu + b\nu - 2R)(\nu - R)}}, \tag{28}$$

$$g = \frac{3b\nu^2(\nu - 2R)R}{(\nu + b\nu - 2R)^2(\nu - R)}, \tag{29}$$

$$\kappa = 2, \tag{30}$$

where we have denoted $R = Dk_c^2 = \nu(Db+1)/2$. The linear coefficient of Eq. (16) is given by Eq. (15).

3D SC-lattice

In the SC-lattice the base vectors are independent and given as $\vec{k}_1 = (1,0,0)$, $\vec{k}_2 = (0,1,0)$ and $\vec{k}_3 = (0,0,1)$. There are no resonant modes. Following the ideas above in the first case we find

(i) $m_2 = k_2$ and $m_3 = -k_2$ with $|m_2 + m_3|^2 = 0$,
(ii) $m_2 = k_1$ and $m_3 = k_2$ with $|m_2 + m_3|^2 = 2$,
(iii) $m_2 = k_1$ and $m_3 = -k_2$ with $|m_2 + m_3|^2 = 2$,

which defines the coefficient $g\kappa = -2F(0) - 4F(2)$ in Eq. (16). The second case yields the permutations

(i) $m_2 = k_1$ and $m_3 = -k_1$ with $|m_2 + m_3|^2 = 0$,
(ii) $m_2 = -k_1$ and $m_3 = k_1$ with $|m_2 + m_3|^2 = 0$,
(iii) $m_2 = k_1$ and $m_3 = k_1$ with $|m_2 + m_3|^2 = 4$,

which gives $g = -2F(0) - F(4)$ for Eq. (16).

For the amplitude equations of the three-dimensional SC-lattice (Eq. (17)) the coefficients are given by

$$g = \frac{-b\nu^2(C^2(8\nu - 23R) - 27R)(\nu - 2R)}{9(\nu + b\nu - 2R)^2(\nu - R)}, \tag{31}$$

$$\kappa = \frac{18(C^2(8\nu - 7R) - 3R)}{C^2(8\nu - 23R) - 27R}, \tag{32}$$

where we have again denoted $R = Dk_c^2 = \nu(Db+1)/2$ and the linear coefficient of Eq. (17) is given by Eq. (15).

3D BCC-lattice

In the BCC-lattice the base vectors given by $\vec{k}_1 = (1,1,1)/\sqrt{3}$, $\vec{k}_2 = (1,1,-1)/\sqrt{3}$, $\vec{k}_3 = (1,-1,1)/\sqrt{3}$ and $\vec{k}_4 = (1,-1,-1)/\sqrt{3}$ are not linearly independent. To find the coefficient of the resonant contribution to Eq. (18) one must consider the possible combinations of the two last terms within the sum $k_2 + k_3 - k_4 = k_1$. These are given by

(i) $m_2 = k_3$ and $m_3 = -k_4$ with $|m_2 + m_3|^2 = \frac{4}{3}$,
(ii) $m_2 = -k_4$ and $m_3 = k_2$ with $|m_2 + m_3|^2 = \frac{4}{3}$,
(iii) $m_2 = k_2$ and $m_3 = k_3$ with $|m_2 + m_3|^2 = \frac{4}{3}$,

which yields the resonant coupling coefficient $\Gamma = 6F(\frac{4}{3})$ for Eq. (18).

The two other coefficient are determined using the same reasoning as above. In the first case one gets

(i) $m_2 = k_2$ and $m_3 = -k_2$ with $|m_2 + m_3|^2 = 0$,
(ii) $m_2 = k_1$ and $m_3 = -k_2$ with $|m_2 + m_3|^2 = \frac{4}{3}$,
(iii) $m_2 = k_1$ and $m_3 = k_2$ with $|m_2 + m_3|^2 = \frac{8}{3}$,

which results in the coefficient $g\kappa = -2F(0) - 2F(\frac{4}{3}) - 2F(\frac{8}{3})$ for Eq. (18). In the second case we get the permutations

(i) $m_2 = k_1$ and $m_3 = -k_1$ with $|m_2 + m_3|^2 = 0$,
(ii) $m_2 = -k_1$ and $m_3 = k_1$ with $|m_2 + m_3|^2 = 0$,
(iii) $m_2 = k_1$ and $m_3 = k_1$ with $|m_2 + m_3|^2 = 4$,

which yields $g = -2F(0) - F(4)$ for the Eq. (18).

The coefficients of the amplitude equations of the three-dimensional BCC-lattice (Eq. (18)) can be written as

$$\Gamma = \frac{6b\nu^2(\nu - 2R)(3C^2(8\nu - 7R) - R)}{(\nu + b\nu - 2R)^2(\nu - R)}, \tag{33}$$

$$g = \frac{-b\nu^2(C^2(8\nu - 23R) - 27R)(\nu - 2R)}{9(\nu + b\nu - 2R)^2(\nu - R)}, \tag{34}$$

$$\kappa = \frac{18(C^2(648\nu - 583R) - 75R)}{25(C^2(8\nu - 23R) - 27R)}, \tag{35}$$

where $R = \nu(Db + 1)/2$ and the linear coefficient λ_c of Eq. (18) is defined by Eq. (15).

4.3. *Stability of different structures*

After we have obtained the system of coupled amplitude equations written with respect to the parameters of the original reaction-diffusion equations, we may now employ linear analysis to study the amplitudes. First, one has to determine the stationary states W_c of the amplitude system, which depends on the symmetry under study (Eq. (16), (17) or (18)). After that one can linearize the system (as in Eq. (6)) and construct the corresponding Jacobian linear matrix $\mathbf{A}$, which is determined by

$$\{\mathbf{A}\}_{ij} = \left|\frac{df_i}{d|W_j|}\right|_{(W_1^c, W_2^c, W_3^c)}, \tag{36}$$

where f_i denotes the right-hand side of the corresponding amplitude equation i and the element is evaluated at the stationary state $W_c = (W_1^c, W_2^c, W_3^c)$. Based on this one can plot the bifurcation diagram, i.e., the eigenvalues of the linearized system $dW/dt = \mathbf{A}W$ as a function of the parameter C in Eq. (3), which contributes to the morphology selection in the generic Turing model. The parameters of the generic Turing model that we have used in the analysis presented here correspond to the mode $k_c = 0.85$ (see Fig. 2).

2D hexagonal lattice

In the case of two-dimensional patterns we are interested in the stability of stripes ($W_c = (W_1^c, 0, 0)^T$) and hexagonally arranged spots ($W_c = (W_1^c, W_2^c, W_3^c)^T$ with $W_1^c = W_2^c = W_3^c$). For the stability analysis of rhombic patterns and anisotropic mixed amplitude states we refer the reader elsewhere [55]. The system of amplitude equations for a two-dimensional hexagonal lattice can be written based on Eq. (16) as

$$\begin{aligned}
\frac{dW_1}{dt} &= \lambda_c W_1 + \Gamma W_2^* W_3^* - g[|W_1|^2 + \kappa(|W_2|^2 + |W_3|^2)]W_1, \\
\frac{dW_2}{dt} &= \lambda_c W_2 + \Gamma W_1^* W_3^* - g[|W_2|^2 + \kappa(|W_1|^2 + |W_3|^2)]W_2, \\
\frac{dW_3}{dt} &= \lambda_c W_3 + \Gamma W_1^* W_2^* - g[|W_3|^2 + \kappa(|W_1|^2 + |W_2|^2)]W_3,
\end{aligned} \tag{37}$$

where the coefficients λ_c, Γ, g and κ are given by Eqs. (15), (28), (29) and (30), respectively. In the case of stripes, $W_2^c = W_3^c = 0$ and the system reduces to only one equation. Now the stationary states defined by the zeros of the right-hand side of (37) can easily be shown to be $W_1^c = \sqrt{\frac{\lambda_c}{g}} e^{i\phi_1}$. In the case of hexagonal spots we have three equations and by choosing a solution such that $W_c = W_1^c = W_2^c = W_3^c$ we obtain the two stationary states defined by

$$|W_{\pm}^c| = \frac{|\Gamma| \pm \sqrt{\Gamma^2 + 4\lambda_c g[1+2\kappa]}}{2g(1+2\kappa)}. \tag{38}$$

In the case of stripes the eigenvalues of the linearized matrix (Eq. (36)) are given as $\mu_1^s = -2\lambda_c$, $\mu_2^s = -\Gamma\sqrt{\frac{\lambda_c}{g}} + \lambda_c(1-\kappa)$ and $\mu_3^s = \Gamma\sqrt{\frac{\lambda_c}{g}} + \lambda_c(1-\kappa)$. Noticing that $\mu_1^s < 0$ and $\mu_3^s > \mu_2$ follows that the stability of stripes is determined by the sign of μ_3^s. The stripes are unstable for $\mu_3^s > 0$ and stable for $\mu_3^s < 0$.

In the case of the hexagonally arranged spots the eigenvalues of the system are given as $\mu_{1,2}^h = \lambda_c - W_c^{\pm}(\Gamma + 3gW_c^{\pm})$ and $\mu_3^h = \lambda_c + W_c^{\pm}(2\Gamma - 3g(2\kappa+1)W_c^{\pm})$, where $W_c^{\pm}$ is defined by Eq. (38). Since there are two stationary states corresponding to hexagonal symmetry, one must analyze the stability of both of them. For stability all the eigenvalues must be negative, i.e., $\mu_{1,2}^h < 0$ and $\mu_3^h < 0$. After writing the eigenvalues in terms of the original parameters (Eqs. (15), (28)–(30)) one can plot the eigenvalues as a function of the nonlinear coefficient C, which is known to adjust the competition between stripes and spots [39]. The result is shown in Fig. 5,

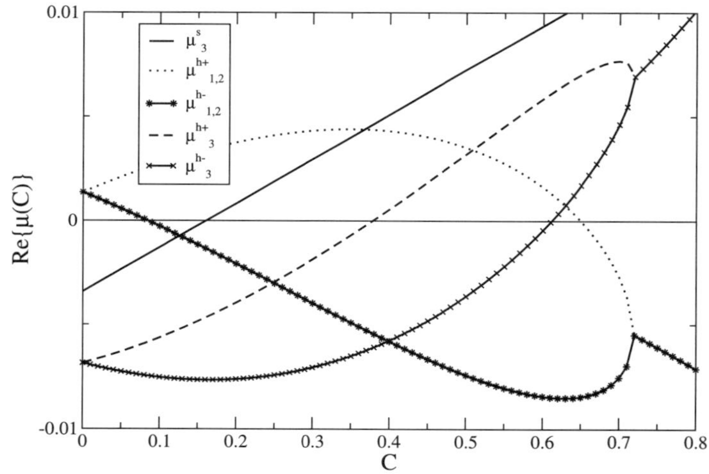

Fig. 5. The real part of the eigenvalues $\mu(C)$ of the linearized amplitude system of the two-dimensional hexagonal symmetry as a function of the parameter C. Eigenvalue μ_3^s determines the stability of the stripes, $\mu_{1,2}^{h+}$ and μ_3^{h+} determine the stability of one hexagonal branch, and $\mu_{1,2}^{h-}$ and μ_3^{h-} determine the stability of the other hexagonal branch. The morphology is stable if the corresponding $\mu(C) < 0$.

from where one can determine the parameter regimes for which a given pattern is stable, i.e., $\mu(C) < 0$.

Figure 5 implies that the hexagonal branch corresponding to W_c^+ is always unstable. Thus there is only one isotropic hexagonal solution to the equations that is stable within certain parameter regime. The analysis predicts that stripes are stable for $C < 0.161$ while using the parameters of mode $k_c = 0.85$. On the other hand, the other hexagonal branch is predicted to be stable for $0.084 < C < 0.611$. The most important information obtained from Fig. 5 is the region of bistability, which is predicted to be between $0.084 < C < 0.161$. Since the bifurcation analysis is based on weakly nonlinear approximation of the dynamics, it can be expected that it fails, when a strong nonlinear action is present. For example, based on the result of the numerical simulation presented in Fig. 3 one can see that the hexagonal spot pattern exists for $C = 1.57$. The bifurcation analysis, however, predicts that hexagons are unstable for all $C > 0.611$. This discrepancy is due to the approximations of the bifurcation theory, which hold only for weak nonlinearities.

3D SC-lattice

In the case of three-dimensional simple cubic lattice there are three possibilities for the structure. One may get planar structures ($W_c = (W_1^c, 0, 0)^T$), cylindrical structures ($W_c = (W_1^c, W_2^c, 0)^T$) or spherical droplet structures ($W_c = (W_1^c, W_2^c, W_3^c)^T$). The amplitude equations of a three-dimensional SC-lattice are based on Eq. (17) and the system is given as

$$\begin{aligned}\frac{dW_1}{dt} &= \lambda_c W_1 - g[|W_1|^2 + \kappa(|W_2|^2 + |W_3|^2)]W_1,\\ \frac{dW_2}{dt} &= \lambda_c W_2 - g[|W_2|^2 + \kappa(|W_1|^2 + |W_3|^2)]W_2,\\ \frac{dW_3}{dt} &= \lambda_c W_3 - g[|W_3|^2 + \kappa(|W_1|^2 + |W_2|^2)]W_3,\end{aligned} \tag{39}$$

where the coefficients λ_c, g and κ are given by Eqs. (15), (31) and (32), respectively.

The stationary state corresponding to the planar lamellae is given as $|W_1^c| = \sqrt{\frac{\lambda_c}{g}}$. For the cylindrical structure we get $|W_1^c| = |W_2^c| = \sqrt{\frac{\lambda_c}{g(\kappa+1)}}$ and for the isotropic stationary state of SC-droplets $|W_1^c| = |W_2^c| = |W_3^c| = \sqrt{\frac{\lambda_c}{g(2\kappa+1)}}$. In the case of planar lamellae the eigenvalues of the linearized

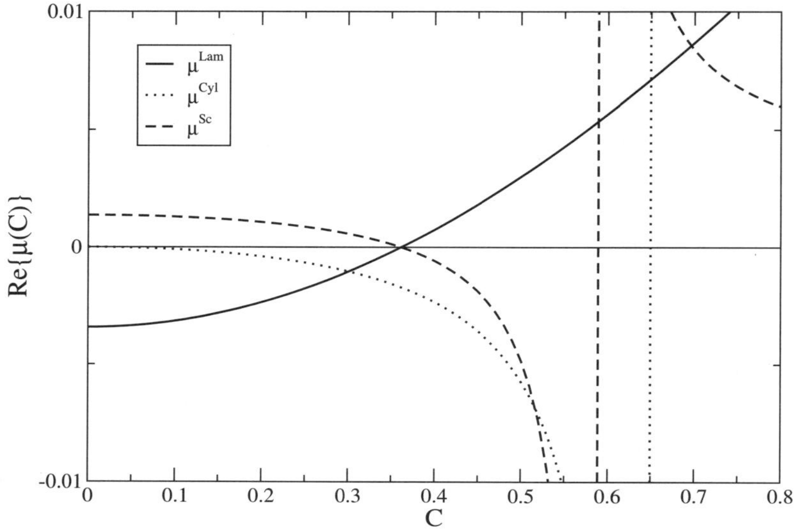

Fig. 6. The real part of the eigenvalues $\mu(C)$ of the linearized amplitude system of the three-dimensional SC-lattice as a function of the parameter C. Eigenvalue μ^{Lam} determines the stability of the planar lamellae, μ^{Cyl} determines the stability of cylindrical structures, and μ^{Sc} determines the stability of the spherical droplets organized in a SC-lattice. The morphology is stable if the corresponding $\mu(C) < 0$.

matrix (Eq. (36)) are given by $\mu_1^{\mathrm{Lam}} = -2\lambda_c$ and $\mu_{2,3}^{\mathrm{Lam}} = \lambda_c(1-\kappa)$. Noticing that $\mu_1^{\mathrm{Lam}} < 0$ follows that the stability of the planar structures is determined by $\mu_{2,3}^{\mathrm{Lam}}$. Repeating the same treatment for the cylindrical structures we find that the real part of the dominant eigenvalue is $\mu_{2,3}^{\mathrm{Cyl}} = \lambda_c - 3gW_c^2$. For the SC-droplets the stability determining eigenvalue is given by $\mu_{2,3}^{\mathrm{Sc}} = \lambda_c - 3gW_c^2$. The real parts of the eigenvalues are presented in Fig. 6.

Based on Fig. 6 it can be reasoned that the bifurcation analysis does not predict a bistability between planes and spherical shapes in three dimensions, but the stability of those structures is exclusive. The planes are predicted to be stable for $C < 0.361$ and the spherical shapes stable for $0.361 < C < 0.589$. The square packed cylinders, however, are predicted to be stable for all $C < 0.650$. It can again be noticed that the bifurcation analysis fails for strong nonlinear interaction, i.e., high values of parameter C.

3D BCC-lattice

In the case of the three-dimensional BCC-lattice there are numerous possibilities for the structure [52]. Here we analyze only the stability of lamellar structures ($W_c = (W_1^c, 0, 0, 0)^T$) and spherical droplets organized in a BCC-lattice ($W_c = (W_1^c, W_2^c, W_3^c, W_4^c)^T$). The amplitude equations of a three-dimensional BCC-lattice are defined by Eq. (18) and the system is given by

$$\frac{dW_1}{dt} = \lambda_c W_1 + \Gamma W_2^* W_3^* W_4^* - g\big[|W_1|^2 + \kappa\left(|W_2|^2 + |W_3|^2\right) + |W_4|^2\big]W_1,$$

$$\frac{dW_2}{dt} = \lambda_c W_2 + \Gamma W_1^* W_3^* W_4^* - g\big[|W_2|^2 + \kappa\left(|W_1|^2 + |W_3|^2\right) + |W_4|^2\big]W_2,$$

$$\frac{dW_3}{dt} = \lambda_c W_3 + \Gamma W_1^* W_2^* W_4^* - g\big[|W_3|^2 + \kappa\left(|W_1|^2 + |W_2|^2\right) + |W_4|^2\big]W_3,$$

$$\frac{dW_4}{dt} = \lambda_c W_4 + \Gamma W_1^* W_2^* W_3^* - g\big[|W_4|^2 + \kappa\left(|W_1|^2 + |W_2|^2\right) + |W_3|^2\big]W_4, \tag{40}$$

where the coefficients λ_c, Γ, g and κ are given by Eqs. (15), (33)–(35), respectively.

The stationary state corresponding to planar lamellae is defined by $|W_1^c| = \sqrt{\frac{\lambda_c}{g}}$, with $|W_2^c| = |W_3^c| = |W_4^c| = 0$. For the stationary state of BCC-droplets with constraint $|W_1^c| = |W_2^c| = |W_3^c| = |W_4^c|$ we get $|W_i^c| = \sqrt{\frac{\lambda_c}{g(3\kappa+1)-\Gamma}}$. For planar lamellae the eigenvalues of the linearized matrix (Eq. (36)) are given by $\mu_1^{\text{Lam}} = -2\lambda_c$ and $\mu_{2,3,4}^{\text{Lam}} = \lambda_c(1-\kappa)$. Noticing that $\mu_1^{\text{Lam}} < 0$ follows that the stability of the planar structures is again determined by $\mu_{2,3,4}^{\text{Lam}}$. For the BCC-droplets, on the other hand, the stability determining eigenvalue is given by $\mu_{2,3,4}^{\text{Sc}} = \lambda_c - 3(\Gamma + g(K+3))W_c^2$. The real parts of these eigenvalues are presented in Fig. 7 as a function of C.

From Fig. 7 one can see that there is a bistability between planes and BCC droplet structures for $0.181 < C < 0.204$. The planes are predicted to be stable for $C < 0.204$ and the spherical shapes to be stable for $0.181 < C < 0.255$. It can again be observed that the bifurcation analysis fails already at a reasonable low nonlinear interaction predicting that the spherical structures become unstable at $C < 0.255$. The stability of the other possible structures in the BCC-lattice remains to be studied in the case of the generic Turing model.

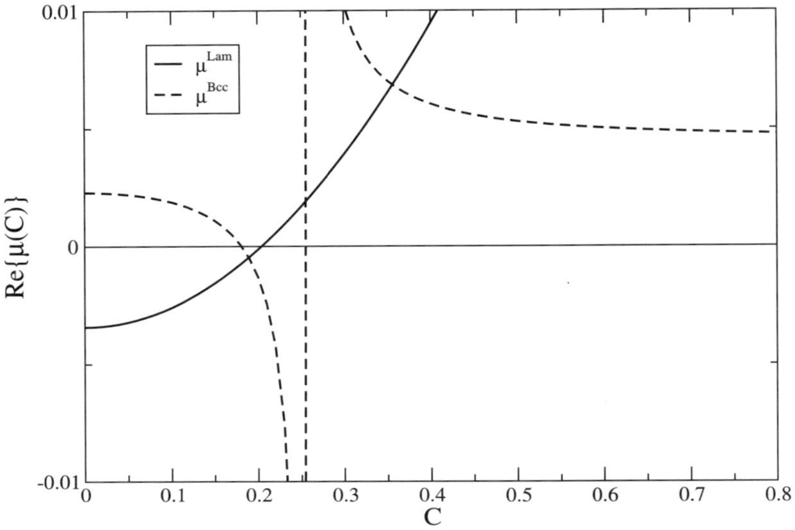

Fig. 7. The real part of the eigenvalues $\mu(C)$ of the linearized amplitude system of the three-dimensional BCC-lattice as a function of the parameter C. Eigenvalue μ^{Lam} determines the stability of the planar lamellae and μ^{Bcc} determines the stability of the spherical droplets organized in a BCC-lattice. The morphology is stable if the corresponding $\mu(C) < 0$.

5. Conclusions

Turing pattern formation is nowadays of great interest. The observation of real chemical patterns some 14 years ago confirmed that the theoretical ideas hypotetized by Alan Turing almost 40 years earlier were not only mathematical formulations, but a pioneering contribution to the theory of nonlinear dynamics. The contribution of Alan Turing to bioinformation technology and biology still remains controversial, although Turing models have been shown to be able to imitate many biological patterns found in animals [11, 13, 56], skeptics argue that more evidence is needed and the exact morphogens that behave according to the Turing mechanism have to be named based on experimental studies by developmental biologists. However, there is a seed of truth in the cautiousness, since a numerical Turing model has even been shown to be able to exhibit patterns resembling the letters of alphabet if some heavy manipulation of the dynamics is carried out [57].

Irrespective of the biological relevance of the Turing systems, they are also of great interest from the physicist's point of view. Nowadays symmetry-breaking and self-organization are all in a day's work for physicists and the knowledge of other fields of physics may be applied to Turing systems and *vice versa*. The difficulty is, however, that making fundamental theoretical contributions to the theory of Turing pattern formation seems to be a penultimate challenge and thus most of the work in the field relies at least in part on experimental data [7, 58, 59]. The use of numerical simulations in studying the Turing pattern formation seems promising since computationally one may study systems that are beyond the reach of experiments and the numerical data is more accurate and easier to analyze as compared to experimental data [60, 61].

This article reviewed some rather theoretical methods that are generally used in analyzing the dynamical behavior of reaction-diffusion systems. The linear analysis is efficient in predicting the presence of instability and the characteristic length of the resulting patterns. However, linear analysis does not reveal anything about the morphology of the resulting pattern. To study the pattern or structure selection we employed the nonlinear bifurcation analysis, which approximately predicts the stability of different symmetries, i.e., the parameter regime that results in certain Turing structures. If one uses homogeneous random initial conditions there is no way to predict the selection of the phase of the resulting pattern (the positions and alignment of stripes or spots). A further difficulty arises if one has to study the pattern selection of a morphologically bistable system: if both stripes and spots are stable, there is no general way to determine which of the states the system will choose. These inadequacies of the theory of pattern formation are in connection with the fundamental problem of non-equilibrium thermodynamics and remain to be answered both in the context of Turing systems and also in the more general framework of non-equilibrium physics.

Acknowledgments

This work was supported by the Finnish Academy of Science and Letters and the Jenny and Antti Wihuri foundation.

References

[1] Ball, P., *The Self-Made Tapestry: Pattern Formation in Nature* (Oxford University Press, Oxford, 2001).
[2] Cross, M.C. and Hohenberg, P.C., *Rev. Mod. Phys.* **65** (1993) 851.
[3] De Groot, S.R. and Mazur, P., *Non-equilibrium Thermodynamics* (North Holland, Amsterdam, 1962).

[4] Nicolis, G. and Prigogine, I., *Self-Organisation in Non-Equilibrium Chemical Systems* (Wiley, New York, 1977).
[5] Prigogine, I., *Bull. Acad. Roy. Belg. Cl. Sci.* **31** (1945) 600.
[6] Landauer, R., *Phys. Rev. A* **12** (1975) 636.
[7] Kapral, R. and Showalter, K. (eds.), *Chemical Waves and Patterns* (Kluwer Academic Publishers, Dordrecht, 1995).
[8] Kapral, R., *Physica D* **86** (1995) 149.
[9] Rashevsky, N., *Mathematical Biophysics* (University of Chicago Press, Chicago, 1938).
[10] Turing, A.M., *Phil. Trans. R. Soc. Lond.* **B237** (1952) 37.
[11] Murray, J.D., *Mathematical Biology* (Springer-Verlag, Berlin, 1989).
[12] Murray, J.D., *Mathematical Biology II: Spatial Models and Biomedical Applications* (Springer-Verlag, Berlin, 2003).
[13] Koch, A.J. and Meinhardt, H., *Rev. Mod. Phys.* **66** (1994) 1481.
[14] Lengyel, I., Rabai, G. and Epstein, I.R., *J. Am. Chem. Soc.* **112** (1990) 4606; *ibid.* **112** (1990) 9104.
[15] Vastano, J.A., Pearson, J.E., Horsthemke, W. and Swinney, H.L., *J. Chem. Phys.* **88** (1988) 6175.
[16] Lengyel, I. and Epstein, I.R., in *Chemical Waves and Patterns*, eds. Kapral, R. and Showalter, K. (Kluwer Academic Publishers, Dordrecht, 1995), p. 297.
[17] Castets, V., Dulos, E., Boissonade, J. and De Kepper, P., *Phys. Rev. Lett.* **64** (1990) 2953.
[18] Ouyang, Q. and Swinney, H.L., *Nature* **352** (1991) 610.
[19] Leppänen, T., Karttunen, M., Kaski, K. and Barrio, R.A., *Int. J. Mod. Phys. B* **17** (2003) 5541.
[20] Astrov, Y., Ammelt, E., Teperick, S. and Purwins, H.-G., *Phys. Lett. A* **211** (1996) 184.
[21] Falta, J., Imbihl, R. and Henzler, M., *Phys. Rev. Lett.* **64** (1990) 1409.
[22] Temmyo, J., Notzel, R. and Tamamura, T., *Appl. Phys. Lett.* **71** (1997) 1086.
[23] Tlidi, M., Mandel, P. and Haelterman, M., *Phys. Rev. E* **56** (1997) 6524.
[24] Walgraef, D. and Ghoniem, N.M., *Phys. Rev. B* **67** (2003) 064103.
[25] Barrio, R.A., Aragon, J.L., Varea, C., Torres, M., Jimenez, I. and Montero de Espinosa, F., *Phys. Rev. E* **56** (1997) 4222.
[26] Vastano, J.A., Pearson, J.E., Horsthemke, W. and Swinney, H.L., *Phys. Lett. A* **124** (1987) 320.
[27] Gray, P. and Scott, S.K., *Chem. Eng. Sci.* **38** (1983) 29.
[28] Gray, P. and Scott, S.K., *Chem. Eng. Sci.* **39** (1984) 1087.
[29] Gray, P. and Scott, S.K., *J. Phys. Chem.* **89** (1985) 22.
[30] Pearson, J.E., *Science* **261** (1993) 189.
[31] Gierer, A. and Meinhardt, H., *Kybernetik* **12** (1972) 30.
[32] Hunding, A., *J. Chem. Phys.* **72** (1980) 5241.
[33] Vance, W. and Ross, J., *J. Phys. Chem. A* **103** (1999) 1347.
[34] Dufiet, V. and Boissonade, J., *J. Chem. Phys.* **96** (1991) 664.
[35] Dufiet, V. and Boissonade, J., *Physica A* **188** (1992) 158.
[36] Borckmans, P., De Wit, A. and Dewel, G., *Physica A* **188** (1992) 137.

[37] Lengyel, I. and Epstein, I.R., *Science* **251** (1991) 650.
[38] Lengyel, I. and Epstein, I.R., *Proc. Natl. Acad. Sci.* **89** (1992) 3977.
[39] Barrio, R.A., Varea, C., Aragón, J.L. and Maini, P.K., *Bull. Math. Biol.* **61** (1999) 483.
[40] Nicolis, G., *Introduction to Nonlinear Science* (Cambridge University Press, Cambridge, 1995).
[41] Strogatz, S., *Nonlinear Dynamics and Chaos* (Perseus, USA, 1994).
[42] Leppänen, T., Karttunen, M., Kaski, K., Barrio, R.A. and Zhang, L., *Physica D* **168 & 169** (2002) 35.
[43] Boissonade, J., Dulos, E. and De Kepper, P., in *Chemical Waves and Patterns*, eds. Kapral, R. and Showalter, K. (Kluwer Academic Publishers, Dordrecht, 1995), p. 221.
[44] Newell, A.C., Passot, T. and Lega, J., *Annu. Rev. Fluid Mech.* **25** (1993) 399.
[45] Crawford, J.D., *Rev. Mod. Phys.* **63** (1991) 991.
[46] Manneville, P., *Dissipative Structures and Weak Turbulence* (Academic Press, USA, 1990).
[47] Walgraef, D., *Spatio-Temporal Pattern Formation* (Springer-Verlag, USA, 1997).
[48] Rovinsky, A. and Menzinger, M., *Phys. Rev. A* **46** (1992) 6315.
[49] Ashcroft, N.W. and Mermin, N.D., *Solid State Physics* (Harcourt Brace College Publishers, USA, 1976).
[50] Callahan, T.K. and Knobloch, E., *Phys. Rev. E* **53** (1996) 3559.
[51] Callahan, T.K. and Knobloch, E., *Nonlinearity* **10** (1997) 1179.
[52] Callahan, T.K. and Knobloch, E., *Physica D* **132** (1999) 339.
[53] Dionne, B., Silber, M. and Skeldon, A.C., *Nonlinearity* **10** (1997) 321.
[54] Kreyszig, E., *Advanced Engineering Mathematics*, 7th edn. (Wiley, USA, 1993).
[55] Borckmans, P., Dewel, G., De Wit, A. and Walgraef, D., in *Chemical Waves and Patterns*, eds. Kapral, R. and Showalter, K. (Kluwer Academic Publishers, Dordrecht, 1995), p. 323.
[56] Kondo, S. and Asai, R., *Nature* **376** (1995) 678.
[57] Kawczynski, A.L. and Legawiec, B., *Phys. Rev. E* **64** (2001) 056202.
[58] Borckmans, P., Dewelm, G., De Wit, A., Dulos, E., Boissonade, J., Gauffre, F. and De Kepper, P., *Int. J. Bif. Chaos* **12** (2002) 2307.
[59] Berenstein, I., Yang, L., Dolnik, M., Zhabotinsky, A.M. and Epstein, I.R., *Phys. Rev. Lett.* **91** (2003) 058302.
[60] De Wit, A., Borckmans, P. and Dewel, G., *Proc. Natl. Acad. Sci.* **94** (1997) 12765.
[61] Leppänen, T., Karttunen, M., Kaski, K. and Barrio, R.A., *Prog. Theor. Phys.* (Suppl.) **150** (2003) 367.

CHAPTER 12

THE DIOCTADECYLAMINE MONOLAYER: NON-EQUILIBRIUM PHASE DOMAINS

A. Flores, E. Corvera-Poiré*, C. Garza and R. Castillo
Instituto de Fisica, UNAM,
P.O. Box 20-364, D. F., 01000 Mexico
**Facultad de Quimica, UNAM,*
Cd. Universitaria, D. F., 45010 Mexico
**rolandoc@fisica.unam.mx*

The phase diagram of the dioctadecylamine Langmuir monolayer is determined from pressure–area isotherms and from direct observations of the monolayer using Brewster angle microscopy. In the $L - S_1$ transition, the domains of the emerging phase are outstanding six-fold dendrites. We study the factors that modify the form of growth of these domains, and we are able to detect that in certain specific conditions of undercooling, the domains have a fingering instability and after some growing time present tip-splitting. If undercooling persists, the domains undergo a transition to side-branching.

1. Introduction

Amphiphilic molecules that are nearly insoluble in water can form Langmuir monolayers (LMs) at the air–water interface. The most common way for studying LMs has been through measurements of the pressure-area isotherms, $\Pi(A,T) = \gamma_o(T) - \gamma(A,T)$, where T is the temperature, A is the area/molecule, γ and γ_o are the surface tensions of the monolayer and of pure water, respectively. Phase diagrams of LMs have been studied intensively for decades. However, significant advances have been obtained only in the last ten years due to new experimental techniques. Grazing incidence X-ray diffraction gives the most explicit information about monolayer organization [1]. Nevertheless, this kind of experiment is time-consuming and expensive for obtaining an entire phase diagram. Other powerful techniques have been developed to study monolayer organization, such as polarized fluorescence microscopy [2] and Brewster angle microscopy (BAM) [3, 4]. These techniques complement the information given by diffraction experiments, because they survey larger scales ($\sim$200 μm) providing information about homogeneity, textures, structure, and the dynamics of monolayers. These optical techniques are quite sensible for observing very fine details in phase transformations such as molecular tilting. All these new experimental techniques have revealed that singularities in the surface pressure–area isotherms are due to phase changes [5].

Studies on secondary amines are not common in the literature. They are potentially interesting because phases with free rotation of the chains about the long molecular axes should not exist, since the coupling of two tails prevents rotation. The role of translational freedom is small compared to the internal degrees of freedom, due to the fact that lateral motion is hindered because of the two chains. Positions and orientations of head groups can also be involved in ordering since interactions among head groups could be laterally anisotropic.

In this paper, the phase diagram of dioctadecylamine (DODA) LM has been determined between 5°C and 45°C, at $pH = 3$ from pressure–area isotherms and from direct observations of the monolayer using BAM. This amine is a secondary amine, with two long aliphatic tails, which is insoluble in water, i.e., the subphase. The phase diagram obtained is very different from those obtained for single chain amphiphiles [5–7]. We found four phases: G, LE, S_1, and S_2, and the phase transitions between these phases were observed. In the $LE - S_1$ transition, we observed dendritic crystallization, where the domains of the emerging phase form outstanding six-fold dendrites. Here, we present a progress report of the stages followed by the non-equilibrium growing domains at the $LE - S_1$ transition for this monolayer.

Morphological instabilities of growing aggregates is one of the subjects that have received considerable attention over past years. Different growth patterns have been observed when LMs undergo a first-order phase transition from a fluid to a condensed phase, such as dendritic domains [8], fractal-like domains [9] and dense-branched domains [10]. Systems that form these patterns seem to have some features in common: some degree of under-cooling is needed, the surfactant molecules have some kind of hindrance (two or more tails, chiral centers, bent tails through carbon double bonds), usually, there is an important density difference between the fluid and the condensed phases, as well as a quite different compressibility, and probably diffusion and the Marangoni effect play important roles.

2. Experimental Details

Dioctadecylamine (DODA), 2C18NH, (⩾99%) was purchased from Fluka Chemie (Switzerland); it was used without any further purification. The amine was spread onto a subphase of ultrapure water (nanopure-UV, 18.3M Ω) at several values of *pH*. The spreading solution was made with chloroform (Aldrich USA, HPLC) at a concentration of 1 mg/ml. HCL (Merck, Mexico) and H_2SO_4 (Merck, Mexico) were used to modify the *pH*.

Two Nima LB troughs (models: TKB2410*A* and 601 BAM, Nima Technology Ltd., England) were used to obtain the LMs; in both cases a Wilhelmy plate was used to measure the lateral pressure, $\Pi(A, T)$. One trough was used to obtain the isotherms and to develop Langmuir–Blodgett transferred monolayers. The other trough was enclosed in a 1 m^3 box to avoid air convection and it was used for the observations of dendritic domains. This trough was isolated from vibrations with a vibration isolation system (model 2*S*, Halcyonics GmbH, Germany). The speed of compression for obtaining the isotherms was of the order of 50 cm^2/min (~7.2 Å molec^{-1}min^{-1}), and for observing dendritic domains it was of 25 cm^2/min (~5.3 Å molec^{-1}min^{-1}). All experiments were carried out in a dust-free environment.

BAM observations during the development of the isotherms were performed in a BAM1 plus (Nanofilm Technologie GmbH, Germany) with a spatial resolution of ca. 4 μm. The experiments for observing the dendritic domains were made with the aid of an Elli2000 imaging ellipsometer (Nanofilm Technologie GmbH, Germany) in the BAM mode (spatial resolution of ca. 2 μm and 1 μm using the 10× and the 20× objectives, respectively), which allowed us to have the whole field of view in focus due to its movable objective lens.

Transferred monolayers of DODA were scanned with a scanning probe microscope (JSTM-4200 JEOL, Japan) with an 80 × 80 μm scanner. Intermittent contact and phase lag modes were used to obtain topographic and phase images; tips with a typical force constant of 40 Nm^{-1} were used.

3. Results and Discussion

3.1. *Phase transitions*

Figure 1 shows the Π–*T* phase diagram for DODA LM for the range of temperatures worked in this study (5–45°C); *pH* was fixed at $pH = 3$ with H_2SO_4. The coexistence lines were obtained from the temperatures and the pressures where phase changes do occur, as detected in the Π–*A* isotherms

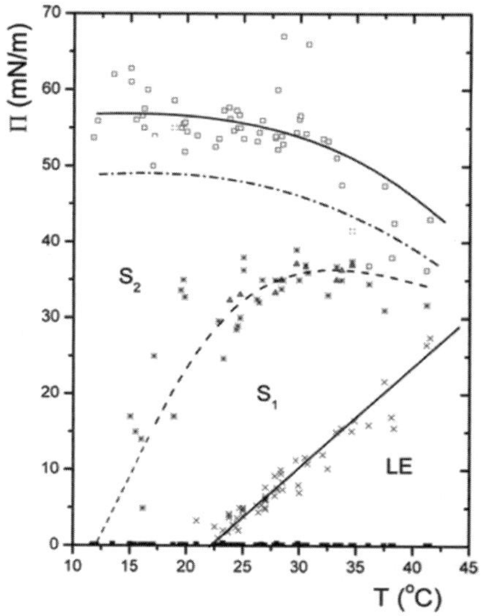

Fig. 1. Surface pressure–temperature phase diagram for dioctadecylamine. The phase changes were obtained from the Π–A isotherms (□, △, ■) and from BAM observations (×, *).

and with BAM observation. Most of the time, the phase changes occur before a change of slope in the $\Pi - A$ isotherms (kink). We found four different phases, these phases were named according to the apparent textures observed with BAM, namely: liquid expanded (LE) phase, solid 1 (S_1) and solid 2 (S_2). There is also a region, at very low lateral pressures ($\Pi \leq 0.2\,\mathrm{mN/m}$), where condensed phases are in coexistence with the gas phase (G). The monolayer collapses above Π~55 mN/m. More details about the DODA phase diagram, pH influence, etc. can be found elsewhere [11]. The S_1 phase grows from LE phase as star-shaped domains or dendritic domains, as it will be shown below. As far as we know, there are no GIXRD studies for this monolayer. Thus, the actual molecular arrangement of the different phases in the diagrams is not known.

The LE phase is found in a temperature range of 22–45°C and at lateral pressures ~0.2–30 mN/m. This phase does not present any kind of domains or shades of gray when observed with BAM; it is optically isotropic. Defects show that the monolayer is not rigid; however, the relative positions of defects do not change easily. Therefore, the LE phase behaves as a liquid phase, although not very fluid. In the LE phase, molecules should be not far apart from each other (~9–10 Å), since typical area density values for this

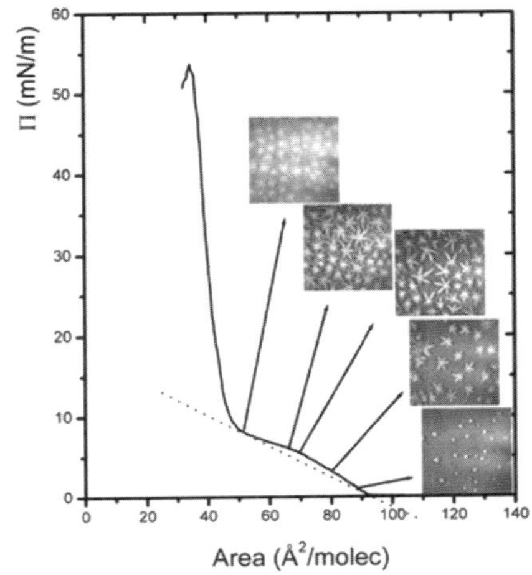

Fig. 2. BAM observations of the dioctadecylamine monolayer along the course of a normal compression to obtain an isotherm (compression rate $50\,\mathrm{cm^2 min^{-1}}$). The dotted line is an aid to the eye to see the wide shoulder in the isotherm.

phase are around 80–100 $\mathrm{\AA}^2$/molec; correlation between the tails of these molecules should be of short range.

At higher lateral pressures, the monolayer in the *LE* phase reaches a phase transition. Here, small light-gray dot domains of the S_1 phase, randomly distributed in the contrastless dark-gray liquid phase, suddenly appear in the BAM images of the monolayer. At the very beginning, the dot domains seem to be round at the microscope resolution. However, as far as the compression goes, the dot domains grow in the form of six-pointed star-shaped domains — all apparently with the same shade of gray. During the growth of the star domains, apparently no new dot domains are formed. If pressure decreases, dot and star domains disappear and they reappear as pressure increases again. As the lateral pressure increases, the points of the stars domains grow as dendrites (see Fig. 2).

Because of compression, star domains grow and become closer to each other, until they are distorted. If the analyzer of the microscope is rotated, there is an angle where the light gray hue of dot and star domains can be exchanged with the dark hue of the contrastless layer. This test reveals that the reflectivity of the monolayer is related to the tilting characteristic of the phases and not to multilayering. Later on, the distorted domains join together until the points of the star domains (dendrites) amalgamate and fade away slowly. Most of the times, these events coincide with the shoulder of the isotherms, which appear ca. 60–70 Å^2/molec. The texture of the S_1 phase seems like a speckled surface with irregularly shaped domains with different shades of gray. This phase is quite rigid; it seems to be like a solid phase. Some of the small bright domains are remains of the dendritic domains. As the pressure increases the size of the domains decreases, although, they remain irregular and some contrast is lost. The $LE - S_1$ transition is reversible for cycles of compression and decompression, and the cycles present hysteresis. Therefore, all seems to indicate that the $LE - S_1$ is a first-order phase transition. Linear fitting of the coexistence line allows us to obtain the two-dimensional Clausius–Clapeyron equation. Our $\frac{dP}{dT} = 1.26\,\text{mN/mK}^{-1}$ is close to the value obtained for other two-tail amphiphile monolayers, as DMPC ($2.35\,\text{mN/mK}^{-1}$) [12], DMPA ($1.05\,\text{mN/mK}^{-1}$) [12], and DPPC ($1.42\,\text{mN/mK}^{-1}$) [13]. The latent heat of transition for $LE - S_1$ transition at 23°C, is $l = 89.86\,\text{KJ/mole}$, which is of the same order of the two-tail DPPC monolayer (114 KJ/mole) [13] at 20°C.

The S_2 phase is located above the S_1 phase in the Π–A phase diagram. S_2 is contrastless using BAM, and it is apparently quite rigid; so it seems to be a solid phase. The $S_1 - S_2$ phase transition is detected most of the time only with BAM, when the contrast is lost as pressure increases. However, the contrast fades away continuously, suggesting that the transition is a second-order phase transition. This makes the localization of the $S_1 - S_2$ border very difficult and not precise, for this reason, we indicated the border as a dotted line in Fig. 1. However, the transition is reversible. The S_2 phase presents large areas almost with the same shade of gray. In this phase, the amine molecules are quite close (~7 Å), since the area density is ca. 45–60 Å^2/molec. If some analogy with fatty acids can be made, probably in this S_2 phase the tails are fully stretched (all-trans) and without tilting to be optically isotropic. On compressing S_2 phase, the monolayer collapses at area densities of the order of 35–50 Å^2/molec.

3.2. *Dendritic domains and domain growth*

At the $LE - S_1$ coexistence line, the transition starts with the formation of light gray dot domains, which grow as six-pointed star-shaped domains (dendrites) as pressure increases at constant temperature. The growth of these star domains is quite slow compared to the compression time. Even at the smallest compression rate used here, the transition occurs, but not at constant pressure. Therefore, a small overpressure is produced in the system before the star domains invade the whole monolayer. This overpressure is not big at low temperatures, but it can reach values ca. 5–7 mN/m at 30–40°C. Consequently, since the system is not in thermal equilibrium, the isotherms are not horizontal along this transition; probably we are coming into a metastable region (supercooled liquid). In Fig. 2, we present an isotherm and the structures observed in the monolayer with BAM, along a normal compression. Taking this behavior into account, it is now clear why the $LE - S_1$ phase transition was determined mainly using BAM instead of isotherms for determining the phase diagrams of Fig. 1.

Compressing the monolayer at lower compression rates than those used for obtaining the isotherms presented above, we reached a lateral pressure where LE and S_1 phases are in coexistence. The lateral pressure was maintained constant with the aid of the servomechanism of the trough. We observed the monolayers for long times. In these experiments, we tried to keep the air above the subphase motionless and at a temperature at most 1°C below the subphase temperature. In these experiments, we observed star domains that grow quite slowly (in the range of hours) and most of them are in fact six-fold dendrites; some examples can be seen in Fig. 3.

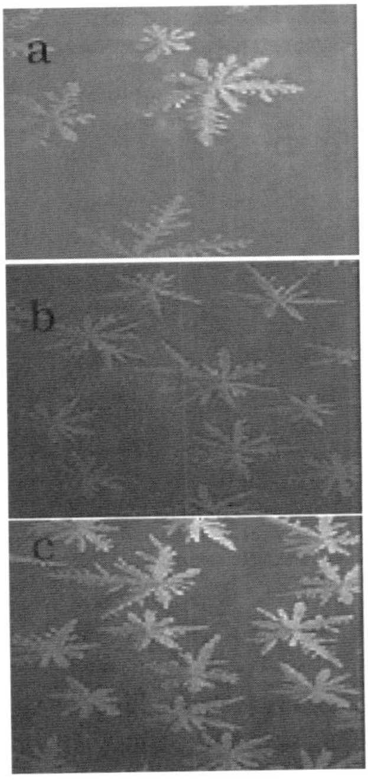

Fig. 3. Images of dendrites of S_1 phase that appear at the LE/S_1 phase transition in the dioctadecylamine monolayer as observed with BAM.

We do not think that these dendritic patterns are formed due to DODA impurities. If impurities were responsible for this form of growth, they would grow at a higher speed, in the range of half a minute, and they would form thin fractal structures, as has been observed in other systems [9]. In the denditric growth regime, the growth direction is correlated to the crystallographic direction of the lattice due to the line tension anisotropy. Therefore, the six-pointed star dendrites, with an angle of 60° on average between adjacent dendritic legs, are probably a consequence of a hexagonal lattice. The weak tilt indicated by the low anisotropy contrast found with BAM probably plays a minor role in the main growth direction, in agreement with what is found in systems with strongly tilted aliphatic chains [14]. In addition, it is clear from our observations that when a star domain does not have a neighbor star domain in some specific direction the dendritic crystallization growth will produce a larger dendritic point in that direction. In mature star domains, it is possible to see that the points have a slightly different shade of gray, revealing a small difference in tilting of the diamine tails in the dendritic points of the star domains. The contrast between domains comes from the different tilting of the hydrocarbon tails. Each shade of gray corresponds to a different azimuthal tilt direction in a domain.

One interpretation of our results is that the two-dimensional disks of S_1 phase grow in a supercooled liquid matrix showing a fingering-like instability similar to those observed in three dimensions, although the fields involved are not yet clear. In three dimensions, dendritic and fingering morphologies in crystallization are due to either the production of latent heat at the moving interface or to the expulsion of chemical impurities from the solid phase at the interface [15–17]. Diffusion of either the excess heat or excess impurities away from the interface proceeds more efficiently for a modulated interface, through what is called the Mullins–Sekerka instability. Studies of morphology of growing phases in LMs were based on polarized fluorescence microscopy before the introduction of BAM. That technique required the use of dyes [9], thus, diffusion of the dye could explain the modulated interface. However, in the case of pure monolayers, the heat build-up at the interface has been ruled out in our community, since LMs rest on a large body of water interface that acts as an isothermal reservoir, and the heat liberated is very small. In our specific case, using the Clayperon equation given above for the $LE - S_1$ phase transition, the released heat per $100\,\mu\text{m}^2$ of S_1 monolayer is 2.4×10^{-11} J, which is quite small.

To obtain the velocity of dendritic growth is not an easy task. A direct measurement of the growth as a function of time is not possible, since the

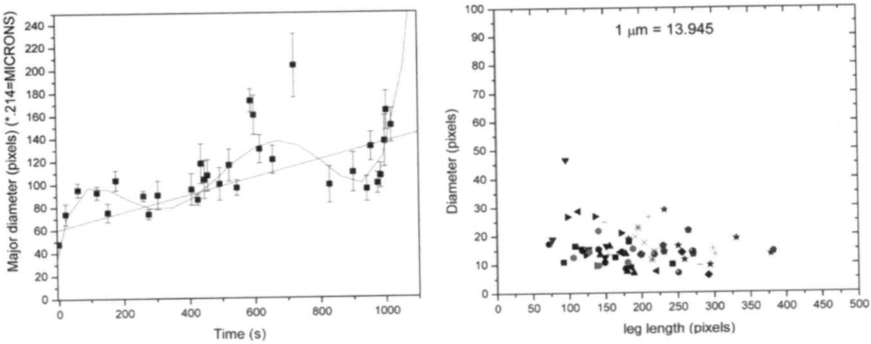

Fig. 4. (a) Size of the dendrites as a function of time. Lines are guides to the eye. (b) Diameter of dendrite tip points versus their point length for many dendrites in many different experiments.

monolayer is continuously moving. Consequently, domains are passing by along the field of view and the dendrites are observed in focus for less than a minute in a small strip of the field of view, due to the inclined position of the BAM beam; this restricts the observation of specific domains for long periods of time. To get a rough estimate, we measured in different areas of the monolayer, as they were passing through the field of view, the dendrite sizes as a function of time. Since they are asymmetric, we circumscribed each dendrite to an ellipsis. Figure 4(a) shows the mayor axis of this ellipsis as a function of time. At first glance, there is a strange oscillation mounted in a linear size increase versus elapsed time; however, this oscillation reveals the lever rule in a phase transition. It is clear that if there are some areas in the monolayer with too many dendrites, they have to be smaller to preserve the same area ratio as in another area with a small number of dendrites that are larger. In this form, our estimate is that the dendrite points grow with a rate between 0.22–0.02 μm/s. We measured the points' length against the radius of the tip, for many dendrites as can be seen in Fig. 4(b), which reveals that, on average, the radius of the tip is more or less the same, no matter how large the point is.

LMs are indeed a three-dimensional system, taking into account the subphase, where molecules forming the monolayer are the only ones which are restricted to move in two dimensions. Quite recently, a hydrodynamic mechanism was proposed [18], based on Marangoni flow, which describes the growth instabilities of liquid-condensed islands in the supercooled liquid-expanded phase. According to the authors of [18], this Marangoni instability seems to be intrinsic to LMs. In this model, the hydrodynamic transport of the insoluble surfactants seems to overwhelm the passive diffusion and to provide a mechanism for fingering instabilities. In this scheme, the authors predict a characteristic subphase flow below the monolayer. However, we used microspheres in the subphase, which can easily be observed with the BAM instrument to get the relation between the subphase velocity and the monolayer velocity. Up to now, everything seems to indicate that the monolayer velocity and the subphase velocity are not correlated. We can easily observe the microspheres below the air/water interface down to $\sim 10\,\mu$m. They can go in any direction below the monolayer and their speed does not seem to have any relation with the monolayer speed or direction of movement.

We performed experiments to determine the non-equilibrium stages followed by the domain growth in certain conditions. In these experiments, we slowly compressed the LE phase up to a certain specific pressure and waited for equilibration (~4–5 hours) until faceted nearly-spherical domains were observed. Then, the lateral pressure was increased by a certain amount to undercool the system and the domains were allowed to evolve freely at constant temperature and pressure. Mostly, the evolution was rapid and masked by the rapid movement of the whole monolayer. To reduce this problem, we added glycerin to the subphase (10–15%), increasing the subphase viscosity, and we also made the trough depth shallower; at these glycerin concentrations, the Π versus A isotherms are just slightly modified. The results were remarkable, because it was possible to observe all the steps involved in the evolution of the domains. Depending on the undercooling, the evolution can start from needles or fat fingers. At low undercooling, the S_1 domains develop fat fingers that show tip-splitting and when a certain size is achieved, there is a transition where a needle grows where the finger has split; after some evolution time the needles develop side branching. At higher undercooling, in the progression which started with needles growing from the S_1 domains, the needles increase in size and after some time there is a transition to side-branching as well. A general evolution scheme of the patterns is shown in Fig. 5. Here, in 5(a) and 5(b), we show how the needles grow from the S_1 domains. In 5(c), we observe some domains with tip-splitting. The additional figures 5(d)–(g) show different stages along the domain growth that evolved after the tip-splitting or after the needles started to side-branch.

The shape of the patterns can be observed with atomic force microscopy in LB transferred monolayers as shown in Fig. 6. The height difference between the top of the dendrites of S_1 phase and the LE phase is

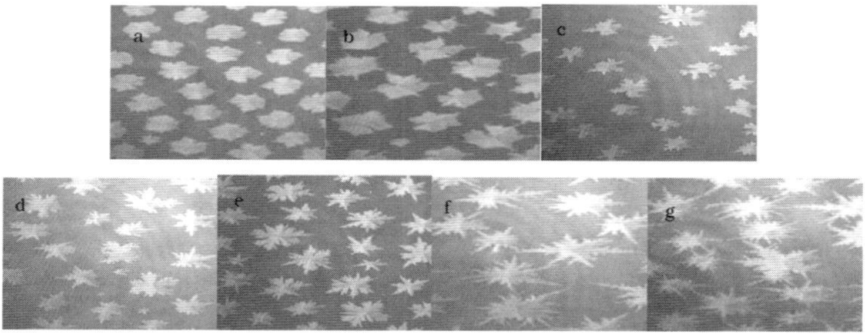

Fig. 5. BAM images showing the sequence that follows the S_1 domains emerging from LE phase ca. 23.5°C and 7–9 mN/m. Depending on the undercooling we observe roughly two different sequences of growth. Starting with the system very close to equilibrium, at low undercooling, the domains present a tip-splitting instability leading to a sequence of the form (c)–(d)–(e)–(f)–(g). If the undercooling is little bit larger, the sequence is (a)–(b)–(d)–(e)–(f)–(g).

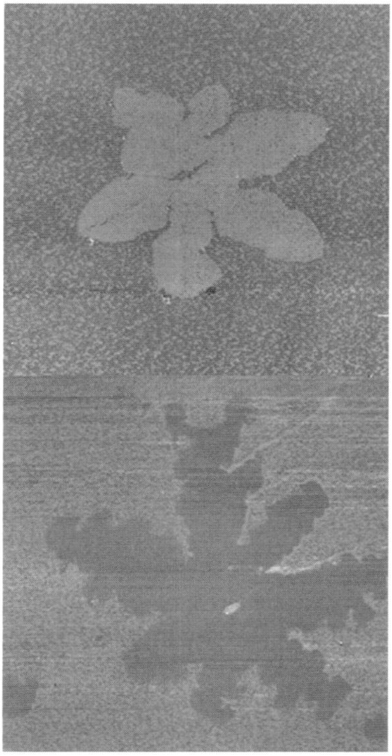

Fig. 6. AFM phase lag images of transferred monolayers of DODA, showing S_1 domains. Upper panel 20 × 20 μm and lower panel 60 × 60 μm.

approximately 0.8 nm due to the tail tilt difference; this difference allows us to estimate the tilt angle change in the LE/S_1 phase transition which is ca. 40°.

In summary, we have obtained the phase diagram of an amine LM with two long aliphatic tails. This system is peculiar because it forms LM for which the phase diagram is very different from those obtained for single chain amphiphiles, such as fatty acids, esters, alcohols, etc., or from two-chain phospholipids. We did not find a mosaic of irregular shaped domains as rich as those obtained for single chain amphiphiles, due to the different azimuthal tilt directions of the all-trans hydrocarbon tails. In our case, even the contrast between domains of different phases is low, although, clearly visible. We found four phases, although more phases could be undetected because of the low order in this LM due to the disorder introduced by the aliphatic tails that cannot align parallel near the interconnecting nitrogen. This system presents a dendritic crystallization, and we found the evolution of patterns during domain growth in certain specific conditions, which show a tip-splitting to side-branching transition common in the study of morphological instabilities [19, 20]. However, the origin of the instabilities that allows this kind of behavior is not clear yet. Details of the mechanism of how these domains grow and which fields drive this kind of instability are under study. Monolayer morphological instabilities are an interesting challenge for promoting theoretical work.

Acknowledgments

We acknowledge partial support of CONACYT and DGAPAUNAM grants (36680-E and IN-113601, IN-117802-2).

References

[1] Als-Nielsen, J., Jacquemain, D., Kjaer, K., Leveiller, F., Lahav, M. and Leiserowitz, L., *Phys. Rep.* **246** (1994) 251.

[2] Knobler, C.M., in *Advances in Chemical Physics*, eds. Prigogine, I. and Rice, S.A., Vol. LXXVII (Wiley, New York, 1990), p. 397.

[3] Höning, D. and Möbius, D., *J. Phys. Chem.* **95** (1991) 4590.

[4] Henon, S. and Meunier, J., *Rev. Sci. Instrum.* **62** (1991) 936.

[5] Kaganer, V.M., Möhwald, H. and Dutta, P., *Rev. Mod. Phys.* **71** (1999) 779.

[6] Ramos, S. and Castillo, R., *J. Chem. Phys.* **110** (1999) 7021.

[7] Riviere, S., Henon, S., Meunier, J., Schwartz, D.K., Tsao, M.W. and Knobler, C.M., *J. Chem. Phys.* **101** (1994) 10045.

[8] Iimura, K.I., Yamauchi, Y., Tsuchiya, Y., Kato, T. and Suzuki, M., *Langmuir* **17** (2001) 4602.

[9] Miller, A. and Möhwald, H., *J. Chem. Phys.* **86** (1987) 4258.
[10] Wiedemann, G. and Vollhardt, D., *Langmuir* **13** (1997) 1623.
[11] Flores, A., Ize, P., Ramos, S. and Castillo, R., *J. Chem. Phys.* **119** (2003) 5644.
[12] Albrecht, O., Gruler, H. and Sackmann, E., *J. Phys. (Paris)* **39** (1978) 301.
[13] Krasteva, N., Vollhardt, D., Brezesinski, G. and Möhwald, H., *Langmuir* **17** (2001) 1209.
[14] Gehlert, U. and Vollhardt, D., *Langmuir* **13** (1997) 277.
[15] Langer, J.S., *Rev. Mod. Phys.* **52** (1980) 1.
[16] Fogedby, H.C., Schwartz Sorensen, E. and Mouritsen, O.G., *J. Chem. Phys.* **87** (1987) 6706.
[17] Mouritsen, O.G., *Int. J. Mod. Phys. B* **4** (1990) 1925.
[18] Bruinsma, R., Rondelez, F. and Levine, A., *Eur. Phys. J. E* **6** (2001) 191.
[19] Kondic, L., Shelley, M.J. and Palffy-Mohoray, P., *Phys. Rev. Lett.* **80** (1998) 1433.
[20] Provatas, N., Wang, Q., Haataja, M. and Grant, M., *Phys. Rev. Lett.* **91** (2003) 155502-1.

PART 3

ELECTRICAL AND MAGNETIC PROPERTIES

CHAPTER 13

MULTIPLE SCATTERING EFFECTS IN THE SECOND HARMONIC GENERATION OF LIGHT REFLECTION FROM RANDOMLY ROUGH METAL SURFACE

A.A. Maradudin*, T.A. Leskova*, M. Leyva-Lucero[†,‡] and E.R. Méndez[†]

Department of Physics and Astronomy, and Institute for Surface and Interface Science, University of California, Irvine, California, 92697, USA

[†]*División de Física Aplicada, Centro de Investigación Científica y de Educación Superior de Ensenada, Apartado Postal 2732, Ensenada, Baja California, Mexico*

[‡]*Escuela de Ciencias Físico-Matematicas, Universidad Autonoma de Sinaloa, Ciudad Universitaria, C.P. 80000, Culiacan, Sinaloa, Mexico*

We present perturbative studies of the second harmonic generation of light in reflection from a one-dimensional, randomly rough, metal surface, when the plane of incidence is perpendicular to the generators of the surface. The random surface is characterized by a power spectrum that is nonzero only in the ranges of wave numbers that can couple the incident light into the surface electromagnetic waves supported by the metal surface at the fundamental and/or harmonic frequencies. The numerical results obtained perturbatively are compared with experimental results and those obtained by means of rigorous numerical simulations. As do the results of the numerical simulations, our perturbative results show that the angular distribution of the intensity of the incoherent component of the scattered light at the harmonic frequency displays either well-defined peaks or dips in the retroreflection direction and in the direction normal to the mean plane of the surface. The mechanisms responsible for the formation of the peaks or dips are discussed.

1. Introduction

Experimental and theoretical studies of second harmonic generation (SHG) of light in reflection from a metal surface go back at least three decades (see, e.g., [1, 2]). However, although considerable attention has been paid to the generation of second harmonic light by surfaces with a periodic profile [3–5] in view of the possibility of obtaining an enhancement of the second harmonic intensity through the excitation of surface plasmon polaritons, comparatively little attention has been devoted to second harmonic generation in reflection from a randomly rough metal surface. Nevertheless, the possiblity of an enhancement of the second harmonic intensity through the excitation of surface plasmon-polaritons via surface roughness has also received some attention [6, 7].

Interest in this problem has increased in the last several years due to the growing interest in interference effects occurring in the multiple scattering of electromagnetic waves from randomly rough metal surfaces and the related backscattering enhancement phenomenon. The enhanced backscattering of light from weakly rough metal surfaces has been of special interest because it had been predicted that it is the multiple scattering of surface plasmon polaritons that is responsible for the enhancement in this case [8]. Recently, experimental observations of the enhanced backscattering of light from a weakly rough random metal surface caused by the interference of surface plasmon polaritons have been reported [9]. As in the case of volume disordered media, where the interplay of nonlinearity and disorder leads to a number of novel effects (see, e.g., [10]), it has been expected that the nonlinear optical interactions at a randomly rough metal surface should produce new features owing to interference effects in the multiple scattering of electromagnetic waves. Especially interesting phenomena could be expected when the nonlinear interaction at a weakly rough random surface leads to the excitation of surface plasmon polaritons of a frequency different from the frequency ω of the incident light, as in the case of second harmonic generation, where surface plasmon polaritons of both the fundamental and second harmonic frequencies are excited. The results of a perturbative calculation carried out by McGurn *et al.* [11] predicted that enhanced second harmonic generation of light at a weakly rough, clean, metal surface occurs not only in the retroreflection direction but also in the direction normal to the mean scattering surface. The multiple scattering of surface plasmon polaritons supported by the vacuum-metal interface, excited by the incident and generated second harmonic light through the roughness of the interface, plays the decisive role in the appearance of both peaks in this theory.

This work stimulated several subsequent experimental studies of second-harmonic generation in the scattering of light from metal surfaces [12–17], and enhanced second harmonic generation peaks in the direction normal to the mean surface and in the retroreflection were observed [12–17]. In these experiments, however, the scattering system was not a clean random interface between vacuum and a semi-infinite metal but the random interface with a dielectric or vacuum of a thin metal film deposited on the planar base of a dielectric prism through which the light was incident (the Kretschmann attenuated total reflection geometry [18]). In the experiments of [12, 13, 15] the scattering system was the random interface between a silver film and a nonlinear quartz crystal, so that the nonlinear interaction occurred in the quartz crystal rather than at the significantly more weakly nonlinear silver surface. A well-defined peak of the second harmonic generation in the direction normal to the mean interface was observed in [12]. When the experiment was carried out with long-range surface polaritons [13], peaks of the enhanced second harmonic generation were detected both in the retroreflection direction and in the direction normal to the surface. In [14, 16, 17] attempts to detect the peaks of the enhanced second harmonic generation at a silver film–vacuum interface were made. A well-defined peak in the direction normal to the mean surface was observed in [14, 17], while only a broad depolarized background, but no peak in the direction normal to the mean surface, was observed in [16]. Theoretical treatments of second harmonic generation from randomly rough metal films in the Kretschmann ATR geometry of the type employed in the experiments of [12–17] can be found in [19, 20], and the reader is referred to these papers for the details of the calculations.

The first experimental studies of multiple-scattering effects in the second harmonic generation of light scattered from a clean one-dimensional vacuum–metal interface were carried out in a series of papers by O'Donnell and his colleagues [21–23]. In the experiments [21, 22] the random surfaces were fabricated in a special way to produce a strong excitation of surface plasmon polaritons of either the fundamental or harmonic frequencies. This allowed separating and identifying different processes of coherent interference occurring in the nonlinear scattering of light from the rough surface. It was found that for both weakly [21, 22] and strongly [23] rough surfaces a dip is present in the retroreflection direction in the angular dependence of the intensity of the scattered second harmonic light rather than the peak that occurs in scattering at the fundamental frequency. No peak in the direction normal to the mean surface was observed in these experiments. Many of the features of second harmonic scattering found in these papers were reproduced by the rigorous numerical simulations of second harmonic generation from random surfaces carried out by Leyva-Lucero *et al.* [24, 25]. There remain, however, some differences that may be attributed to uncertainties in the phenomenological nonlinear surface susceptibilities.

In this paper we present results of perturbative calculations of the second harmonic generation of light in reflection from a weakly rough random metal surface. Such a treatment is useful for interpreting the results of the numerical simulations as well as the experimental data, since it yields analytic expressions for the various contributions to the process of second harmonic generation, whose content is readily understood and, thus, allows clarifying the origin of different processes of nonlinear interaction and coherent interference. Our analysis is based on the well known fact that homogeneous and isotropic metals possess inversion symmetry, so that the dipole contribution to the bulk nonlinear polarization is absent ($\chi^{(2)} = 0$). The presence of the surface breaks the inversion symmetry, and since both the electromagnetic fields and material constants vary rapidly at the surface, their gradients give rise to the optical nonlinearity of the surface. The second harmonic radiation we are interested in is generated in a metal–vacuum interface layer that has a finite thickness on the microscopic scale. Consequently, the nonlinear polarization can be taken into account through the boundary conditions for the second harmonic fields.

In our analysis we will neglect the small contribution to the nonlinearity coming from the bulk and the possible anisotropy of the material constants. In this formulation of the problem both the fundamental and harmonic fields satisfy Helmholtz equations above and below the interface. We first solve the linear problem of the scattering of light of the fundamental frequency ω, and use its solution to determine the surface nonlinear polarization at the harmonic frequency 2ω. In solving the scattering problem for the harmonic fields we will use the nonlinear boundary conditions. The form of these boundary conditions is known to depend on the particular model for the nonlinear surface polarization assumed, and has been under discusssion for quite a long time [20, 26–33].

However, the experimental studies of SHG from a planar metal surface are fundamentally restricted and cannot provide enough information to distinguish among different theoretical models for the nonlinear response. In contrast, the analysis of the linear and nonlinear processes that occur at a rough metal surface presented in this paper allows us to separate the mechanisms leading to the experimentally observed features and, as a result, provides a means of choosing from among different models of the nonlinear optical response.

The outline of this paper is as follows. In Sec. 2 we describe the scattering system. In Sec. 3 we describe the perturbative calculations of the fundamental and second harmonic fields. The results of our calculations presented in Sec. 4 are discussed in Sec. 5. Finally, in Sec. 6 we present the conclusions drawn from the results obtained in this work.

2. Formulation of the Scattering Problem

The physical system we consider consists of an isotropic metal characterized by a complex, frequency-dependent dielectric function $\epsilon(\omega) = \epsilon_1(\omega)+i\epsilon_2(\omega)$ in the region $x_3 < \zeta(x_1)$ and a vacuum in the region $x_3 > \zeta(x_1)$. The surface profile function $\zeta(x_1)$ is assumed to be a single-valued function of x_1 that is differentiable as many times as is necessary, and to constitute a zero-mean, stationary, Gaussian random process defined by

$$\langle\zeta(x_1)\rangle = 0, \tag{1a}$$

$$\langle\zeta(x_1)\zeta(x_1')\rangle = \delta^2 W(|x_1 - x_1'|). \tag{1b}$$

The angle brackets in Eqs. (1) denote an average over the ensemble of realizations of the surface profile function, and $\delta = \langle\zeta^2(x_1)\rangle^{\frac{1}{2}}$ is the RMS height of the surface.

We also introduce the Fourier integral representation of the surface profile function,

$$\zeta(x_1) = \int_{-\infty}^{\infty} \frac{dQ}{2\pi}\hat{\zeta}(Q)e^{iQx_1}. \tag{2}$$

The Fourier coefficient $\hat{\zeta}(Q)$ is a zero-mean Gaussian random process defined by

$$\langle\hat{\zeta}(Q)\rangle = 0, \tag{3a}$$

$$\langle\hat{\zeta}(Q)\hat{\zeta}(Q')\rangle = 2\pi\delta(Q+Q')\delta^2 g(|Q|), \tag{3b}$$

where $g(|Q|)$, the power spectrum of the surface roughness, is given by

$$g(|Q|) = \int_{-\infty}^{\infty} dx_1 W(|x_1|)e^{-iQx_1}. \tag{4}$$

In this paper we will present numerical results calculated for a random surface characterized by a power spectrum of the form

$$g(|k|) = \frac{\pi h_1}{k_{\max}^{(1)} - k_{\min}^{(1)}} \Big[\Theta\Big(k - k_{\min}^{(1)}\Big) \Theta\Big(k_{\max}^{(1)} - k\Big) + \Theta\Big(-k - k_{\min}^{(1)}\Big) \Theta\Big(k_{\max}^{(1)} + k\Big) \Big] + \frac{\pi h_2}{k_{\max}^{(2)} - k_{\min}^{(2)}} \Big[\Theta\Big(k - k_{\min}^{(2)}\Big) \Theta\Big(k_{\max}^{(2)} - k\Big) + \Theta\Big(-k - k_{\min}^{(2)}\Big) \Theta\Big(k_{\max}^{(2)} + k\Big) \Big], \tag{5}$$

where $\Theta(z)$ is the Heaviside unit step function, $k_{\min}^{(1)} < k_{\rm sp}(\omega) < k_{\max}^{(1)}$, $k_{\min}^{(2)} < k_{\rm sp}(2\omega) < k_{\max}^{(2)}$, $k_{\rm sp}(\Omega) = \mathrm{Re}\left[(\Omega/c)[\epsilon(\Omega)/(\epsilon(\Omega)+1)]^{1/2}\right]$ is the wave number of surface plasmon polaritons of frequency Ω, and Ω stands for ω or 2ω, while $h_1 + h_2 = 1$. Surfaces characterized by a power spectrum of this type have been used in recent experimental studies of light scattering from weakly rough random metal surfaces [9, 21, 22].

In our treatment of second harmonic generation we neglect the influence of the nonlinearity on the fundamental fields, as was done in [24, 25, 33] in rigorous numerical simulations of second harmonic generation from random surfaces. To solve the scattering equations at the harmonic frequency, the Maxwell equations satisfied by the electromagnetic fields in regions $x_3 > \zeta(x_1)$ and $x_3 < \zeta(x_1)$ have to be supplemented by boundary conditions at the rough metal interface $x_3 = \zeta(x_1)$. At the fundamental frequency ω these boundary conditions express the continuity of the tangential components of the magnetic and electric fields across the interface. The nonlinear boundary conditions for the harmonic fields were obtained in several studies [20, 27–33] by integrating Maxwell's equations for these fields across the interface layer, and then passing to the limit of a vanishing layer thickness. We will use them in the form presented in [33]

$$H^{>}(x_1|2\omega) - H^{<}(x_1|2\omega) = 4\pi \frac{2ic}{\omega} \frac{\chi_{ttz}^{\rm s}}{\phi^2(x_1)} L^{>}(x_1|\omega) \frac{d}{dx_1} H^{>}(x_1|\omega), \tag{6}$$

$$L^{>}(x_1|2\omega) - \frac{1}{\epsilon(2\omega)} L^{<}(x_1|2\omega) = 4\pi \frac{2ic}{\omega} \frac{d}{dx_1} \left\{ \frac{1}{\phi^2(x_1)} \left[\chi_{zzz}^{\rm s} \left(\frac{d}{dx_1} H^{>}(x_1|\omega) \right)^2 + \chi_{ztt}^{\rm s} \left(L^{>}(x_1|\omega) \right)^2 \right] \right\}, \tag{7}$$

where the superscripts $>$ and $<$ denote the fields in the vacuum and in the metal, respectively, and χ^{s}_{ijk} is the second order surface susceptibility tensor whose components are the constants of proportionality relating the amplitude of the components of the nonlinear surface polarization to the fundamental field amplitudes at the surface. In Eqs. (6) and (7) we have introduced the source functions

$$H^{>,<}(x_1|2\omega) = H_2^{>,<}(x_1, x_3|2\omega)\big|_{x_3=\zeta(x_1)}, \tag{8a}$$

$$L^{>,<}(x_1|2\omega) = \frac{\partial}{\partial N}\, H_2^{>,<}(x_1, x_3|2\omega)\big|_{x_3=\zeta(x_1)}, \tag{8b}$$

where

$$\frac{\partial}{\partial N} = -\zeta'(x_1)\frac{\partial}{\partial x_1} + \frac{\partial}{\partial x_3}$$

is a derivative along the normal to the interface at each point directed from the metal into the vacuum. When the incident light is s-polarized the nonlinear sources give a nonzero contribution only to the p-polarized second harmonic field. In this case the nonlinear boundary conditions have the form

$$H^{>}(x_1|2\omega) - H^{<}(x_1|2\omega) = 0, \tag{9}$$

$$L^{>}(x_1|2\omega) - \frac{1}{\epsilon(2\omega)}L^{<}(x_1|2\omega) = -4\pi\frac{2i\omega}{c}\chi^{\mathrm{s}}_{ztt}\frac{d}{dx_1}\left[E^{>}(x_1|\omega)\right]^2, \tag{10}$$

where

$$E^{>}(x_1|\omega) = E_2^{>}(x_1, \zeta(x_1)|\omega). \tag{11}$$

We first solve a linear scattering problem, and with the solution of the linear scattering problem in hand we determine the fields of frequency ω and their normal derivatives evaluated on the rough surface. To simplify the expressions for the nonlinear driving term we will calculate the field and its derivative at the surface inside the medium and will then make use of the linear boundary conditions at the rough surface to obtain the source functions $H^{(>)}(x_1|\omega)$, $H^{(>)}(x_1|\omega)$, and $L^{(>)}(x_1|\omega)$.

By matching the harmonic fields at $x_3 = \zeta(x_1)$ with the use of the nonlinear boundary conditions we derive an equation for the scattering amplitude at frequency 2ω analogous to the reduced Rayleigh equation in the linear theory. On solving this equation we determine the intensity of second harmonic generation by averaging the intensity of the harmonic light over the ensemble of realizations of the surface profile function and normalizing it by the square of the intensity of the incident light.

To describe the most prominent features of the angular dependence of the mean differential intensity of the generated light using the small amplitude perturbation theory it is neccessary to calculate the intensity up to, at least, sixth order in the surface profile function. As is known [34], the use of small-amplitude perturbation theory in the linear problem of light scattering leads to both quantitatively and qualitatively good results. However, in the problem of nonlinear scattering this might be not the case, since the higher order processes of scattering can, and actually do, give the dominant contribution to the intensity of the generated light. The second reason for the small-amplitude perturbation theory to break down is that although the surfaces we study are weakly rough, i.e., the inequalities $\delta \ll \lambda$ and $|\zeta'(x_1)| \ll 1$ are satisfied, the influence of roughness on the surface plasmon polaritons could be extremely strong due to the specific form of the power spectrum of the surface roughness assumed in this work. The experiments of [21] and [22] were done with surfaces which were fabricated so that the surface plasmon polaritons of frequencies 2ω [21] or ω [22] were dominantly converted into volume waves in the vacuum above the surface rather than scattered into the surface plasmon polaritons. This means that the influence of roughness on the propagation of surface plasmon polaritons is strong, although the surface is weakly rough in a conventional sense.

To avoid the problems that might arise when using the small-amplitude perturbation theory we will use the Green's function approach usually used in a many-body perturbation theory.

3. Perturbative Solution

3.1. *Fundamental fields*

We assume that a p- or s-polarized plane wave of frequency ω is incident on the surface $x_3 = \zeta(x_1)$ from the vacuum side. The angle of incidence, measured counterclockwise from the normal to the mean surface is θ_0, and the plane of incidence is the x_1x_3-plane. We assume that the surface is weakly rough and satisfies the condition for the validity of the Rayleigh hypothesis, $|\zeta'(x_1)| \ll 1$ [35–37]. In this case we can seek the x_2-component of the magnetic (p-polarization) or electric (s-polarization) fields in the vacuum in the form of a sum of the fields of the incident and scattered waves,

$$\mathcal{F}^{>}(x_1, x_3|\omega) = \mathcal{F}_0 \left[e^{ikx_1} e^{-i\alpha_0(k,\omega)x_3} + \int_{-\infty}^{\infty} \frac{dq}{2\pi} R_\omega(q|k) e^{iqx_1} e^{i\alpha_0(q,\omega)x_3} \right], \tag{12a}$$

while in the medium

$$\mathcal{F}^{<}(x_1, x_3|\omega) = \mathcal{F}_0 \int_{-\infty}^{\infty} \frac{dq}{2\pi} T_\omega(q|k) e^{iqx_1} e^{-i\alpha(q,\omega)x_3}, \tag{12b}$$

where $\mathcal{F}(x_1, x_3|\omega)$ stands for either $H_2(x_1, x_3|\omega)$ or $E_2(x_1, x_3|\omega)$, $\mathcal{F}_0$ is the amplitude of the incident field, $R_\omega(q|k)$ and $T_\omega(q|k)$ are the scattering and transmission amplitudes, respectively, and $k = (\omega/c)\sin\theta_0$ and $\alpha_0(k,\omega) = (\omega/c)\cos\theta_0$ are the tangential and normal components of the wave vector of the incident light, $\alpha_0(q,\omega) = \sqrt{(\omega^2/c^2) - q^2}$, $\mathrm{Re}(\alpha_0(q,\omega)) > 0$, $\mathrm{Im}(\alpha_0(q,\omega)) > 0$, and $\alpha(q,\omega) = \sqrt{\epsilon(\omega)(\omega^2/c^2) - q^2}$, $\mathrm{Re}(\alpha(q,\omega)) > 0$, $\mathrm{Im}(\alpha(q,\omega)) > 0$.

To calculate the nonlinear source functions entering the nonlinear boundary conditions (6) and (7) it is neccessary to know only the field and its normal derivative on the surface. In the linear problem the field and its normal derivative on the surface in the vacuum and in the metal are related through the Maxwell boundary conditions, $\mathcal{F}^{>}(x_1|\omega) = \mathcal{F}^{<}(x_1|\omega)$ and $\mathcal{L}^{>}(x_1|\omega) = \kappa(\omega)\mathcal{L}^{<}(x_1|\omega)$, where $\mathcal{F}^{>,<}(x_1|\omega) = \mathcal{F}^{>,<}(x_1, \zeta(x_1)|\omega)$ and $\mathcal{L}^{>,<}(x_1|\omega) = \partial\mathcal{F}^{>,<}(x_1, x_3|\omega)/\partial N|_{x_3=\zeta(x_1)}$, while $\kappa(\omega) = 1/\epsilon(\omega)$ for a p-polarized field, and $\kappa(\omega) = 1$ for an s-polarized field. One can therefore solve the scattering problem for the fields of frequency ω either in the vacuum or in the metal. In our study it is more convenient to work with the fields in the metal. First, the scattering theory is more straightforward, because in this case the scattering potential is a function rather than a solution of an integral equation. Second, the nonlinear source functions can be written in a more compact form, because the explicit expressions for $\mathcal{F}^{<}(x_1, \zeta(x_1)|\omega)$ and $\mathcal{L}^{<}(x_1, \zeta(x_1)|\omega)$ are much simpler than those for $\mathcal{F}^{>}(x_1, \zeta(x_1)|\omega)$ and $\mathcal{L}^{>}(x_1, \zeta(x_1)|\omega)$.

An equation analogous to the reduced Rayleigh equation for the transmission amplitude can be derived in a standard manner [38] and has the form

$$\int_{-\infty}^{\infty} \frac{dq}{2\pi} N(p, q|\omega) T_\omega(q|k) = -2i\alpha_0(k,\omega) 2\pi\delta(p-k), \tag{13}$$

where

$$N(p, k|\omega) = \frac{n(p, k|\omega)}{\alpha_0(p,\omega) - \alpha(k,\omega)} I(\alpha(k,\omega) - \alpha_0(p,\omega)|p-k), \tag{14a}$$

with

$$n(p, q|\omega) = i\left(1 - \frac{1}{\epsilon(\omega)}\right)(pq + \alpha_0(p,\omega)\alpha(q,\omega)), \tag{14b}$$

for p-polarized incident light,

$$n(p,q|\omega) = i\left(\epsilon(\omega) - 1\right)\frac{\omega^2}{c^2}, \tag{14c}$$

for s-polarized incident light, and

$$I(\gamma|Q) = \int_{-\infty}^{\infty} dx_1 e^{-iQx_1} e^{-i\gamma\zeta(x_1)}. \tag{15}$$

Following the procedure developed in [39] in solving the reduced Rayleigh equation for the scattering amplitude we will seek the solution of Eq. (13) in the form

$$T_\omega(q|k) = -2iG_\omega(q|k)\alpha_0(k,\omega), \tag{16}$$

where we have introduced the Green's function $G_\omega(q|k)$ associated with the randomly rough interface between the vacuum and the metal. We define it as the solution of the equation

$$G_\omega(q|k) = 2\pi\delta(q-k)G_0(k,\omega) + \int_{-\infty}^{\infty}\frac{dp}{2\pi}G_\omega(q|p)V(p,k|\omega)G_0(k,\omega), \tag{17}$$

where

$$G_0(q,\omega) = \frac{i}{\alpha_0(q,\omega) + \kappa(\omega)\alpha(q,\omega)} \tag{18}$$

is the Green's function associated with a planar surface. An equation for the scattering potential $V(p,k|\omega)$ can be derived in a following way. We substitute Eq. (16) into Eq. (13) with the result:

$$\int_{-\infty}^{\infty}\frac{dq}{2\pi}N(p,q|\omega)G_\omega(q|k) = 2\pi\delta(p-k). \tag{19}$$

We now multiply Eq. (17) for the Green's function from the left by $N(p,q|\omega)$ and integrate the result over q, to obtain

$$\int_{-\infty}^{\infty}\frac{dq}{2\pi}N(p,q|\omega)G_\omega(q|k) = N(p,k|\omega)G_0(k,\omega) + \int_{-\infty}^{\infty}\frac{dq}{2\pi}\int_{-\infty}^{\infty}\frac{dr}{2\pi}N(p,q|\omega) \times G_\omega(q|r)V(r|k)G_0(k,\omega). \tag{20}$$

With the aid of Eq. (19) the integral equation (20) reduces to the algebraic equation

$$2\pi\delta(p-k) = N(p,k|\omega)G_0(k,\omega) + V(p,k|\omega)G_0(k,\omega). \tag{21}$$

If we further use the identity $I(\gamma|Q) = 2\pi\delta(Q) + J(\gamma|Q)$, where

$$J(\gamma|Q) = \int_{-\infty}^{\infty} dx_1 e^{-iQx_1} \left(e^{-i\gamma\zeta(x_1)} - 1 \right), \tag{22}$$

we can represent $N(p, k|\omega)$ as

$$N(p, k|\omega) = 2\pi\delta(q-k)G_0^{-1}(k,\omega) + \tilde{N}(q, k|\omega), \tag{23}$$

where

$$\tilde{N}(q, p|\omega) = \frac{n(q, p|\omega)}{\alpha_0(q,\omega) - \alpha(p,\omega)} J(\alpha(p,\omega) - \alpha_0(q,\omega)|q-p), \tag{24}$$

whereupon Eq. (21) takes the simple form

$$V(q, p|\omega) = -\tilde{N}(q, p|\omega). \tag{25}$$

Due to the stationarity of $\zeta(x_1)$ the averaged Green's function $\langle G_\omega(q|k)\rangle$ is given by

$$\langle G_\omega(q|k)\rangle = 2\pi\delta(q-k)\frac{1}{G_0^{-1}(k,\omega) - M(k,\omega)} \tag{26a}$$

$$\equiv 2\pi\delta(q-k)G(k,\omega), \tag{26b}$$

where $M(k,\omega)$ is the averaged proper self-energy. The latter is given by

$$\langle M_\omega(q|k)\rangle = 2\pi\delta(q-k)M(k,\omega), \tag{27}$$

where the (unaveraged) proper self-energy $M_\omega(q|k)$ is the solution of [39]

$$M_\omega(q|k) = V(q, k|\omega) + \int_{-\infty}^{\infty} \frac{dp}{2\pi} M_\omega(q|p)G(p,\omega)W(p, k|\omega), \tag{28}$$

and we have introduced the notation

$$W(q, k|\omega) = V(q, k|\omega) - \langle M_\omega(q|k)\rangle. \tag{29}$$

In order to incorporate into the calculation of the fields the averaged Green's function $G(p,\omega)$ instead of the unperturbed Green's function $G_0(p,\omega)$, we rewrite the equation for the Green's function, Eq. (17), in the form [39]

$$G_\omega(q|k) = G(q,\omega)2\pi\delta(q-k) + G(q,\omega)t_\omega(q|k)G(k,\omega). \tag{30}$$

In Eq. (30) the operator $t_\omega(p|r)$ was introduced to satisfy

$$\int_{-\infty}^{\infty} \frac{dp}{2\pi} W(q,p|\omega)G_\omega(p|k) = t_\omega(q|k)G(k,\omega), \tag{31}$$

and is the solution of the equation

$$t_\omega(q|k) = W(q,k|\omega) + \int_{-\infty}^{\infty} \frac{dp}{2\pi} \int_{-\infty}^{\infty} \frac{dr}{2\pi} W(q,p|\omega)G(p,\omega)t_\omega(p|k). \tag{32}$$

From Eq. (30) it follows that $\langle t_\omega(q|k)\rangle = 0$. The transmission amplitude $T_\omega(q|k)$ in terms of the averaged Green's function $G(p,\omega)$ and the operator $t_\omega(q|k)$ is then given by

$$T_\omega(q|k) = -2i\left[2\pi\delta(q-k) + G(q,\omega)t_\omega(q|k)\right]G(k,\omega)\alpha_0(k,\omega). \tag{33}$$

Thus, we have reduced the problem of light scattering to a standard scattering problem [39], that can be solved by a many-body perturbation theory approach. The Green's function, $G(q,\omega)$, present in the expression for the scattered field, plays an essential role in the scattering theory, since it describes the excitation of surface plasmon polaritons when the scattering of p-polarized light is studied. In what follows it will enable us to extract the surface polariton-related processes of scattering and nonlinear interaction.

We can now calculate the source functions $\mathcal{F}^>(x_1|\omega) = \mathcal{F}^<(x_1|\omega)$ and $\mathcal{L}^>(x_1|\omega) = \kappa(\omega)\mathcal{L}^<(x_1|\omega)$ by setting $x_3 = \zeta(x_1)$ in Eq. (12b) and in its normal derivative. As a result we obtain

$$\begin{aligned} \mathcal{F}^>(x_1|\omega) &= -2i\mathcal{F}_0 \int_{-\infty}^{\infty} \frac{dq}{2\pi} e^{iqx_1 - i\alpha(q,\omega)\zeta(x_1)} \left[2\pi\delta(p-k)\right. \\ &\quad \left. + G(q,\omega)t_\omega(q|k)\right] G(k,\omega)\alpha_0(k,\omega), \end{aligned} \tag{34a}$$

$$\begin{aligned} \mathcal{L}^>(x_1|\omega) &= -2\kappa(\omega)\mathcal{F}_0 \int_{-\infty}^{\infty} \frac{dq}{2\pi} \left[\alpha(q,\omega) + q\zeta'(x_1)\right] e^{iqx_1 - i\alpha(q,\omega)\zeta(x_1)} \\ &\quad \times \left[2\pi\delta(p-k) + G(q,\omega)t_\omega(q|k)\right] G(k,\omega)\alpha_0(k,\omega). \end{aligned} \tag{34b}$$

The Fourier coefficients of the source functions then are given by

$$\begin{aligned} \widehat{\mathcal{F}}(q) &= -2i\mathcal{F}_0 \int_{-\infty}^{\infty} \frac{dp}{2\pi} I(\alpha(p,\omega)|q-p) \left[2\pi\delta(p-k)\right. \\ &\quad \left. + G(p,\omega)t_\omega(p|k)\right] G(k,\omega)\alpha_0(k,\omega), \end{aligned} \tag{35a}$$

and

$$\mathcal{L}(q) = -2\mathcal{F}_0 \int_{-\infty}^{\infty} \frac{dp}{2\pi} u(q|p) I(\alpha(p,\omega)|q-p) \left[2\pi\delta(p-k) + G(p,\omega)t_\omega(p|k)\right] G(k,\omega)\alpha_0(k,\omega), \tag{35b}$$

where

$$u(q|k) = \kappa(\omega)\frac{\epsilon(\omega)(\omega^2/c^2) - qk}{\alpha(k,\omega)}. \tag{36}$$

3.2. *Fields of frequency 2ω*

We now turn to the treatment of the fields of frequency 2ω. As was shown in [20, 26–33] both p- and s-polarized incident light give rise to a surface nonlinear polarization that radiates only p-polarized waves of frequency 2ω. Since the field of the generated waves satisfies Helmholtz equations in the vacuum and in the metal, and the surface roughness is such that use of the Rayleigh method is allowed, we will write the magnetic field at frequency 2ω in the form of the Fourier integrals

$$H_2^>(x_1,x_3|2\omega) = \int_{-\infty}^{\infty} \frac{dq}{2\pi} R_{2\omega}(q|2k)e^{iqx_1}e^{i\alpha_0(q,2\omega)x_3}, \quad x_3 > \zeta(x_1), \tag{37a}$$

$$H_2^<(x_1,x_3|2\omega) = \int_{-\infty}^{\infty} \frac{dq}{2\pi} T_{2\omega}(q|2k)e^{iqx_1}e^{-i\alpha(q,2\omega)x_3}, \quad x_3 < \zeta(x_1). \tag{37b}$$

We derive an integral equation for the scattering amplitude $R_{2\omega}(q|2k)$ analogous to the reduced Rayleigh equation in a standard manner. To do this we first substitute the expressions given by Eqs. (37a) and (37b) into the nonlinear boundary conditions at $x_3 = \zeta(x_1)$, Eqs. (6) and (7), and obtain a pair of coupled integral equations for the amplitudes $R_{2\omega}(q|2k)$ and $T_{2\omega}(q|2k)$. To eliminate the function $T_{2\omega}(q|2k)$ we multiply the equation obtained from the boundary condition for the tangential component of the magnetic field by $(i/\epsilon(2\omega))[-i\alpha(p,2\omega) + ip\zeta'(x_1)]\exp[-i\alpha(p,2\omega)\zeta(x_1) - ipx_1]$, and the equation obtained from the boundary condition for the tangential component of the electric field by $\exp[-i\alpha(p,2\omega)\zeta(x_1) - ipx_1]$, integrate the resulting equations with respect to x_1, and add them. In this way we obtain the reduced Rayleigh equation for $R_{2\omega}(q|2k)$:

$$\int_{-\infty}^{\infty} \frac{dq}{2\pi} N(p,q|2\omega)R_{2\omega}(q|2k) = -Q(p|2k), \tag{38}$$

where

$$N(p,q|2\omega) = i\left(1 - \frac{1}{\epsilon(2\omega)}\right)\frac{pq + \alpha(p,2\omega)\alpha_0(q,2\omega)}{\alpha(p,2\omega) - \alpha_0(q,2\omega)} \times I(\alpha(p,2\omega) - \alpha_0(q,2\omega)|p-q). \tag{39}$$

The driving term $Q(p|2k)$, which describes the nonlinear surface source, is given by

$$Q(p|2k) = -\frac{2c}{\omega}\int_{-\infty}^{\infty} dx_1 e^{-ipx_1 - i\alpha(p,2\omega)\zeta(x_1)} \times \left\{\chi^{s}_{ttz}\frac{\alpha(p,2\omega) - p\zeta'(x_1)}{\epsilon(2\omega)}\phi^{-2}(x_1)L(x_1|\omega)\frac{dH(x_1|\omega)}{dx_1} - i\frac{d}{dx_1}\phi^{-2}(x_1)\left[\chi^{s}_{zzz}\left(\frac{dH(x_1|\omega)}{dx_1}\right)^2 + \chi^{s}_{ztt}L^2(x_1|\omega)\right]\right\} \tag{40}$$

in the case of p-polarized incident light, and by

$$Q(p|2k) = -\frac{2i\omega}{c}\chi^{s}_{ztt}\int_{-\infty}^{\infty} dx_1 e^{-ipx_1 - i\alpha(p,2\omega)\zeta(x_1)}\frac{d}{dx_1}E^2(x_1|\omega) \tag{41}$$

in the case of s-polarized incident light.

A more convenient form of Eq. (38) is obtained by the use of the identity $I(\gamma|Q) = 2\pi\delta(Q) + J(\gamma|Q)$,

$$R_{2\omega}(p|2k) = G_0(p,2\omega)Q(p|2k) + G_0(p,2\omega)\int_{-\infty}^{\infty}\frac{dq}{2\pi}\tilde{N}(p,q|2\omega)G_0(q,2\omega)R_{2\omega}(q|2k), \tag{42}$$

where

$$\tilde{N}(p,q|2\omega) = i\left(1 - \frac{1}{\epsilon(\omega)}\right)\frac{pq + \alpha(p,2\omega)\alpha_0(q,2\omega)}{\alpha(p,2\omega) - \alpha_0(q,2\omega)}J(\alpha(p,2\omega) - \alpha_0(q,2\omega)|p-q). \tag{43}$$

We now introduce a new unknown function $S(q|2k)$ by the equation

$$R_{2\omega}(p|2k) = G_0(p,2\omega)Q(p|2k) + G_0(p,2\omega)\int_{-\infty}^{\infty}\frac{dq}{2\pi}S(p|q)G_0(q,2\omega)Q(q|2k). \tag{44}$$

From Eq. (38) it follows that the function $S(q|p)$ satisfies the equation

$$S(q|p) = \tilde{N}(q, p|2\omega) + \int_{-\infty}^{\infty} \frac{dr}{2\pi} \tilde{N}(q, r|2\omega) G_0(r, 2\omega) S(r|p), \tag{45a}$$

$$= \tilde{N}(q, p|2\omega) + \int_{-\infty}^{\infty} \frac{dr}{2\pi} S(q|r) G_0(r, 2\omega) \tilde{N}(r, p|2\omega). \tag{45b}$$

Therefore, $S(q|p)$ is the transition matrix for the problem of the scattering of light of frequency 2ω, and the scattering potential $V(q, p|2\omega)$ is

$$V(q, p|2\omega) = \tilde{N}(q, p|2\omega). \tag{46}$$

We can now introduce the Green's function $G_{2\omega}(q|k)$ associated with the randomly rough interface between the vacuum and the scattering medium through the equation

$$G_{2\omega}(p|k) = G_0(p, 2\omega) 2\pi\delta(p - k) + G_0(p, 2\omega) \int_{-\infty}^{\infty} \frac{dq}{2\pi} V(p, q|2\omega) G_{2\omega}(q|k), \tag{47a}$$

$$= G_0(p, 2\omega) 2\pi\delta(p - k) + G_0(p, 2\omega) S(p|k) G_0(k, 2\omega), \tag{47b}$$

and obtain a simple relation between the scattering amplitude $R_{2\omega}(q|2k)$ and the Green's function $G_{2\omega}(q|p)$:

$$R_{2\omega}(p|2k) = \int_{-\infty}^{\infty} \frac{dq}{2\pi} G_{2\omega}(p|q) Q(q|2k). \tag{48}$$

As in the case of linear scattering, the Green's function satisfies the equation

$$G_{2\omega}(q|k) = 2\pi\delta(q - k) G(q, 2\omega) + G(q, 2\omega) t_{2\omega}(q|k) G(k, 2\omega), \tag{49}$$

where $G(q, 2\omega)$ is the averaged Green's function, $\langle G_{2\omega}(q|k)\rangle = 2\pi\delta(q - k) G(k, 2\omega)$, and is given by

$$G(k, 2\omega) = \frac{1}{G_0^{-1}(k, 2\omega) - M(k, 2\omega)}. \tag{50}$$

In Eq. (50) $M(k, 2\omega)$ is the averaged proper self-energy, $\langle M_{2\omega}(q|k)\rangle = 2\pi\delta(q - k) M(k, 2\omega)$, where the (unaveraged) proper self-energy $M_{2\omega}(q|k)$

is the solution of [39]

$$M_{2\omega}(q|k) = V(q,k|2\omega) + \int_{-\infty}^{\infty} \frac{dp}{2\pi} M_{2\omega}(q|p)G(p,2\omega)W(p,k|2\omega), \tag{51}$$

and

$$W(q,k|2\omega) = V(q,k|2\omega) - \langle M_{2\omega}(q|k)\rangle. \tag{52}$$

As in the preceding subsection the operator $t_{2\omega}(p|r)$ was introduced to satisfy

$$\int_{-\infty}^{\infty} \frac{dp}{2\pi} W(q,p|2\omega)G_{2\omega}(p|k) = t_{2\omega}(q|k)G(k,2\omega), \tag{53}$$

and is the solution of the equation

$$t_{2\omega}(q|k) = W(q,k|2\omega) + \int_{-\infty}^{\infty} \frac{dp}{2\pi} W(q,p|2\omega)G(p,2\omega)t_{2\omega}(p|k). \tag{54}$$

From Eq. (49) it follows that $\langle t_{2\omega}(q|k)\rangle = 0$.

Finally, the Eq. (48) for the scattering amplitude of frequency 2ω, $R_{2\omega}(q|2k)$ can be rewritten in terms of the operator $t_{2\omega}(q|k)$ as

$$R_{2\omega}(q|2k) = G(q,2\omega)Q(q|2k) + G(q,2\omega)\int_{-\infty}^{\infty} \frac{dp}{2\pi} t_{2\omega}(q|p)G(p,2\omega)Q(p|2k). \tag{55}$$

3.3. *The nonlinear source function $Q(q|2k)$*

In terms of the Fourier coefficients of the source functions of the linear scattering of light of frequency ω, the nonlinear source term $Q(q|2k)$, given by Eqs. (40) and (41), has the form

$$\begin{aligned} Q(q|2k) = &-\frac{2c}{\omega}\int_{-\infty}^{\infty}\frac{dp}{2\pi}\int_{-\infty}^{\infty}\frac{dr}{2\pi}\int_{-\infty}^{\infty}\frac{dt}{2\pi} I(\alpha(q,2\omega)|q-p-r-t) \\ &\times \Bigg[i\chi^{\mathrm{s}}_{ttz}\frac{\alpha^2(q,2\omega)+q(q-p-r-t)}{\alpha(q,2\omega)\epsilon(2\omega)} pH(p)L(r) \\ &- (p+r+t)\big(\chi^{\mathrm{s}}_{zzz}pH(p)rH(r) - \chi^{\mathrm{s}}_{ztt}L(p)L(r)\big)\Bigg]\Phi(t), \end{aligned} \tag{56}$$

in the case of p-polarized incident light, and

$$Q(q|2k) = \frac{2\omega}{c}\chi^{\mathrm{s}}_{ztt}\int_{-\infty}^{\infty}\frac{dp}{2\pi} I(\alpha(q,2\omega)|q-p-r)(p+r)E(p)E(r), \tag{57}$$

in the case of s-polarized incident light. In Eq. (56) we have introduced the function

$$\Phi(t) = \int_{-\infty}^{\infty} dx_1 e^{-itx_1} \phi^{-2}(x_1). \tag{58}$$

If we now substitute the expressions for the Fourier coefficients of the field of frequency ω and its normal derivative evaluated on the rough surface, Eq. (35), into Eqs. (56) and (57), we obtain the expression for the nonlinear driving term in the case of p-polarized incident light

$$\begin{aligned} Q(q|2k) = A_p \int_{-\infty}^{\infty} \frac{dp}{2\pi} \int_{-\infty}^{\infty} \frac{dp'}{2\pi} \int_{-\infty}^{\infty} \frac{dr}{2\pi} \int_{-\infty}^{\infty} \frac{dr'}{2\pi} \int_{-\infty}^{\infty} \frac{dt}{2\pi} \\ \times I(\alpha(q, 2\omega)|q - p - r - t)\Phi(t) \\ \times \Gamma_p(q, p, p', r, r', t) I(\alpha(p', \omega)|p - p') \\ \times I(\alpha(r', \omega)|r - r') A(p'|k) A(r'|k), \end{aligned} \tag{59}$$

and in the case of s-polarized incident light

$$\begin{aligned} Q(q|2k) = A_s \int_{-\infty}^{\infty} \frac{dp}{2\pi} \int_{-\infty}^{\infty} \frac{dp'}{2\pi} \int_{-\infty}^{\infty} \frac{dr}{2\pi} \int_{-\infty}^{\infty} \frac{dr'}{2\pi} I(\alpha(q, 2\omega)|q - p - r) \\ \times \Gamma_s(p, r) I(\alpha(p', \omega)|p - p') I(\alpha(r', \omega)|r - r') A(p'|k) A(r'|k), \end{aligned} \tag{60}$$

where the amplitudes $A_{p,s}$ are

$$A_p = (8c/\omega)\mathcal{F}_0^2 \alpha_0^2(k) G^2(k, \omega), \tag{61a}$$

$$A_s = -(8\omega/c)\mathcal{F}_0^2 \alpha_0^2(k) G^2(k, \omega). \tag{61b}$$

The effective nonlinear coefficients in Eqs. (59) and (60) are given by

$$\begin{aligned} \Gamma_p(q, p, p', r, r', t) = \chi_{ttz}^{\mathrm{s}} \frac{\alpha^2(q, 2\omega) + q(q - p - r - t)}{\epsilon(2\omega)\alpha(q, 2\omega)} pu(r|r') \\ - (p + r + t)\left[\chi_{zzz}^{\mathrm{s}} pr + \chi_{ztt}^{\mathrm{s}} u(p|p')u(r|r')\right] \end{aligned} \tag{62a}$$

in the case of p-polarized incident light and

$$\Gamma_s(p, r) = (p + r)\chi_{ztt}^{\mathrm{s}} \tag{62b}$$

in the case of s-polarized incident light, and the field amplitude $A(p|k)$ is

$$A(q|k) = 2\pi\delta(q - k) + G(q, \omega)t_\omega(q|k). \tag{63}$$

We can now analyze different contributions to the nonlinear source function $Q(q|2k)$ given by Eqs. (59) and (60). The functions $I(\gamma|Q)$ and $\Phi(Q)$ entering the expressions account for the fact that the fields of frequency ω are evaluated on the local surface, the δ-function in the amplitude $A(q|k)$ stands for the specular component of the fundamental field, and the term $G(q,\omega)t_\omega(q|k)$ accounts for the scattered field of frequency ω. In the case of p-polarized incident light the Green's function $G(q,\omega)$ has simple poles associated with the excitation of surface plasmon polaritons of frequency ω.

Thus, if the product of the amplitudes $A(p'|k)A(r'|k)$ is written down explicitly, the nonlinear source function $Q(q|2k)$ is the sum of three terms which have different physical meanings. The part of $Q(q|2k)$ which contains only the product of the δ-functions, $(2\pi)^2\delta(p'-k)\delta(r'-k)$, describes the nonlinear mixing of the fields of frequency ω which would be specular if the surface were planar. The part of $Q(q|2k)$ that contains the product of the δ-function and the term with the Green's function, $2\pi\delta(p'-k)G(r',2\omega)t_{2\omega}(r'|k)+2\pi\delta(r'-k)G(p',2\omega)t_{2\omega}(p'|k)$, describes the interaction of the "specular" and scattered fields, including the nonlinear mixing of the excited surface plasmon polaritons with the incident light. Finally, the part of $Q(q|2k)$ that contains the product of the Green's functions, $G(p',2\omega)t_{2\omega}(p'|k)G(r',2\omega)t_{2\omega}(r'|k)$, describes the nonlinear mixing of the scattered fields, and includes the mixing of co- and contrapropagating surface plasmon polaritons. We, therefore, subdivide the nonlinear source function $Q(q|2k)$ into three contributions according to the classification we have just described:

$$\begin{aligned} Q(q|2k) = A_{p,s}\Bigg[& Q_v(q|2k) + \int_{-\infty}^{\infty}\frac{dp}{2\pi}Q_s(q,p,k)G(p,\omega)t_\omega(p|k) \\ & + \int_{-\infty}^{\infty}\frac{dp}{2\pi}\int_{-\infty}^{\infty}\frac{dp'}{2\pi}Q_{ss}(q,p,p')G(p,\omega)t_\omega(p|k)G(p',\omega)t_\omega(p'|k)\Bigg], \end{aligned} \tag{64}$$

where the expressions for the functions $Q_v(q|2k)$, $Q_s(q,p,k)$, and $Q_{ss}(q,p,p')$ are obtained from Eqs. (59) and (60):

$$\begin{aligned} Q_v(q|2k) = & \int_{-\infty}^{\infty}\frac{dp}{2\pi}\int_{-\infty}^{\infty}\frac{dr}{2\pi}\int_{-\infty}^{\infty}\frac{dt}{2\pi}I(\alpha(q,2\omega)|q-p-r-t)\Phi(t) \\ & \times\Gamma_p(q,p,k,r,k,t)I(\alpha(k,\omega)|p-k)I(\alpha(k,\omega)|r-k), \end{aligned} \tag{65a}$$

$$Q_s(q,p,k) = \int_{-\infty}^{\infty} \frac{dr}{2\pi} \int_{-\infty}^{\infty} \frac{dr'}{2\pi} \int_{-\infty}^{\infty} \frac{dt}{2\pi} I(\alpha(q,2\omega)|q-r-r'-t)\Phi(t) \\ \times [\Gamma_p(q,r,k,r',p,t) + \Gamma_p(q,r',p,r,k,t)] \\ \times I(\alpha(k,\omega)|r-k) I(\alpha(p,\omega)|r'-p), \tag{65b}$$

and

$$Q_{ss}(q,p,p) = \int_{-\infty}^{\infty} \frac{dr}{2\pi} \int_{-\infty}^{\infty} \frac{dr'}{2\pi} \int_{-\infty}^{\infty} \frac{dt}{2\pi} I(\alpha(q,2\omega)|q-r-r'-t)\Phi(t) \\ \times \Gamma_p(q,r,p,r',p',t) I(\alpha(p,\omega)|r-p) I(\alpha(p',\omega)|r'-p'), \tag{65c}$$

in the case of p-polarized incident light and

$$Q_v(q|2k) = \int_{-\infty}^{\infty} \frac{dp}{2\pi} \int_{-\infty}^{\infty} \frac{dr}{2\pi} I(\alpha(q,2\omega)|q-p-r)\Gamma_s(p,r) \\ \times I(\alpha(k,\omega)|p-k) I(\alpha(k,\omega)|r-k), \tag{66a}$$

$$Q_s(q,p,k) = \int_{-\infty}^{\infty} \frac{dr}{2\pi} \int_{-\infty}^{\infty} \frac{dr'}{2\pi} I(\alpha(q,2\omega)|q-r-r') [\Gamma_s(r,r') + \Gamma_s(r',r)] \\ \times I(\alpha(k,\omega)|r-k) I(\alpha(p,\omega)|r'-p), \tag{66b}$$

and

$$Q_{ss}(q,p,p) = \int_{-\infty}^{\infty} \frac{dr}{2\pi} \int_{-\infty}^{\infty} \frac{dr'}{2\pi} I(\alpha(q,2\omega)|q-r-r') \\ \times \Gamma_s(r,r') I(\alpha(p,\omega)|r-p) I(\alpha(p',\omega)|r'-p'), \tag{66c}$$

in the case of s-polarized incident light.

3.4. *The mean intensity of second harmonic light*

In terms of $R_{2\omega}(q|2k)$ the mean normalized intensity of second harmonic light defined in Sec. 3 has the form

$$\langle I(\theta_s|2\omega)\rangle_{\text{incoh}} = \frac{8\omega}{L_1 c^2 |\mathcal{F}_0|^4} \frac{\cos^2\theta_s}{\cos^2\theta_0} \left[\langle |R_{2\omega}(q|2k)|^2\rangle - |\langle R_{2\omega}(q|2k)\rangle|^2\right], \tag{67}$$

where $q = (2\omega/c)\sin\theta_s$ and $k = (\omega/c)\sin\theta_0$. To evaluate this expression we need the average $\langle |R_{2\omega}(q|2k)|^2\rangle_{\text{incoh}} = \left[\langle |R_{2\omega}(q|2k)|^2\rangle - |\langle R_{2\omega}(q|2k)\rangle|^2\right]$. We could calculate this average by first calculating the scattering amplitude

$R_{2\omega}(q|2k)$ by solving the reduced Rayleigh equation, Eq. (38), in the framework of small-amplitude perturbation theory. However, to account for the processes of nonlinear mixing of multiply scattered surface plasmon polaritons of frequency ω we have to calculate the source functions of the field of frequency ω through terms of at least the seventh order in the surface profile function. Instead we will calculate the mean normalized intensity of the second harmonic light starting from $R_{2\omega}(q|2k)$ given by Eq. (55). With the use of the property $\langle t_{2\omega}(q|p)\rangle = 0$, we obtain

$$\langle |R_{2\omega}(q,2k)|^2\rangle_{\rm incoh} = |A_{p,s}|^2 |G(q,2\omega)|^2 \Bigg\{ \langle |Q(q|2k)|^2\rangle_{\rm incoh} + L_1\tau_{2\omega}(q|2k)|G(2k,2\omega)|^2|Q(2k|2k)|^2 + L_1\tilde{\tau}_{2\omega}(q|2k) + 2{\rm Re}\int_{-\infty}^{\infty}\frac{dp}{2\pi}\langle t_{2\omega}(q|p)G(p,2\omega)Q(p|2k)Q^*(q|2k)\rangle_c \Bigg\}, \tag{68}$$

where $\langle\cdot\rangle_c$ denotes the cumulant average [40]. In writing Eq. (68) we have introduced the notation

$$\langle t_\Omega(q|p)t^*_\Omega(q|p')\rangle = 2\pi\delta(p-p')L_1\tau_\Omega(q|p), \tag{69}$$

for the averaged reducible vertex function $\tau_\Omega(q|p)$ in the problem of the linear scattering of light from a rough surface, and the specular component of the field of frequency 2ω generated at the rough surface, $Q(2k|2k)$, is determined by

$$\langle Q(p|2k)\rangle = \tilde{Q}(2k|2k)2\pi\delta(p-2k). \tag{70}$$

The function $\tilde{\tau}_{2\omega}(q|2k)$ appearing in Eq. (68) is the analog of the reducible vertex function in the problem of nonlinear scattering and is given by

$$\tilde{\tau}_{2\omega}(q|2k) = \int_{-\infty}^{\infty}\frac{dp}{2\pi}\int_{-\infty}^{\infty}\frac{dp'}{2\pi}\Bigg[\frac{1}{L_1}\langle t_{2\omega}(q|p)G(p,2\omega)Q(p|2k)t^*_{2\omega}(q|p')G^*(p',2\omega) \times Q^*(p'|2k)\rangle_c + \tau_{2\omega}(q|p)|G(p,2\omega)|^2\langle|Q(p|2k)|^2\rangle_c + \frac{1}{L_1}\langle t_{2\omega}(q|p) \times G(p,2\omega)Q^*(p'|2k)\rangle_c\langle t_{2\omega}(q|p')G^*(p',2\omega)Q(p|2k)\rangle_c\Bigg]. \tag{71}$$

The function $\langle|Q(q|2k)|^2\rangle_{\text{incoh}} = \langle|Q(q|2k)|^2\rangle - |\langle Q(q|2k)\rangle|^2$ is given by

$$\begin{aligned}\langle|Q(q|2k)|^2\rangle_{\text{incoh}} &= \langle|Q_v(q|2k)|^2\rangle_{\text{incoh}} + \chi_1(q|k) + \chi_2(q|k) + \chi_3(q|k) \\ &\quad + \chi_4(q|k) + 2\text{Re}\left\langle Q_v^*(q|2k) \int_{-\infty}^{\infty} \frac{dp}{2\pi} \right. \\ &\quad \left. \times Q_s(q,p,k)G(p,\omega)t_\omega(p|k) \right\rangle_c \\ &\quad + 2\text{Re}\left\langle Q_v^*(q|2k) \int_{-\infty}^{\infty} \frac{dp}{2\pi} \int_{-\infty}^{\infty} \frac{dp'}{2\pi} \right. \\ &\quad \left. \times Q_s(q,p,p')G(p,\omega)t_\omega(p|k)G(p',\omega)t_\omega(p'|k) \right\rangle_c, \end{aligned} \tag{72}$$

where

$$\begin{aligned}\chi_1(q|k) = \tau_\omega(q-k|k)|G(q-k,\omega)|^2 \times \left|\left\langle Q_s(q|q-k|k) \right.\right. \\ \left.\left. + \int_{-\infty}^{\infty} \frac{dp}{2\pi} \widehat{Q}_{ss}(q,q-p,p)t(p|k) \right\rangle_0 \right|^2, \end{aligned} \tag{73a}$$

with

$$\widehat{Q}_{ss}(q,p,p') = Q_{ss}(q,p,p') + Q_{ss}(q,p',p). \tag{73b}$$

The function $\chi_2(q|k)$ is a sum of two contributions,

$$\chi_2(q|k) = \chi_2^{(L)}(q|k) + \chi_2^{(MC)}(q|k), \tag{73c}$$

where

$$\chi_2^{(L)} = \int_{-\infty}^{\infty} \frac{dp}{2\pi} \langle|Q_s(q,p,k)|^2\rangle_0 |G(p,\omega)|^2 \tau_\omega(p|k), \tag{73d}$$

and

$$\begin{aligned}\chi_2^{(MC)} = \int_{-\infty}^{\infty} \frac{dp}{2\pi} \langle Q_s(q,p,k)t_\omega^*(q-p|k)\rangle_0 G(p,\omega) \\ \times G^*(q-p,\omega)\langle Q_s^*(q,q-p,k)t_\omega(p|k)\rangle_0. \end{aligned} \tag{73e}$$

Finally, the function $\chi_3(q|k)$ and $\chi_4(q|k)$ are given by

$$\begin{aligned}\chi_3(q|k) = \int_{-\infty}^{\infty} \frac{dp}{4\pi} |\langle \widehat{Q}_{ss}(q,p,q-p)\rangle_0|^2 \langle|G(q-p,\omega)G(p,\omega)|^2 \\ \times \tau_\omega(q-p|k)\tau_\omega(p|k), \end{aligned} \tag{73f}$$

and

$$\chi_4(q|k) = \int_{-\infty}^{\infty} \frac{dp}{4\pi} \int_{-\infty}^{\infty} \frac{dp'}{2\pi}$$
$$\times \left\{ \frac{1}{2} \langle |\widehat{Q}_{ss}(q,p,p')|^2 \rangle_0 |G(p,\omega)G(p',\omega)|^2 \tau_\omega(p|k)\tau_\omega(p'|k) \right.$$
$$+ \langle \widehat{Q}_{ss}(q,p,p') t^*_\omega(q+k-p-p'|k) \rangle_0 |G(p,\omega)|^2$$
$$\times G(p')G^*(q+k-p-p',\omega)$$
$$\left. \times \langle \widehat{Q}_{ss}(q,q+k-p-p',p) t^*_\omega(p'|k) \rangle^*_0 \tau_\omega(p|k) \right\}. \tag{73g}$$

The four terms on the right-hand side of Eq. (68) have different physical meanings. The contribution $\langle |Q(q|2k)|^2 \rangle_{\mathrm{incoh}}$ contains only those processes in which the multiply-scattered waves of frequency ω interact nonlinearly giving rise to volume waves of frequency 2ω in the vacuum above the surface. The second term is proportional to $\tau_{2\omega}(q|2k)$, and describes the multiple scattering of the generated volume waves of frequency 2ω. The third term, that is $\tilde{\tau}_{2\omega}(q|2k)$, describes the multiple scattering of the diffusely generated waves. The last term is nonresonant and contributes only to the background intensity.

4. Results

To illustrate the results presented in the preceding sections we have calculated the mean differential intensity of the second harmonic light generated in reflection from a one-dimensional, random silver surface characterized by the power spectrum (5). In our calculations the wavelength of the incident light was chosen to be 1.064 μm, as in the experiments of [21–23], so that the wavelength of the generated light is 0.532 μm. The dielectric constants of silver at the fundamental and harmonic frequencies are then $\epsilon(\omega) = -56.25 + i0.60$ and $\epsilon(2\omega) = -11.56 + i0.37$, respectively [41], which ensures that surface plasmon polaritons exist at both frequencies. The real parts of their wave numbers are $k_{\mathrm{sp}}(\omega) = 1.009008(\omega/c)$ and $k_{\mathrm{sp}}(2\omega) = 1.0462234(2\omega/c)$, respectively.

To illustrate the specific effects of the rectangular power spectrum, in Fig. 1 we present the results of linear scattering calculations at the fundamental frequency ω for three angles of incidence $\theta_0 = 0°$ (Fig. 1(a)), $\theta_0 = 8°$ (Fig. 1(b)), and $\theta_0 = 10°$ (Fig. 1(c)) for the case where the power spectrum is centered at the wave number $k_{\mathrm{sp}}(\omega)$ of the surface plasmon

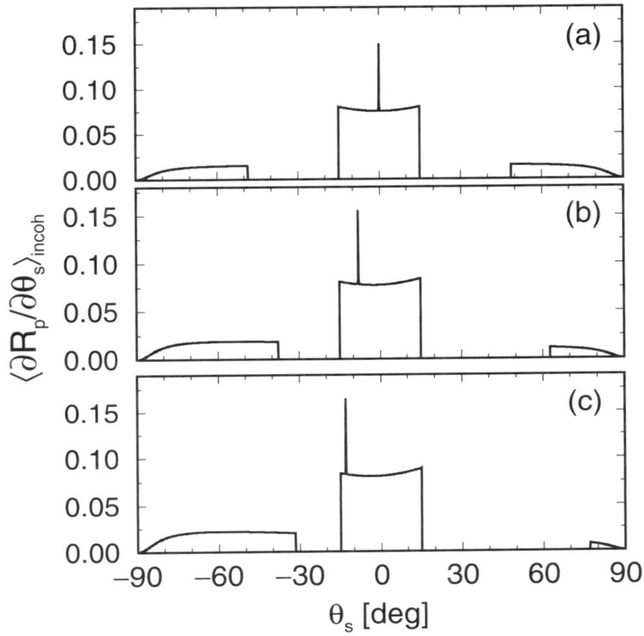

Fig. 1. The incoherent component of the mean differential reflection coefficient as a function of the scattering angle θ_s for the scattering of p-polarized light of wavelength $\lambda = 1.064\,\mu$m from a random silver surface characterized by the rectangular power spectrum (5) centered at the wavenumber of surface plasmon polaritons of wavelength λ, with roughness parameters $\delta = 10.8$ nm and $\theta_{max} = 15°$, and a dielectric constant $\epsilon(\omega) = -56.25 + i0.60$. The angles of incidence are (a) $\theta_0 = 0°$, (b) $\theta_0 = 8°$, and (c) $\theta_0 = 13°$.

polaritons at frequency ω, and has a halfwidth equal to $(\omega/c)\sin\theta_{max}$. The RMS height of the surface roughness is $\delta = 10.8$ nm, and the characteristic angle θ_{max} is $\theta_{max} = 15°$. In this case the light whose angle of incidence is within the range $-\theta_{max} < \theta_0 < \theta_{max}$ is converted effectively into surface plasmon polaritons of frequency ω, and the surface plasmon polaritons of frequency ω are converted effectively into the light that propagates into the vacuum within the range of scattering angles $-\theta_{max} < \theta_s < \theta_{max}$. The plots display single-scattering wings at large angles of scattering and an almost rectangular distribution coming from the double-scattering processes mediated by the surface plasmon polaritons of frequency ω. The height of the enhanced backscattering peak in Fig. 1 is exactly twice the background intensity, as is expected in the linear scattering theory.

In our calculations of the second harmonic generation we use nonlinear constants calculated on the basis of the free-electron model [1, 26–28] because it leads to simple algebraic expressions:

$$\chi^{\rm s}_{zzz} = -\frac{2}{3}\beta\left[\frac{(\epsilon(\omega)-1)(\epsilon(\omega)-3)}{2\epsilon^2(\omega)} - \frac{2}{3}\ln\left(\frac{\epsilon(\omega)}{\epsilon(2\omega)}\right)\right], \tag{74a}$$

$$\chi^{\rm s}_{ztt} = 0, \tag{74b}$$

and

$$\chi^{\rm s}_{ttz} = \beta\left(\frac{\epsilon(\omega)-1}{\epsilon(\omega)}\right), \tag{74c}$$

where $\beta = e/(8\pi m\omega^2)$.

Apart from the sign, these nonlinear susceptibilities coincide with those obtained by Mendoza and Mochán [30] by a different approach. The numerical values of the nonlinear coefficients are $\chi^{\rm s}_{zzz} = (0.2384987 \times 10^{-14} + i6.384 \times 10^{-17})CGSE$, and $\chi^{\rm s}_{ttz} = (0.6818093 \times 10^{-14} + i1.316 \times 10^{-18})CGSE$. Note, that in the free-electron model $\chi^{\rm s}_{ztt} = 0$. A nonzero value of $\chi^{\rm s}_{ztt}$ arises when the surface nonlinear polarization has the form discussed by Agranovich and Darmanyan in [29]. We can estimate this constant assuming that the values of the nonlinear coefficients $\chi^{\rm s}_{zzz}$ and $\chi^{\rm s}_{ttz}$ coincide with those in the free-electron model and expressing the phenomenological constants entering the expression for the surface nonlinear polarization in the Agranovich and Darmanyan model in terms of the parameters appearing in the free-electron model. Then, we obtain an expression for $\chi^{\rm s}_{ztt}$ in the form

$$\chi^{\rm s}_{ztt} = \frac{2}{3}\beta\ln\left[\frac{\epsilon(\omega)}{\epsilon(2\omega)}\right], \tag{75}$$

so that $\chi^{\rm s}_{ztt} = (0.7732004 \times 10^{-14} + i9.871 \cdot 10^{-17})CGSE$.

As was done in the experiments of [21] and [22], to separate different mechanisms for the interplay of the nonlinearity and roughness of the surface we have calculated the mean intensity of the generated light under different scattering conditions imposed by the power spectrum of the surface roughness. In Figs. 2(a)–(c) we present the mean intensity of the second harmonic light, calculated when p-polarized light is incident on a one-dimensional, random silver surface, characterized by the power spectrum (5) with $h_1 = 0$ and $h_2 = 1$, an RMS height $\delta = 11.1\,\text{nm}$ and $\theta_{\max} = 12.2°$. The power spectrum is centered at the wave numbers $\pm k_{\rm sp}(2\omega)$ of the surface plasmon polariton at frequency 2ω, and has

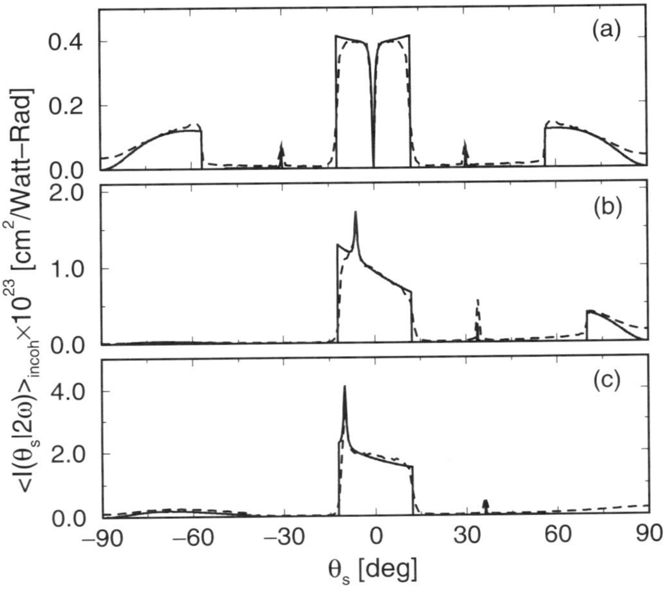

Fig. 2. The mean differential intensity of the second harmonic light as a function of the scattering angle θ_s for the scattering of p-polarized light from a randomly rough silver surface whose roughness is characterized by the rectangular power spectrum Eq. (5), centered at the wavenumbers of surface plasmon polaritons of frequency 2ω, with $\delta = 11.1\,\text{nm}$ and $\theta_m = 12.2°$. The nonlinear coefficients are given by the free-electron model, Eqs. (74). The angles of incidence are (a) $\theta_s = 0°$, (b) $\theta_s = 6°$, and (c) $\theta_s = 10°$. The solid lines represent the results of the perturbative calculations, the dashed lines represent the numerical results.

a halfwidth equal to $(2\omega/c)\sin\theta_{\max}$. Therefore, we have $k_{\min}^{(2)} = k_{\text{sp}}(2\omega) - (2\omega/c)\sin\theta_{\max}$ and $k_{\max}^{(2)} = k_{\text{sp}}(2\omega) + (2\omega/c)\sin\theta_{\max}$. In this case the excitation of surface polaritons of frequency ω is strongly suppressed, while a strong conversion of surface plasmon polaritons of frequency 2ω into volume waves of frequency 2ω in the vacuum is ensured. The nonlinear parameters used in the calculations, are given by the free-electron model. For the sake of comparison the results obtained by means of rigorous numerical simulations, representing the averages of results obtained from 3,000 realizations of the random surface, are plotted by dashed lines [24, 32].

The plots in Figs. 2(a)–(c) look similar to the plots of the differential reflection coefficient in the case of the linear scattering of light from surfaces whose roughness is characterized by the rectangular power spectrum (5) [9]. They display the single-scattering wings at large angles of scattering and an almost rectangular distribution coming from the double-scattering processes mediated by the surface plasmon polaritons of frequency 2ω. However, in contrast to the results for linear scattering, the mean intensity of the second harmonic light displays a dip in the retroreflection direction when the fundamental light is incident normally on the surface, $\theta_0 = 0°$ (Fig. 2(a)), and a peak in the retroreflection direction for larger angles of incidence $\theta_0 = 6°$ (Fig. 2(b)) and $\theta_0 = 10°$ (Fig. 2(c)). Two weak peaks positioned at $\theta_s = \pm 30.3°$ when the angle of incidence is 0°, at $\theta_s = 33.83°$ and $\theta_s = -26.89°$ when the angle of incidence is 6°, and at $\theta_s = 36.25°$ and $\theta_s = -24.69°$ when the angle of incidence is 10°, are also displayed.

Although the results presented in Fig. 2 are in quite good quantitative agreement with the experimental results of [21], they disagree with them qualitatively, since only a dip in the retroreflection direction was observed in the experiments of [21] for all angles of incidence of the fundamental light.

In Figs. 3(a)–(c) we present plots of the mean intensity of second harmonic of light generated in reflection from the same surface used in the

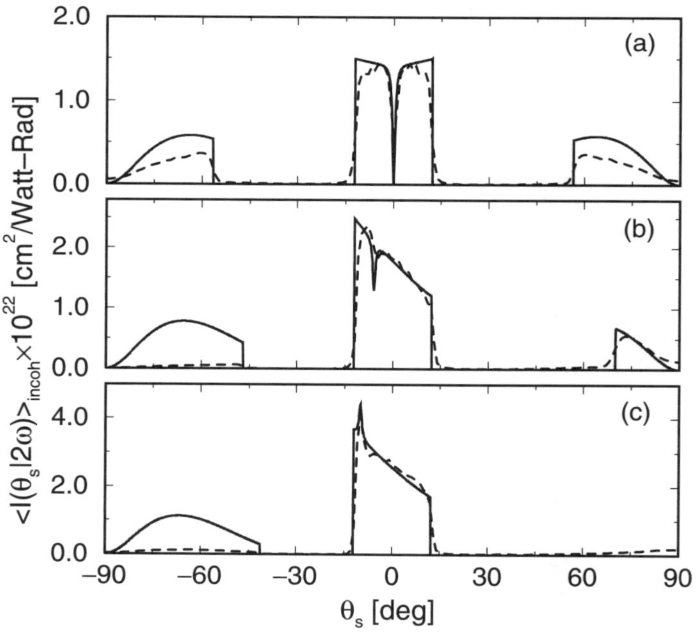

Fig. 3. The same as Fig. 2, except that the nonlinear coefficient χ^s_{ztt} is given by Eq. (75) rather than by Eq. (74b).

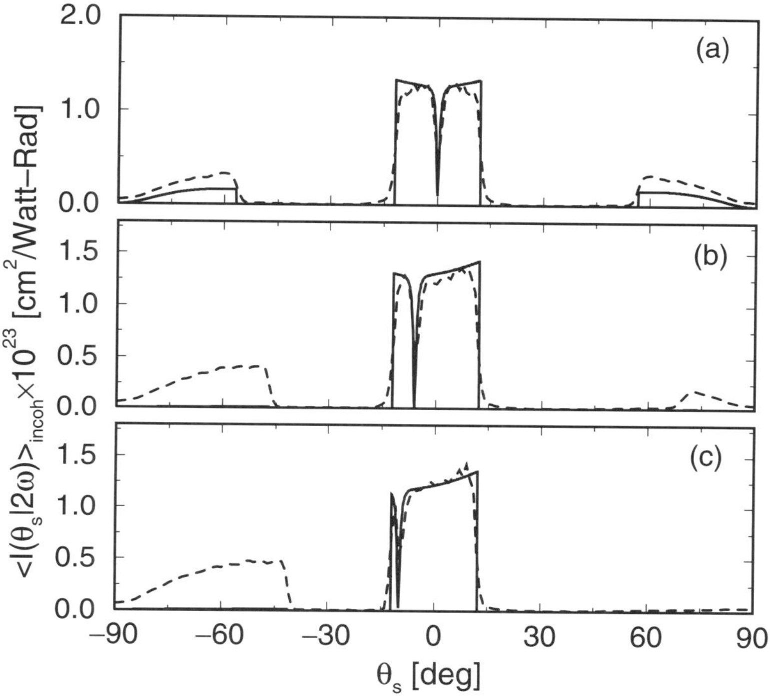

Fig. 4. The same as Fig. 3, except that the nonlinear coefficients χ^{s}_{tzz} and χ^{s}_{ztt} are given by Eq. (74c) and Eq. (75), respectively, while the coefficient $\chi^{\mathrm{s}}_{zzz} = 0$.

calculations of the results presented in Fig. 2, but with the surface nonlinear polarization [29] that give rise to nonzero value of χ^{s}_{ztt}, Eq. (75).

To illustrate the importance of the model of the surface nonlinear polarization assumed in the calculation, in Figs. 4(a)–(c) we present plots of the intensity of the second harmonic light for the case when the nonlinear coefficients χ^{s}_{ztt} and χ^{s}_{ttz} are the same as those used in the calculations of the results presented in Fig. 3, while $\chi^{\mathrm{s}}_{zzz} = 0$.

In Figs. 5(a)–(c) we present the mean differential intensity of the second harmonic light when *s*-polarized light is incident on the same surface used in the calculations of the results presented in Fig. 2, at three angles of incidence $\theta_0 = 0°$ (Fig. 5(a)), $\theta_0 = 6°$ (Fig. 5(b)), and $\theta_0 = 10°$ (Fig. 5(c)). The results obtained by means of numerical simulations [32] are shown by dashed lines. In this case the surface nonlinear polarization is given by the Agranovich and Darmanyan model [29] with χ^{s}_{ttz} given by Eq. (75). The plots are similar

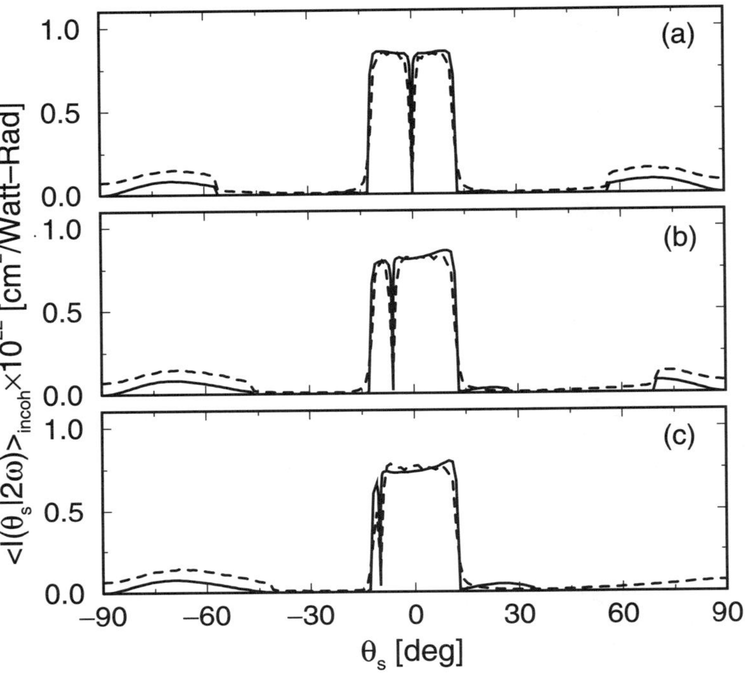

Fig. 5. The mean differential intensity of the second harmonic light as a function of the scattering angle θ_s for the scattering of s-polarized light from a randomly rough silver surface whose roughness is characterized by the rectangular power spectrum Eq. (5), centered at the wavenumbers of surface plasmon polaritons of frequency 2ω, with $\delta = 11.1\,\text{nm}$ and $\theta_m = 12.2°$. The nonlinear coefficient χ^s_{ztt} is given by Eq. (75). The angles of incidence are (a) $\theta_s = 0°$, (b) $\theta_s = 6°$, and (c) $\theta_s = 10°$. The solids lines represent the results of the perturbative calculations, the dashed lines represent the numerical results.

to those presented in Fig. 4. They display the central, almost rectangular, distribution coming from the double-scattering processes mediated by the surface plasmon polaritons of frequency 2ω, and a dip in the retroreflection direction for all angles of incidence.

In Figs. 6(a)–(c) we present the mean intensity of the second harmonic light when p-polarized light is incident on a one-dimensional, random silver surface at three angles of incidence $\theta_0 = 3°$ (Fig. 6(a)), $\theta_0 = 8°$ (Fig. 6(b)), and $\theta_0 = 10°$ (Fig. 6(c)). The dashed lines show the results of rigorous numerical simulations [32]. The nonlinear coefficients are given by the free-electron model. The surface roughness is characterized by the power spectrum (5) with $h_1 = 1$ and $h_2 = 0$, an RMS height $\delta = 10.8\,\text{nm}$, and

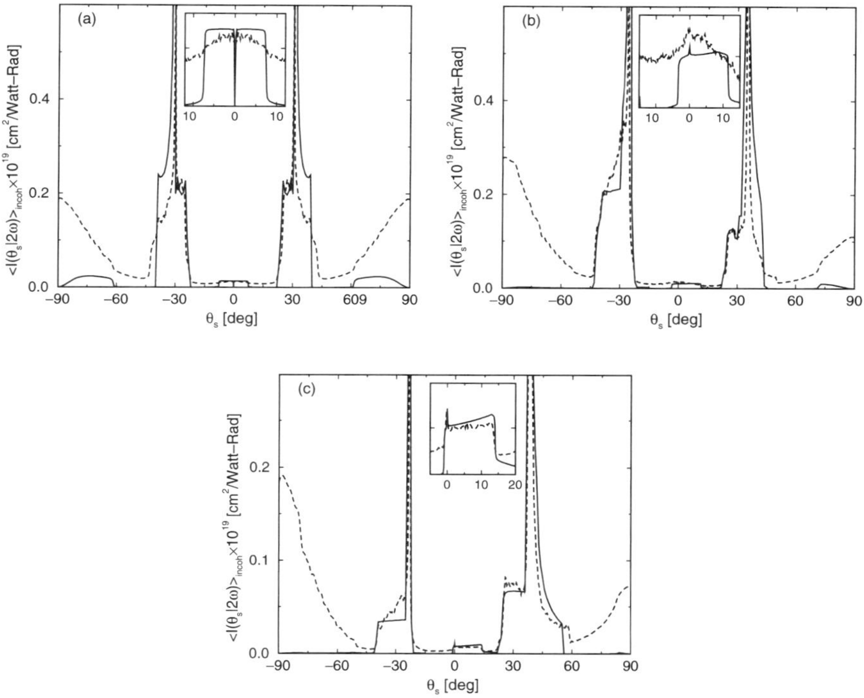

Fig. 6. The mean differential intensity of the second harmonic light as a function of the scattering angle θ_s for the scattering of p-polarized light from a randomly rough silver surface whose roughness is characterized by the rectangular power spectrum Eq. (5), centered at the wavenumbers of surface plasmon polaritons of frequency ω, with $\delta = 10.8\,\text{nm}$ and $\theta_m = 15°$. The nonlinear coefficients are given by the free-electron model. The angles of incidence are (a) $\theta_s = 0°$, (b) $\theta_s = 8°$, and (c) $\theta_s = 13°$. The plots represent the results of the perturbative (solid line) and numerical (dashed line) calculations.

$\theta_{\max} = 15°$. This power spectrum is centered at the wave number $k_{sp}(\omega)$ of the surface plasmon polaritons at frequency ω, and has a halfwidth equal to $(\omega/c)\sin\theta_{\max}$. Therefore, we have $k^{(1)}_{\min} = k_{sp}(\omega) - (\omega/c)\sin\theta_{\max}$ and $k^{(1)}_{\max} = k_{sp}(\omega) + (\omega/c)\sin\theta_{\max}$. In this case the light whose angle of incidence is within the range $-\theta_{\max} < \theta_0 < \theta_{\max}$ is converted effectively into surface plasmon polaritons of frequency ω. Note that the intensity of the light of frequency 2ω in this case is an order of magnitude greater than in the case when the excitation of surface plasmon polaritons of frequency ω is forbidden (see Figs. 2–5). The two peaks at $q = k \pm k_{sp}(\omega)$ are associated with the resonant nonlinear interaction of the excited surface plasmon polariton of frequency ω with the incident light. The most striking feature of the plots in Fig. 6 is a narrow dip at $\theta_s = 0$ that is present in the plots at small angles of incidence and evolves into a peak when the angle of incidence is 10°. A weak dip in the retroreflection direction is displayed in Fig. 6(b). Although the main features of the experimental results of [22] are displayed by the plots presented in Fig. 6, no dips or peaks in the direction normal to the mean surface and in the retroreflection direction were observed experimentally.

In Figs. 7–9 we present plots of the intensity of second harmonic light of light in reflection from the same surface used in the calculations of the results presented in Fig. 6, but with a different set of nonlinear parameters: we present plots of the differential intensity of second harmonic generation for the cases where $\chi^s_{ttz} = \chi^s_{ztt} = 0$, (Fig. 7) where $\chi^s_{zzz} = \chi^s_{ztt} = 0$ (Fig. 8), and where $\chi^s_{zzz} = \chi^s_{ttz} = 0$ (Fig. 9). From the plots we can see that when the fundamental light is incident normally on the surface a narrow

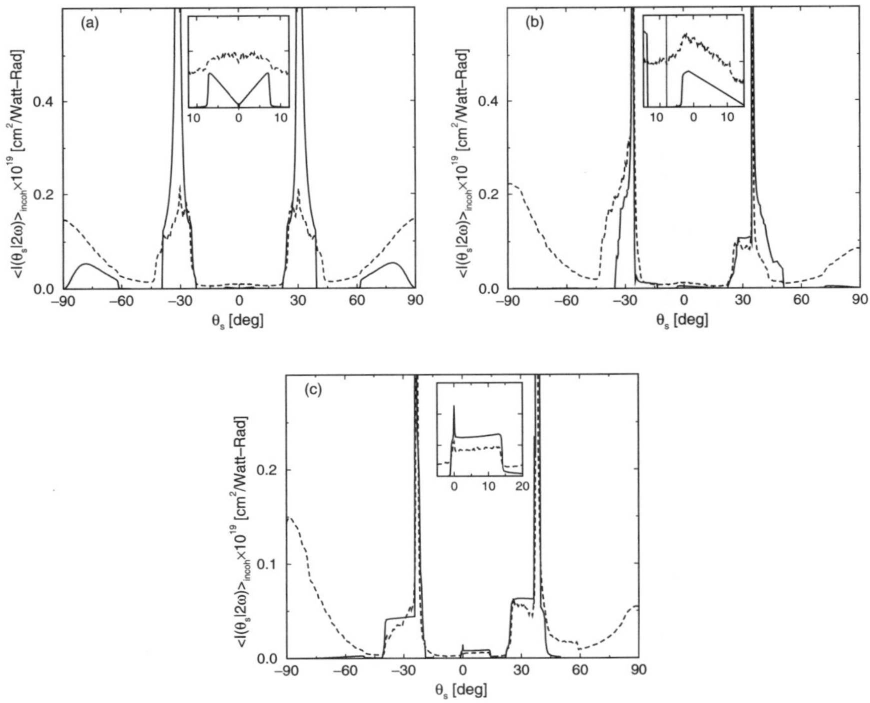

Fig. 7. The same as Fig. 6, except that χ^s_{zzz} is given by Eq. (74a), while $\chi^s_{ttz} = \chi^s_{ztt} = 0$.

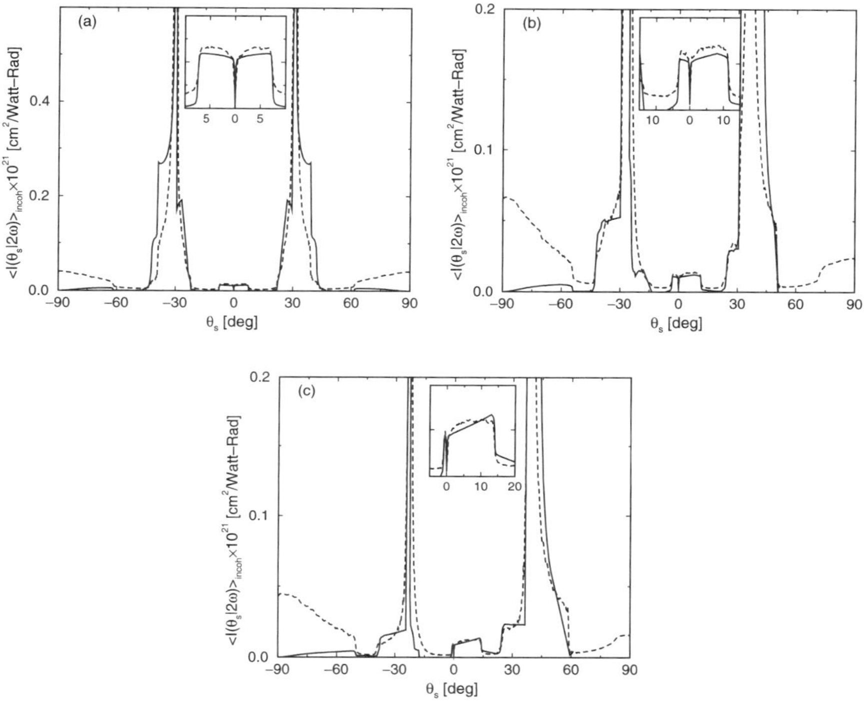

Fig. 8. The same as Fig. 6, except that χ^{s}_{ztt} is given by Eq. (75), while $\chi^{s}_{zzz} = \chi^{s}_{ttz} = 0$.

dip occurs in the direction normal to the mean surface independent of the values of the nonlinear coefficients. With the increase of the angle of incidence the dip evolves into a peak when the nonlinear coefficient χ^{s}_{zzz} is nonzero, while in the case when only $\chi^{s}_{ztt} \neq 0$ or $\chi^{s}_{ttz} \neq 0$ only a dip appears in the angular dependence of the mean intensity. For all three cases a weak dip in the retroreflection direction is displayed when the angle of incidence is $8°$.

The parameters of the surface roughness used in the calculations of the results presented in Figs. 6–9 are the parameters of the surface 1 studied in [22]. Unfortunately, perturbation theory cannot be applied to the other samples studied in [22]. The experimental curves for the mean intensity of the harmonic light presented in [22] also support the conclusion that perturbation theory is applicable only to the case of surface 1, the most weakly rough surface used in the study.

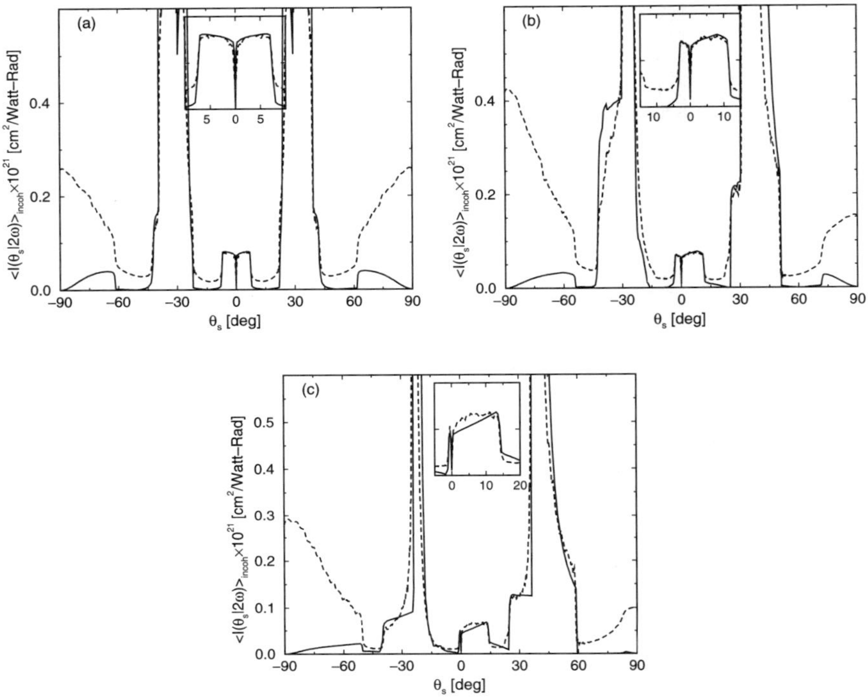

Fig. 9. The same as Fig. 6, except that χ^{s}_{ttz} is given by Eq. (74c), while $\chi^{s}_{zzz} = \chi^{s}_{ztt} = 0$.

5. Discussion

We preface the analysis of the numerical results presented in Figs. 2–9 with a discussion of the processes of scattering and nonlinear interaction that are possible for different models of the surface nonlinear polarization.

The results of a perturbative calculation carried out by McGurn *et al.* [11] predicted that enhanced second harmonic generation of light at a weakly rough, clean, metal surface occurs not only in the retroreflection direction but also in the direction normal to the mean scattering surface. The multiple scattering of surface plasmon polaritons of frequency 2ω is responsible for an enhanced second harmonic generation peak in the retroreflection direction. On the other hand, an enhancement peak of the second harmonic intensity in the direction normal to the mean surface arises due to the nonlinear mixing of the multiply-scattered surface plasmon polaritons of frequency ω. The results of the calculations presented in the preceding section, as well as the experiments of [21] and [22], support the general physical picture of the nonlinear scattering discussed in [11]. However, in contrast to the results of [11], the coherent interference effects mediated by surface plasmon polaritons of frequency 2ω lead to the appearance of a peak or a dip in the retroreflection direction depending on the model of the surface nonlinear polarization and, what is more, in the experiments of [21] only dips in the retroreflection direction were observed. We recall that no particular structure in the direction normal to the mean surface was observed in the experiments of [22].

To understand the origin of the peaks and dips appearing in the angular dependence of the mean intensity of the second harmonic light we will analyze the possible processes of scattering and nonlinear mixing. First we recall the physical origin of the enhanced backscattering phenomenon occurring in the scattering of light from a weakly rough random metal surface that supports surface electromagnetic waves — surface plasmon polaritons [8]. This is the presence of a well-defined peak in the retroreflection direction in the angular dependence of the intensity of the light scattered incoherently from the random surface. The presence of the roughness breaks the translational invariance of the scattering system, so that the incident light can couple into surface plasmon polaritons. The excited surface plasmon polaritons propagate along the surface and, after being scattered several times by the surface roughness, are converted back into volume electromagnetic waves in the vacuum, which propagate away from the surface. In general, all such multiply-scattered optical paths are incoherent due to the random phase introduced by the roughness of the surface. However, in the backscattering direction, for any such path there is a corresponding reciprocal partner, in which the light and the surface plasmon polariton are scattered from the same points on the surface, but in the reverse order. The waves emerging into the vacuum after travelling along these two paths interfere constructively, and give a contribution to the intensity of the scattered light. So, when the direction of observation is opposite to the direction of the incident light these two scattering paths are coherent, and it is necessary to add the amplitudes for these two scattering sequences when calculating their contribution to the intensity of the scattered light. If A_d and A_r are the amplitudes of the direct and reciprocal scattering sequences, the contribution to the intensity of the scattered light from them is therefore $|A_d + A_r|^2$. As the scattering angle departs from the retroreflection direction, a random phase difference with nonzero mean and increasing variance between the two scattering paths develops. As a result, they are no longer coherent and their contribution to the intensity of the scattered light becomes equal to $|A_d|^2 + |A_r|^2$. Thus, within a narrow angular range about the retroreflection direction, the intensity of the scattered light can become twice as large as the background intensity related to the multiple-scattering processes, when or if the amplitudes A_d and A_r are equal, due to the cross terms in $|A_d + A_r|^2$. Since in the scattering of light from a random rough surface the scattering potential $V(q|k)$ is reciprocal, the amplitudes A_d and A_r are equal, and their interference is constructive leading to the appearance of the enhanced backscattering peak. The situation can be quite different when the nonlinear interaction is included in the scattering processes, for example, when we consider the second harmonic generation of light in reflection from a randomly rough surface. The incident field of frequency ω generates a nonlinear polarization at the surface. Due to the gradient form of the nonlinear polarization, the "reflection" coefficient from a planar metal surface is a linear function of the tangential component k of the wave vector of the incident light. Therefore the amplitude of the field of the generated second harmonic waves changes its sign when the sign of the angle of incidence is changed, and the response of the planar nonlinear surface is antireciprocal [25]. This means that the amplitudes of the direct and reciprocal scattering sequences, A_d and A_r, have opposite signs so that the coherent interference in this case is destructive and leads to the appearance of an enhanced backscattering dip [25].

When the surface is rough the translational invariance of the scattering system is broken by the surface roughness, and the nonlinear polarization can generate, along with volume waves of frequency 2ω, surface plasmon polaritons of frequency 2ω. So surface plasmon polaritons of both the frequencies ω and 2ω can be excited by the incident light.

5.1. *Multiple scattering of surface plasmon polaritons of frequency* 2ω

First we analyze the processes of multiple scattering of surface plasmon polaritons of frequency 2ω. As was done in the experiments of [21] and in obtaining the results presented in Figs. 2–5, we assume that the power spectrum of the surface roughness is such that the excitation of surface plasmon polaritons of frequency ω through single-scattering processes is forbidden when the fundamental light is incident at sufficiently small angles of incidence. On the other hand, the strong conversion of the surface plasmon polaritons of frequency 2ω into volume electromagnetic waves radiated into vacuum in the given angular range of the scattering angles is ensured. This is achieved by the use of the rectangular power spectrum, Eq. (5), centered at the wave numbers of surface plasmon polaritons of frequency 2ω, $k_{\mathrm{sp}}(2\omega)$, with a halfwidth equal to $(2\omega/c)\sin\theta_{\max}$. The central distribution of the intensity of the light of frequency 2ω in Figs. 2–5 is then due to the roughness-induced radiation of surface plasmon polaritons of frequency 2ω and is determined by the second, $\tau_{2\omega}(q|2k)|G(2k,2\omega)|^2|Q(2k|2k)|^2$, and third, $\tilde{\tau}_{2\omega}(q|2k)$, terms on the right-hand side of Eq. (68).

Before proceeding we recall that the averaged reducible vertex function $\tau_{\Omega}(q|p)$ describes the processes of scattering of incident waves with the wavenumber p into scattered waves with the wavenumber q which are mediated by surface plasmon polaritons with the wavenumbers $\pm k_{\mathrm{sp}}(\Omega)$. It was calculated in [8, 39] by the use of a pole approximation for the Green's function $G(q,\Omega)$ of the form

$$G(q,\Omega) = \frac{C(\Omega)}{q - k_{\mathrm{sp}}(\Omega) - i\Delta_t(\Omega)} - \frac{C(\Omega)}{q + k_{\mathrm{sp}}(\Omega) + i\Delta_t(\Omega)}, \tag{76}$$

where

$$C(\Omega) = \frac{(|\epsilon_1(\Omega)|)^{3/2}}{\epsilon_1^2(\Omega) - 1} \tag{77a}$$

is the residue of the Green's function at the poles $q = \pm k_{\mathrm{sp}}(\Omega)$. The decay rate $\Delta_t(\Omega)$ of the surface plasmon polaritons is

$$\Delta_t(\Omega) = \Delta_\epsilon(\Omega) + \Delta_{\mathrm{sp}}(\Omega), \tag{77b}$$

where

$$\Delta_\epsilon(\Omega) = \frac{1}{2}k_{\mathrm{sp}}(\omega)\frac{\epsilon_2(\omega)}{(\epsilon_1(\omega)+1)\epsilon_1(\omega)} \tag{77c}$$

is the decay rate of the surface plasmon polaritons due to ohmic losses, and

$$\Delta_{\mathrm{sp}}(\Omega) = C(\Omega)\mathrm{Im}\, M(k_{\mathrm{sp}}(\Omega)), \tag{77d}$$

is the decay rate of surface plasmon polaritons due to their scattering by the surface roughness. In this approximation, and in the limit of a weakly rough surface, the averaged reducible vertex function $\tau_{\Omega}(q|p)$, calculated in [8, 39], has the form

$$\begin{aligned}\tau_{\Omega}(q|p) = K(q,p|\Omega) &+ \frac{C^2(\Omega)}{2\Delta_t(\Omega)}\tau_L(q,p|\Omega) \\ &+ \frac{\Delta_t(\Omega)C^2(\Omega)}{(q+p)^2 + 4\Delta_t^2(\Omega)}\tau_{MC}(q,p|\Omega),\end{aligned} \tag{78}$$

where $C(\Omega)$ and Δ_t are given by Eqs. (77a) and (77b).

The functions $\tau_L(q,p|\Omega)$ and $\tau_{MC}(q,p|\Omega)$ in Eq. (78) are given by [8, 39]

$$\tau_L(q,p|\Omega) = \frac{1}{1-(\Delta_{\rm sp}(\Omega)/\Delta_t(\Omega))^2}\Bigg[K(q,k_{\rm sp}(\Omega)|\Omega)K(k_{\rm sp}(\Omega),p|\Omega)$$
$$+K(q,-k_{\rm sp}(\Omega)|\Omega)K(-k_{\rm sp}(\Omega),p|\Omega)$$
$$+\frac{\Delta_{\rm sp}(\Omega)}{\Delta_t(\Omega)}\left(K(q,k_{\rm sp}(\Omega)|\Omega)K(-k_{\rm sp}(\Omega),p|\Omega)\right.$$
$$\left.+K(q,-k_{\rm sp}(\Omega)|\Omega)K(k_{\rm sp}(\Omega),p|\Omega)\right)\Bigg], \tag{79a}$$

and

$$\tau_{MC}(q,p|\Omega) = \frac{1}{1-(\Delta_{\rm sp}(\Omega)/\Delta_t(\Omega))^2}\Bigg[K(q,k_{\rm sp}(\Omega)|\Omega)K(k_{\rm sp}(\Omega),p|\Omega)$$
$$+K(q,-k_{\rm sp}(\Omega)|\Omega)K(-k_{\rm sp}(\Omega),p|\Omega)$$
$$+K(q,q+p-k_{\rm sp}(\Omega)|\Omega)K(q+p-k_{\rm sp}(\Omega),p|\Omega)$$
$$+K(q,q+p+k_{\rm sp}(\Omega)|\Omega)K(q+p+k_{\rm sp}(\Omega),p|\Omega)$$
$$+\frac{\Delta_{\rm sp}(\Omega)}{\Delta_t(\Omega)}\left(K(q,k_{\rm sp}(\Omega)|\Omega)K(-k_{\rm sp}(\Omega),p|\Omega)\right.$$
$$+K(q,-k_{\rm sp}(\Omega)|\Omega)K(k_{\rm sp}(\Omega),p|\Omega)$$
$$+K(q,q+p-k_{\rm sp}(\Omega)|\Omega)K(q+p+k_{\rm sp}(\Omega),p|\Omega)$$
$$\left.+K(q,q+p+k_{\rm sp}(\Omega)|\Omega)K(q+p-k_{\rm sp}(\Omega),p|\Omega)\right)\Bigg], \tag{79b}$$

where the function $K(q,p|\Omega)$ is related to the scattering potential $V(q,p|\Omega)$ through

$$K(q,p) = \langle |W(q,p|\Omega)|^2\rangle. \tag{79c}$$

It can be seen from Eqs. (78) and (79b), that the averaged reducible vertex function $\tau_\Omega(q|p)$ displays a Lorentzian enhanced backscattering peak whose position is determined by the condition $q+p=0$ and whose half-width at half-maximum equals the decay rate of the surface plasmon polaritons propagating along the rough surface.

Thus, the contribution to the mean intensity of the second harmonic light from the second term on the right-hand side of Eq. (68), $\tau_{2\omega}(q|2k)|G(2k,2\omega)|^2|Q(2k|2k)|^2$, can be viewed as a linear scattering one, with the only difference that the "amplitude" of the incident field, that is $G(2k,2\omega)Q(2k|2k)$, depends on the angle of incidence (the processes

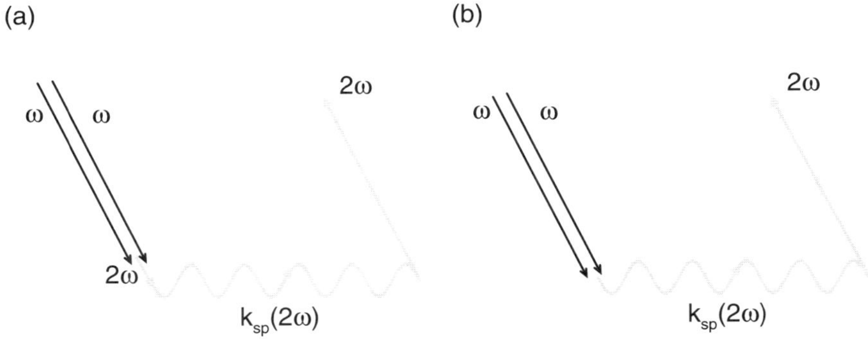

Fig. 10. The diagrams of the multiple scattering processes involving surface plasmon polaritons of frequency 2ω. The solid line arrows represent light of frequency ω, the wavy gray line arrows represent surface plasmon polaritons of frequency 2ω and the solid gray line arrows represent light of frequency 2ω.

associated with this term are illustrated schematically in Fig. 10(a)). Due to the presence of $\tau_{2\omega}(q|2k)$, this contribution displays an enhanced second harmonic generation peak. The position of the peak is determined by the condition $q + 2k = 0$. Since $q = (2\omega/c)\sin\theta_s$, the condition $q = -2k$ is equivalent to $\sin\theta_s = -\sin\theta_0$, i.e., the enhanced second harmonic generation peak occurs in the retroreflection direction. The height of the peak as a function of the angle of incidence is determined by the effective amplitude of the field being scattered, $G(2k, 2\omega)Q(2k|2k)$, i.e., by the coherent component of the nonlinear source function of frequency 2ω. The strongest contribution to it is the specular contribution to the term $Q_v(q|2k)$ (see Eq. (65a)), which is

$$\gamma_0(q,k) = \Gamma_p(q,k,k,k,k,0) = \chi^{\mathrm{s}}_{ttz}\frac{\alpha(q,2\omega)}{\epsilon(2\omega)}k\frac{\alpha(k,\omega)}{\epsilon(\omega)} - 2k\left[\chi^{\mathrm{s}}_{zzz}k^2 - \chi^{\mathrm{s}}_{ztt}\frac{\alpha^2(k,\omega)}{\epsilon^2(\omega)}\right]. \tag{80}$$

This is the nonlinear Fresnel reflection coefficient [25] for the scattering of p-polarized light from a planar surface. Since this coefficient is proportional to k, in contrast to the problem of the linear scattering, the peak of the enhanced second harmonic generation in this case will be absent when light is incident normally on the surface. We note that this contribution to the mean intensity of the generated light always displays a peak in the retroreflection direction. It is just this contribution that was considered in [11] when discussing the effects of the coherent interference in the multiple scattering of surface plasmon polaritons of frequency 2ω. The remaining term, which contains the effects of the coherent interference of the surface plasmon polaritons of frequency 2ω, $\tilde{\tau}_{2\omega}(q|2k)$, has been neglected.

We now turn to an analysis of the third term in Eq. (68), $\tilde{\tau}_{2\omega}(q|2k)$. In a sense, this term is analogous to the contributions from the ladder and maximally-crossed diagrams in the linear problem of the scattering of light from a randomly rough surface. However, in the nonlinear scattering problem the amplitude of the field being scattered is the nonlinear source function $Q(q|2k)$. Therefore, it describes the scattering of the waves of frequency 2ω generated through the nonlinear mixing of the diffusely scattered fundamental radiation. To second order in the nonlinear coefficients the function $\tilde{\tau}_{2\omega}(q|2k)$ has the form

$$\begin{aligned}\tilde{\tau}_{2\omega}(q|2k) = \int_{-\infty}^{\infty} \frac{dp}{2\pi} & \left[\tau_{2\omega}(q|p)|G(p,2\omega)|^2\langle|Q(p|2k)|^2\rangle_0\right. \\ & + \langle t_{2\omega}(q,p)Q^*(q+2k-p|2k)\rangle_0 G(p,2\omega)G^*(q+2k-p,2\omega) \\ & \times \left.\langle t^*_{2\omega}(q,q+2k-p)Q(p|2k)\rangle_0\right], \qquad (81)\end{aligned}$$

where we have introduced the notation $\langle f(q,p)g(p',q')\rangle = 2\pi\delta(q-p+p'-q')$ $\langle f(q,p)g(q'-q+p,q')\rangle_0$.

As was done in [8, 39], to calculate the most important contributions to $\tilde{\tau}_{2\omega}(q|2k)$ we will use the pole approximation for the Green's function $G(q,\Omega)$, Eq. (76). In this approximation we obtain the following expression for $\tilde{\tau}_{2\omega}(q|2k)$:

$$\begin{aligned}\tilde{\tau}_{2\omega}(q|2k) = & \frac{C^2(2\omega)}{\Delta_t(2\omega)}[\tau_{2\omega}(q|k_{\rm sp}(2\omega))\langle|Q(k_{\rm sp}(2\omega)|2k)|^2\rangle_0 \\ & + \tau_{2\omega}(q|-k_{\rm sp}(2\omega))\langle|Q(-k_{\rm sp}(2\omega)|2k)|^2\rangle_0] \\ & + \frac{4\Delta_t(2\omega)C^2(2\omega)}{(q+2k)^2+4\Delta_t^2(2\omega)}{\rm Re}[\mathcal{P}_{2\omega}(q,k_{\rm sp}(2\omega)|q \\ & + 2k-k_{\rm sp}(2\omega),2k)\mathcal{P}^*_{2\omega}(q,q+2k-k_{\rm sp}(2\omega)|k_{\rm sp}(2\omega),2k) \\ & + \mathcal{P}_{2\omega}(q,-k_{\rm sp}(2\omega)|q+2k+k_{\rm sp}(2\omega),2k) \\ & \times \mathcal{P}^*_{2\omega}(q,q+2k+k_{\rm sp}(2\omega)|-k_{\rm sp}(2\omega),2k)] \\ & + \frac{1}{2}C^2(2\omega)\tau_{MC}(q,-q|2\omega)|G(q,2\omega)|^2\langle|Q(-q|2k)|^2\rangle_0, \qquad (82)\end{aligned}$$

where

$$\mathcal{P}_{2\omega}(q,p|p',q') = \langle t_{2\omega}(q,p)Q^*(p'|q')\rangle_0. \qquad (83)$$

From Eq. (82) it is seen that the function $\tilde{\tau}_{2\omega}(q|2k)$ displays a Lorentzian peak centered at $q = -2k$, i.e., in the retroreflection direction. However, the height of the peak is determined by the Fourier component of the nonlinear source through which the excitation of surface plasmon polaritons of frequency 2ω occurs. As a result, the height of the peak depends on the particular trajectory of the scattering path. In our case it is the "incident" surface plasmon polaritons of frequency 2ω, which are excited due to the nonlinear mixing of the fundamental waves, that are multiply scattered. (The schematic illustrations of the processes contained in the term under discussion are presented in Fig. 10(b).) The mean intensities of these "incident" surface plasmon polaritons are determined by the nonlinear source functions, and are proportional to $Q(k_{\mathrm{sp}}(2\omega)|2k)$ and $Q(-k_{\mathrm{sp}}(2\omega)|2k)$. Generally speaking, both the nonlinear excitation and the roughness induced radiation of the surface plasmon polaritons of frequency 2ω are nonreciprocal processes and, as a result, there is no reason to expect a backscattering enhancement peak in the angular distribution of the intensity of the generated light. Depending on the phases acquired in the processes of excitation and radiation, a peak or a dip can occur.

From the expressions for the nonlinear source functions, Eqs. (59) and (60), one can see that the strongest contribution to $Q(\pm k_{\mathrm{sp}}(2\omega)|2k)$ comes from the function $Q_{\mathrm{s}}(\pm k_{\mathrm{sp}}(2\omega), k, k)$. This function is governed by the effective nonlinearity

$$\begin{aligned}
&\Gamma_p(\pm k_{\mathrm{sp}}(2\omega), k, k, \pm k_{\mathrm{sp}}(2\omega) - k, -k, 0) \\
&\quad + \Gamma_p(\pm k_{\mathrm{sp}}(2\omega), \pm k_{\mathrm{sp}}(2\omega) - k, k, k, k, 0) \\
&= i\chi^{\mathrm{s}}_{ttz} \frac{\alpha(k_{\mathrm{sp}}(2\omega), 2\omega)}{\epsilon(2\omega)} \left[k \frac{\alpha(k,\omega) \pm k k_{\mathrm{sp}}(2\omega)}{\epsilon(\omega)\alpha(k,\omega)} + (\pm k_{\mathrm{sp}}(2\omega) - k) \frac{\alpha(k,\omega)}{\epsilon(\omega)} \right] \\
&\quad \mp 2k_{\mathrm{sp}}(2\omega) \left[\chi^{\mathrm{s}}_{zzz} k(\pm k_{\mathrm{sp}}(2\omega) - k) + \chi^{\mathrm{s}}_{ztt} \frac{\alpha(k,\omega)}{\epsilon(\omega)} \frac{\alpha(k,\omega) \pm k k_{\mathrm{sp}}(2\omega)}{\epsilon(\omega)\alpha(k,\omega)} \right],
\end{aligned} \tag{84}$$

in the case of p-polarized incident light and

$$\Gamma_s(k, \pm k_{\mathrm{sp}}(2\omega) - k) + \Gamma_s(\pm k_{\mathrm{sp}}(2\omega) - k, k) = \pm 2k_{\mathrm{sp}}(2\omega)\chi^{\mathrm{s}}_{ztt} \tag{85}$$

in the case of s-polarized incident light.

We can see that in the case of s-polarized incident light the effective nonlinear coefficient, Eq. (85), is proportional to the wave vector of the intermediate excitations, in our case $\pm k_{\mathrm{sp}}(2\omega)$. Therefore, when the surface plasmon polaritons of frequency 2ω propagating in opposite directions are excited in the processes of nonlinear mixing, they will have a phase difference π, and, as a result, only a dip can be formed in the retroreflection direction. This is confirmed by the results presented in Fig. 5. In the case of p-polarized incident light the situation is different. The expression Eq. (84) for the effective nonlinear coefficient in this case can be rewritten as

$$\begin{aligned}
&\Gamma_p(\pm k_{\rm sp}(2\omega), k, k, \pm k_{\rm sp}(2\omega) - k, -k, 0) \\
&\quad + \Gamma_p(\pm k_{\rm sp}(2\omega), \pm k_{\rm sp}(2\omega) - k, k, k, k, 0) \\
&= \pm k_{\rm sp}(2\omega)\left[i\chi^{\rm s}_{ttz}\frac{\alpha(k_{\rm sp}(2\omega), 2\omega)}{\epsilon(2\omega)}\frac{\omega^2/c^2}{\epsilon(\omega)\alpha(k,\omega)} + 2\chi^{\rm s}_{zzz}k^2 - 2\chi^{\rm s}_{ztt}\frac{\alpha^2(k,\omega)}{\epsilon^2(\omega)}\right] \\
&\quad - 2kk^2_{\rm sp}(2\omega)\left[\chi^{\rm s}_{zzz} + \chi^{\rm s}_{ztt}\frac{1}{\epsilon^2(\omega)}\right]. \qquad (86)
\end{aligned}$$

The first term in Eq. (86) is proportional to the wave vector of the intermediate excitations, exactly as in the case of s-polarized incident light and, therefore, leads to the appearance of a dip in the retroreflection direction. The second term in Eq. (86) is quadratic in $k_{\rm sp}(2\omega)$ and, thus, does not introduce the phase difference π in the fields of surface plasmon polaritons of frequency 2ω propagating in opposite directions. It, therefore, can lead to the appearance of a peak of the enhanced second harmonic generation in the retroreflection direction. However, this term is linear in the tangential component of the wave vector of the incident light, k. Therefore, it vanishes at normal incidence of the fundamental light. Thus, at normal incidence only a dip can be formed in the retroreflection direction. With an increase of the angle of incidence the second term in Eq. (86) becomes dominant since $\chi^{\rm s}_{zzz}$ is the largest nonlinear coefficient and, what is more, the terms with $\chi^{\rm s}_{ttz}$ and $\chi^{\rm s}_{ztt}$ are even smaller due to the presence of factors $1/\epsilon(2\omega)$ or $1/\epsilon(\omega)$. As a result, a peak in the retroreflection direction will be formed. The angle of incidence at which the dip disappears and the peak begins to evolve is determined not only by the relative magnitudes of the nonlinear coefficients, but by the dielectric functions of the medium, $\epsilon(\omega)$ and $\epsilon(2\omega)$, as well. In the case when $\chi^{\rm s}_{zzz} = 0$ (see Figs. 4 and 5) the angular distribution of the intensity displays a dip in the retroreflection direction for all angles of incidence. The strong dependence of the intensity of the generated light on the angle of incidence displayed in the plots in Fig. 2 also shows the dominant role of the nonlinear constant $\chi^{\rm s}_{zzz}$ when the calculations are carried out within the framework of the free-electron model.

The last term on the right-hand side of Eq. (68) is nonresonant and contributes to the structureless background intensity. The first term on the right-hand side of Eq. (68), $\langle |Q(q|2k)|^2 \rangle_{\text{incoh}}$, also contributes to the background intensity and, since the power spectrum of the surface roughness is such that the excitation of surface plasmon polaritons of frequency ω through a single scattering process is forbidden, is a structureless function of the scattering angle when the surface is weakly rough, so that only single- and double-scattering processes contribute to the nonlinear source function $Q(q|2k)$. However, in the gap between the central distribution and the single-scattering wings in Fig. 2, as well as in the experimental data of [21], peaks at $q = k \pm k_{\text{sp}}(\omega)$, which are due to the resonant nonlinear mixing of the incident light with the surface plasmon polaritons of frequency ω, are present. For the given power spectrum of the surfaces under study, surface plasmon polaritons of frequency ω can be excited in higher-order scattering processes, in fact in the third event of scattering by the surface roughness. Therefore, the resonant peaks due to the nonlinear mixing of surface plasmon polaritons of frequency ω with the incident light can arise, but only in the higher-order scattering processes. They can be seen in Fig. 2 but are too weak to be seen in Figs. 3 and 4. The processes leading to the appearance of the resonant peaks are described by the second term in Eq. (72), $\chi_1(q|k)$, given by Eq. (73a). In the pole approximation the contribution of lowest order in the surface profile function which describes the resonant peaks has the form

$$\begin{aligned}\chi_1(q|k) &= |\langle Q_s(q, q-k, k)\rangle_0|^2 \delta^4 |G(q-k, \omega)|^2 \frac{C^2(\omega)}{2\Delta_t(\omega)} \\ &\times \left[\tau_\omega(q-k|k_{\text{sp}}(\omega)) \int_{-\infty}^{\infty} \frac{dp}{2\pi} \langle |w(k_{\text{sp}}(\omega), p|\omega)|^2 \rangle_0 \right. \\ &\times |G(p,\omega)|^2 \langle |w(p, k|\omega)|^2 \rangle_0 \\ &\times \tau_\omega(q-k|-k_{\text{sp}}(\omega)) \int_{-\infty}^{\infty} \frac{dp}{2\pi} \langle |w(-k_{\text{sp}}(\omega), p|\omega)|^2 \rangle_0 \\ &\left. \times |G(p,\omega)|^2 \langle |w(p, k|\omega)|^2 \rangle_0 \right]. \end{aligned} \tag{87}$$

5.2. *Multiple scattering of surface plasmon polaritons of frequency ω*

As was done in the experiments of [22] and in obtaining the results presented in Figs. 6–9, we assume that the power spectrum of the surface roughness is now such that the fundamental light incident on the surface at angles of incidence $\theta_0 < \theta_{\max}$ is strongly coupled into surface plasmon polaritons of frequency ω and surface plasmon polaritons of frequency ω are radiated into vacuum in the angular range $|\theta_s| < \theta_{max}$. This is achieved by the use of the rectangular power spectrum, Eq. (5), centered at the wave numbers of surface plasmon polaritons of frequency ω, $k_{sp}(\omega)$, with the halfwidth equal to $(\omega/c)\sin\theta_{max}$.

The second harmonic light generated through the nonlinear mixing of the surface plasmon polaritons of the fundamental frequency emerges into vacuum in the angular range determined by the conditions

$$-\frac{1}{2}(\sin\theta_{max} - \sin\theta_0) < \sin\theta_s < \frac{1}{2}(\sin\theta_{max} + \sin\theta_0), \tag{88a}$$

and

$$\begin{aligned} -\frac{1}{2}(n_{sp}(\omega) + \sin\theta_{max}) < \sin\theta_s < -\frac{1}{2}(n_{sp}(\omega) - \sin\theta_{max}), \\ \frac{1}{2}(n_{sp}(\omega) - \sin\theta_{max}) < \sin\theta_s < \frac{1}{2}(n_{sp}(\omega) + \sin\theta_{max}). \end{aligned} \tag{88b}$$

In contrast, the surface plasmon polaritons of frequency 2ω excited through the nonlinear interaction are converted into vacuum light that radiates within the angular range determined by the conditions

$$\begin{aligned} &\frac{1}{2}(n_{sp}(\omega) - \sin\theta_{max}) - n_{sp}(2\omega) \\ &\quad < \sin\theta_s < \frac{1}{2}(n_{sp}(\omega) + \sin\theta_{max}) - n_{sp}(2\omega), \end{aligned} \tag{89a}$$

and

$$\begin{aligned} &-\frac{1}{2}(n_{sp}(\omega) + \sin\theta_{max}) + n_{sp}(2\omega) \\ &\quad < \sin\theta_s < -\frac{1}{2}(n_{sp}(\omega) - \sin\theta_{max}) + n_{sp}(2\omega). \end{aligned} \tag{89b}$$

In Eqs. (89a) and (89b) $n_{sp}(\Omega) = (k_{sp}(\Omega)c)/\Omega$ is the refractive index of the surface plasmon polaritons of frequency Ω. The single scattering processes give a contribution to the intensity of the second harmonic light only in the angular ranges

$$\begin{aligned} &\frac{1}{2}(n_{sp}(\omega) - \sin\theta_{max}) + \sin\theta_0 \\ &\quad < \sin\theta_s < \frac{1}{2}(n_{sp}(\omega) + \sin\theta_{max}) + \sin\theta_0, \end{aligned} \tag{90a}$$

and

$$\begin{aligned} &-\frac{1}{2}(n_{sp}(\omega) + \sin\theta_{max}) + \sin\theta_0 \\ &\quad < \sin\theta_s < -\frac{1}{2}(n_{sp}(\omega) - \sin\theta_{max}) + \sin\theta_0. \end{aligned} \tag{90b}$$

The plots in Figs. 6–9 indeed display a nonzero intensity of the scattered light of frequency 2ω in only these angular intervals.

The effects of the multiple scattering of surface plasmon polaritons of frequency ω influence all the terms contributing to the mean differential intensity of the second harmonic light. The first term in Eq. (68), $\langle Q(q|2k)|^2\rangle_{\rm incoh}$, describes the second harmonic generation of volume waves of frequency 2ω through the nonlinear mixing of the scattered fundamental light, including the multiply-scattered surface plasmon polariton of frequency ω, while the remaining contributions describe those processes in which the generated waves of frequency 2ω have been scattered by the rough surface. The latter give rise to the structureless background of the intensity of the second harmonic light, because the power spectrum chosen in this case forbids the conversion of surface plasmon polaritons of frequency 2ω into radiative waves in vacuum. The contributions of the first and two last terms to $\langle Q(q|2k)|^2\rangle_{\rm incoh}$, Eq. (72), are also structureless, and contribute to the background intensity. The effects of the multiple scattering of surface plasmon polaritons of frequency ω are contained in the functions $\chi_1(q|k)$, $\chi_2(q|k)$, $\chi_3(q|k)$, and $\chi_4(q|k)$. With the use of the pole approximation (76) for the Green's function $G(q,\omega)$ we can calculate each of these contributions to $\langle Q(q|2k)|^2\rangle_{\rm incoh}$. The function $\chi_1(q|k)$ in this approximation has the form

$$\chi_1(q,k) = |\gamma_1(q,k)|^2|G(q-k,\omega)|^2\tau_\omega(q-k|k), \tag{91}$$

where

$$\begin{aligned}\gamma_1(q,k) = \langle Q_s(q|q-k|k)\rangle_0 &+ iC(\omega)\langle(\widehat{Q}_{ss}(q,k_{\rm sp}(\omega),q-k)t_\omega(k_{\rm sp}(\omega)|k) \\ &+ \widehat{Q}_{ss}(q,-k_{\rm sp}(\omega),q-k)t_\omega(-k_{\rm sp}(\omega)|k))\rangle_0.\end{aligned} \tag{92}$$

This term describes the intensity of the light of frequency 2ω generated by the nonlinear interaction of the incident light with the scattered waves of frequency ω. (Schematic illustrations of these processes are presented in Figs. 11(a)–(c).) In the discussion of the effects of the multiple scattering of surface plasmon polaritons of frequency ω in [11] only this term was kept and all the remaining contributions into $\langle|Q(q|2k)|^2\rangle_{\rm incoh}$ were neglected. The function $\chi_1(q|k)$ contains the product of two highly peaked functions, $|G(q-k,\omega)|^2\tau_\omega(q-k|k)$ and, as a result, displays three peaks of the enhanced second harmonic generation. Two strong resonant peaks at $q = k \pm k_{\rm sp}(\omega)$ are due to the resonant interaction of the incident light with the excited surface plasmon polaritons of frequency ω [7, 11] (Figs. 11(a) and (b)) and appear already in the single scattering processes

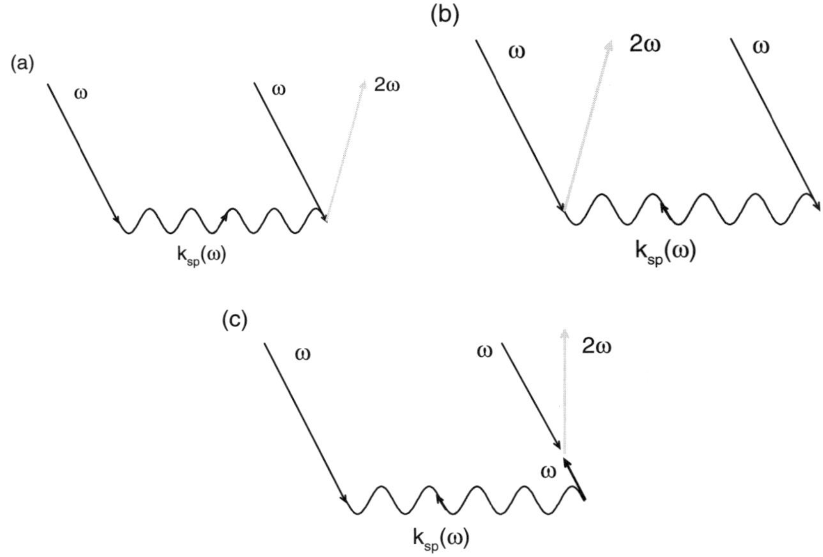

Fig. 11. The diagrams of the multiple scattering processes involving surface plasmon polaritons of frequency ω. The solid line arrows represent light of frequency ω, the wavy line arrows represent surface plasmon polaritons of frequency ω, and the solid gray line arrows represent light of frequency 2ω.

due to the presence of the factor $|G(q-k,\omega)|^2$ [7, 11]. We note here that in the experiments of [12, 14], and [17] in which the Kretchmann ATR geometry was used to excite surface plasmon polaritons of frequency ω, one of the resonant peaks, namely the one whose position is determined by the relation $q = k - k_{\rm sp}(\omega)$, moves to the direction normal to the mean surface, since in this case the tangential component of the wave vector of the incident wave equals the wavenumber of the surface plasmon polaritons of frequency ω, $k = k_{\rm sp}(\omega)$. Therefore, in the Kretchmann ATR geometry any possible features of the coherent interference in the direction normal to the mean surface are masked by this strong resonant peak and, as a result, are hardly observable.

The coherent interference of multiply-scattered surface plasmon polaritons of frequency ω leads to the appearance of a peak in the direction normal to the surface, because the reducible vertex function $\tau_\omega(q-k|k)$ (see Eq. (78)) displays a Lorentzian peak centered at $q - k + k = 0$. The function $\tau_\omega(q-k|k)$ describes the second harmonic generation by the nonlinear mixing of the incident light with volume waves of frequency ω emerging into the vacuum after being multiply scattered by the surface roughness. It displays a peak when $q = 0$, i.e., in the direction normal to the mean surface. The presence of this peak can be understood easily, since it is due to the mixing of the incident light and the light scattered in the retroreflection direction, i.e., in the direction of the enhanced backscattering (Fig. 11(c)).

In this case the contrapropagating beams of volume waves interact nonlinearly giving rise to the waves of frequency 2ω propagating into the vacuum in the direction normal to the surface. The main contribution to the efficiency of the nonlinear mixing comes from $\langle Q_s(q, q-k, k)\rangle_0$ in Eq. (91), i.e., from the nonlinear coefficient $\gamma_1(q, k)$ defined by

$$\begin{aligned}\gamma_1(q,k) &= \langle Q_s(q,q-k,k)\rangle_0 = \Gamma_p(q,q-k,k,k,k,0) + \Gamma_p(q,k,k,q-k,k,0)\\ &= \chi^{s}_{ttz}\frac{\alpha(q,2\omega)}{\epsilon(2\omega)}\left[(q-k)\frac{\alpha(k,\omega)}{\epsilon(\omega)} + ku(q-k|k)\right]\\ &\quad - 2q\left[\chi^{s}_{zzz}k(q-k) + \chi^{s}_{ztt}u(q-k|k)\frac{\alpha(k,\omega)}{\epsilon(\omega)}\right]. \end{aligned} \tag{93}$$

At small values of q, $\gamma_1(q,k) \propto q$; therefore, the efficiency of the nonlinear mixing vanishes when $q = 0$, i.e., in the direction normal to the mean surface.

The processes of nonlinear mixing of the multiply-scattered surface plasmon polaritons of frequency ω with the incident light (we illustrate them in Figs. 12(a) and (b)) are described by the contribution $\chi_2(q|k)$, Eq. (73c). It contains two contributions which come from the ladder, $\chi_2^L(q|k)$, and maximally-crossed, $\chi_2^{MC}(q|k)$, diagrams and have the same physical meaning as the corresponding contributions in the case of linear scattering. Calculated in the pole approximation for $G(q, \omega)$ they have

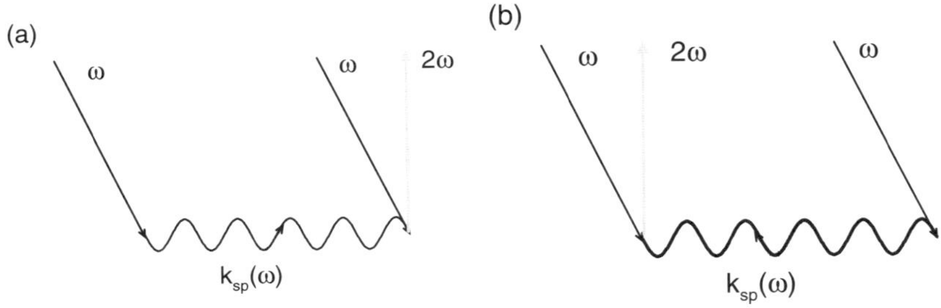

Fig. 12. The diagrams of the multiple scattering processes involving surface plasmon polaritons of frequency ω. The solid line arrows represent light of frequency ω, the wavy line arrows represent surface plasmon polaritons of frequency ω, and the solid gray line arrows represent light of frequency 2ω.

the forms

$$\chi_2^L(q|k) = \frac{C^2(\omega)}{2\Delta_t(\omega)}\Big[\tau_\omega(k_{\rm sp}(\omega)|k)\langle|Q_s(q,k_{\rm sp}(\omega),k)|^2\rangle_0 + \tau_\omega(-k_{\rm sp}(\omega)|k)\langle|Q_s(q,-k_{\rm sp}(\omega),k)|^2\rangle_0 + \frac{1}{2}\tau_{MC}(-k,k|\omega)|G(k,\omega)|^2\langle|Q_s(q,-k,k)|^2\rangle_0\Big], \tag{94a}$$

and

$$\chi_2^{MC}(q,k) = \frac{C^2(\omega)\Delta_t(\omega)}{q^2+4\Delta_t^2(\omega)}\left[\mathcal{P}(q,k_{\rm sp}(\omega),k) + \mathcal{P}(q,-k_{\rm sp}(\omega),k) + \mathcal{P}(q,q-k_{\rm sp}(\omega),k)\mathcal{P}(q,q+k_{\rm sp}(\omega),k)\right], \tag{94b}$$

where

$$\mathcal{P}(q,p,k) = \langle Q_s(q,p,k)t_\omega^*(q-p|k)\rangle_0\langle Q_s^*(q,q-p,k)t_\omega(p|k)\rangle_0. \tag{94c}$$

The contribution from the ladder diagrams yields the intensity of the light generated in the direct and reciprocal processes of multiple scattering, while the contribution from the maximally-crossed diagrams yields the interference between them. Since the surface plasmon polaritons of frequency ω which propagate in opposite directions play the key role in the nonlinear interaction, the peak of the enhanced second harmonic generation in this case occurs in the direction normal to the mean surface.

However, the strongest contributions to $\langle|Q(q|2k)|^2\rangle_{\rm incoh}$ come from the function $\chi_3(q|k)$, given by Eq. (73f), and from $\chi_4(q|k)$, given by Eq. (73g). The function $\chi_3(q,k)$ describes the nonlinear mixing of the multiply scattered surface plasmon polaritons of frequency ω propagating in opposite directions, illustrated in Fig. 13. In the pole approximation for the Green's functions it has the form

$$\begin{aligned}\chi_3(q,k) = {}& \frac{C^2(\omega)}{2\Delta_t(\omega)}\frac{C^2(\omega)}{q^2+4\Delta_t^2(\omega)}[\tau_\omega(k_{\rm sp}(\omega)|k)\tau_\omega(q-k_{\rm sp}(\omega)|k) \\ & \times|\langle\widehat{Q}_{ss}(q,k_{\rm sp}(\omega),q-k_{\rm sp}(\omega))\rangle_0|^2 \\ & + \tau_\omega(-k_{\rm sp}(\omega)|k)\tau_\omega(q+k_{\rm sp}(\omega)|k) \\ & \times|\langle\widehat{Q}_{ss}(q,-k_{\rm sp}(\omega),q+k_{\rm sp}(\omega))\rangle_0|^2] \\ & + C^2(\omega)\tau_{MC}(-k,k|\omega)\tau_\omega(q+k|k)|G(k)G(q+k)|^2 \\ & \times|\langle\widehat{Q}_{ss}(q,-k,q+k)\rangle_0|^2.\end{aligned} \tag{95}$$

As should be expected, $\chi_3(q|k)$ contains a Lorentzian factor centered at $q=0$. However, the efficiency of the nonlinear mixing of the

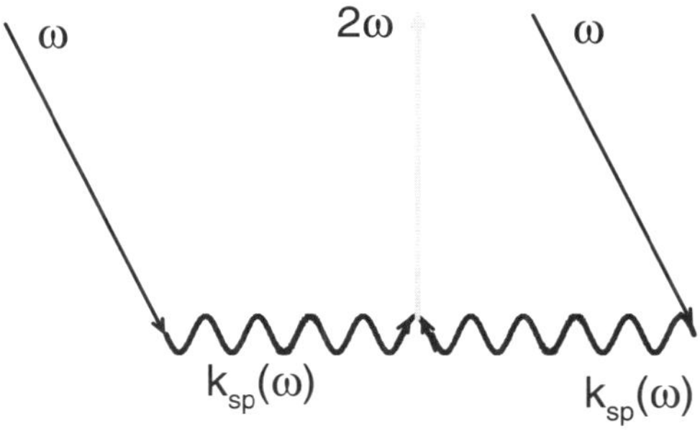

Fig. 13. The diagrams of the multiple scattering processes involving surface plasmon polaritons of frequency ω. The solid line arrows represent light of frequency ω, the wavy line arrows represent surface plasmon polaritons of frequency ω, and the solid gray line arrows represent light of frequency 2ω.

contrapropagating surface plasmon polaritons is determined by the effective nonlinear coefficient $\gamma_2(q, k_{\rm sp}(\omega)) = \langle \widehat{Q}_{ss}(q, -k_{\rm sp}(\omega), q + k_{\rm sp}(\omega))\rangle_0$, the main contribution to which is

$$\gamma_2(q,p) = \chi^{\rm s}_{ttz}\frac{\alpha(q,2\omega)}{\epsilon(2\omega)}\left[p\frac{\alpha(q-p,\omega)}{\epsilon(\omega)} + (q-p)\frac{\alpha(p,\omega)}{\epsilon(\omega)}\right]$$
$$- 2q\left[\chi^{\rm s}_{zzz}p(q-p) + \chi^{\rm s}_{ztt}\mu_2\frac{\alpha(p,\omega)}{\epsilon(\omega)}\frac{\alpha(q-p,\omega)}{\epsilon(\omega)}\right]. \qquad (96)$$

This effective nonlinear coefficient is linear in q for small q due to the symmetry of the surface nonlinear polarization. As is well known the symmetry of the nonlinear polarization of a metal surface forbids such processes [1]. Therefore, this contribution displays a dip rather than a peak in the direction normal to the mean surface. The depth of this dip depends strongly on the values of the material parameters and the angle of incidence of the fundamental light. The last term in the expression for the function $\chi_3(q|k)$ has quite a different origin and describes the resonant nonlinear mixing of the multiply-scattered surface plasmon polaritons of frequency ω with the enhanced backscattered radiation of frequency ω. These processes are illustrated in Figs. 14(a) and (b). They lead to the peaks/dips of the enhanced/suppressed second harmonic generation in the directions determined by the conditions $q = -k \pm k_{\rm sp}(\omega)$, depending on the effective nonlinear coefficient $\gamma_2(q, -k)$

The last resonant contribution to $\langle |Q(q|2k)|^2\rangle_{\rm incoh}$, $\chi_4(q,k)$, Eq. (73g), has a complicated resonant structure and, in the pole approximation is given by

$$\chi_4(q,k) = \left[\chi_4^L(q,k) + \chi_4^{MC}(q,k)\right], \qquad (97a)$$

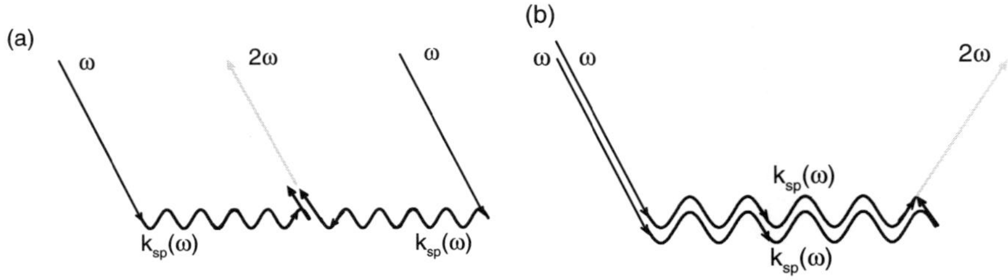

Fig. 14. The diagrams of the multiple scattering processes involving surface plasmon polaritons of frequency ω. The solid line arrows represent light of frequency ω, the wavy line arrows represent surface plasmon polaritons of frequency ω, and the solid gray line arrows represent light of frequency 2ω.

with

$$\begin{aligned}\chi_4^L(q,k) = {} & \frac{C^4(\omega)}{4\Delta_t^2(\omega)}\Big[\tau_\omega^2(k_{\rm sp}(\omega)|k)\langle|\widehat{Q}_{ss}(q,k_{\rm sp}(\omega),k_{\rm sp}(\omega))|^2\rangle_0 \\ & + \tau_\omega^2(-k_{\rm sp}(\omega)|k)\langle|\widehat{Q}_{ss}(q,-k_{\rm sp}(\omega),-k_{\rm sp}(\omega))|^2\rangle_0 \\ & + 2\tau_\omega(k_{\rm sp}(\omega)|k)\tau_\omega(-k_{\rm sp}(\omega)|k)\langle|\widehat{Q}_{ss}(q,k_{\rm sp}(\omega),-k_{\rm sp}(\omega)) \\ & + \widehat{Q}_{ss}(q,-k_{\rm sp}(\omega),k_{\rm sp}(\omega))|r\rangle_0\Big] + \frac{C^4(\omega)}{4\Delta_t(\omega)}\tau_\omega^{(MC)}(-k|k)|G(k,\omega)|^2 \\ & \times\big[\tau_\omega(k_{\rm sp}(\omega)|k)\langle|\widehat{Q}_{ss}(q,-k,k_{\rm sp}(\omega))|^2\rangle_0 \\ & + \tau_\omega(-k_{\rm sp}(\omega)|k)\langle|\widehat{Q}_{ss}(q,-k,-k_{\rm sp}(\omega))|^2\rangle_0\big] + \frac{C^4(\omega)}{8} \\ & \times\Big[\tau_\omega^{(MC)}(-k|k)\Big]^2|G(k,\omega)|^4\langle|\widehat{Q}_{ss}(q,-k,-k)|^2\rangle_0, \end{aligned} \tag{97b}$$

and

$$\begin{aligned}\chi_4^{MC}(q,k) = {} & \frac{1}{2}\frac{C^4(\omega)}{(q+k-k_{\rm sp}(\omega))^2+4\Delta_t^2(\omega)}\tau_\omega(k_{\rm sp}(\omega)|k)\mathcal{F}(q,k_{\rm sp}(\omega),k_{\rm sp}(\omega)) \\ & + \frac{1}{2}\frac{C^4(\omega)}{(q+k+k_{\rm sp}(\omega))^2+4\Delta_t^2(\omega)} \\ & \times\tau_\omega(-k_{\rm sp}(\omega)|k)\mathcal{F}(q,-k_{\rm sp}(\omega),-k_{\rm sp}(\omega)) + \frac{C^2(\omega)}{4\Delta_t(\omega)}\tau_\omega(q+k|k) \\ & \times\ [\mathcal{F}(q,q+k,k_{\rm sp}(\omega)) + \mathcal{F}(q,q+k,-k_{\rm sp}(\omega))]|G(q+k,\omega)|^2 \\ & + \frac{C^4(\omega)\Delta_t}{(q+2k)^2+4\Delta_t^2(\omega)}\mathcal{F}(q,-k,-k)|G(k)|^2. \end{aligned} \tag{97c}$$

The functions $\mathcal{F}(q,p,p')$ entering Eq. (97c) are given by

$$\mathcal{F}(q,p,p') = \langle \widehat{Q}_{ss}(q,p,p')t^*_\omega(q-p-p'+k|k)\rangle_0 \times \langle \widehat{Q}_{ss}(q,q+k-p-p'p)t_\omega(p'|k)\rangle_0. \tag{97d}$$

The function $\chi_4(q,k)$, describes the processes of nonlinear mixing in which at least one of the participating waves is the volume wave emerging into the vacuum after the scattering of surface plasmon polaritons of frequency ω by the surface roughness. In particular, the nonlinear mixing of the enhanced backscattered radiation with the enhanced backscattered surface plasmon polariton beam leads to the appearance of resonant peaks or dips when $q = -k \pm k_{\rm sp}(\omega)$ (see Figs. 14(a) and (b)). In a sense, these processes are analogous to the processes of the resonant mixing of the incident light with the surface plasmon polaritons of frequency ω which lead to the peaks/dips at $q = k \pm k_{\rm sp}(\omega)$ (Figs. 6–9). However, these peaks are weak and are displayed in the plots of Figs. 6–9, as well as in the experimental curves of [22], as a weak structure on the left and right shoulders of the rectangular distributions of the intensity in the angular ranges $-45^\circ < \theta_{\rm s} < -20^\circ$ and $20^\circ < \theta_{\rm s} < 45^\circ$. The function $\chi_4(q|k)$ also contains the nonlinear mixing of the backscattered waves of frequency ω (see Fig. 15(a)). It displays a peak or a dip in the retroreflection direction, since the waves of frequency ω scattered into the retroreflection direction are coherent. No surface plasmon polaritons of frequency 2ω participate in the formation of the peak/dip. The dip is clearly seen in the plots in Figs. 6–9. In addition, the last, and possibly the strongest, processes described by $\chi_4(q|k)$ are the processes of the nonlinear mixing of copropagating surface plasmon

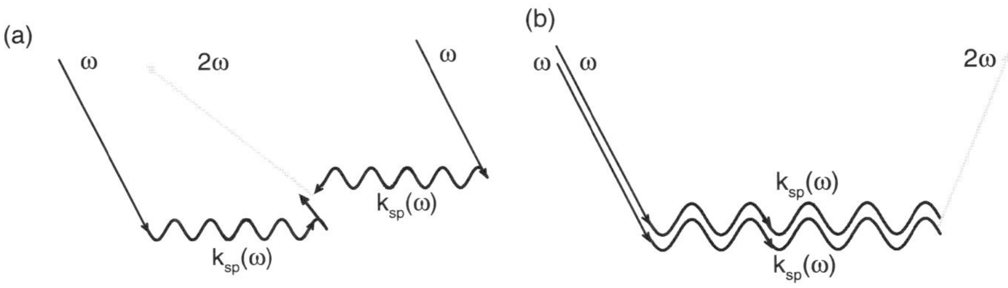

Fig. 15. The diagrams of the multiple scattering processes involving surface plasmon polaritons of frequency ω. The solid line arrows represent light of frequency ω, the wavy line arrows represent surface plasmon polaritons of frequency ω and the solid gray line arrows represent light of frequency 2ω.

polaritons of frequency ω (see Fig. 15(b)). The latter processes are nonresonant since the frequency dispersion of the dielectric function of the metal breaks the phase matching conditions. It should be pointed out that in the processes of the nonlinear mixing described by the function $\chi_4(q,k)$ only the multiply-scattered waves participate, and the resulting angular pattern of the intensity of the generated light does not depend on the angle of incidence. This is why the rectangular distributions of the intensity in the angular ranges $-45° < \theta_s < -20°$ and $20° < \theta_s < 45°$ in Figs. 6–9 do not move with an increase of the angle of incidence.

The nonlinear interaction of the multiply-scattered surface plasmon polaritons of frequency ω can also lead to the excitation of surface plasmon polaritons of frequency 2ω. These types of processes are of a higher order in the surface profile function, and are usually weak. What is more, the particular form of the power spectrum used in our calculations makes them even weaker or totally forbids them. However, the nonlinear mixing of copropagating surface plasmon polaritons of frequency ω can result in the excitation of surface plasmon polaritons of frequency 2ω on the rough surface. These processes can give quite a strong contribution to the intensity of the second harmonic radiation independent of the power spectrum. They are contained in the third term in Eq. (71) for the contribution $\tilde{\tau}_{2\omega}(q|2k)$ to the intensity of the scattered light and have the form

$$\begin{aligned}\tilde{\tau}^{(co)}_{2\omega}(q|2k) = \frac{1}{L_1}\int_{-\infty}^{\infty}\frac{dp}{2\pi}\int_{-\infty}^{\infty}\frac{dp'}{2\pi}\int_{-\infty}^{\infty}\frac{dr}{2\pi} \\ \times \langle t_{2\omega}(q|p)G(p,2\omega)G^*(q-p+r,\omega)t^*_{\omega}(q-p+r|r)\rangle \\ \times \langle t_{2\omega}(q|p')G^*(p',2\omega)G(q-p'+r,\omega)t_{\omega}(q-p'+r|r)\rangle \\ \times |G(p+p'-q-r,\omega)|^2\tau_{\omega}(p-r|k)|G(r,\omega)|^2\tau_{\omega}(r|k). \qquad (98)\end{aligned}$$

The most important contribution to the integral calculated in the pole approximation for the Green's functions has the form

$$\begin{aligned}\tilde{\tau}^{(co)}_{2\omega}(q|2k) = \sum_{\pm}|G(q\pm 2k_{\mathrm{sp}}(\omega),2\omega)|^2|G(q\pm 3k_{\mathrm{sp}}(\omega),\omega)|^2 \\ \times \langle t_{2\omega}(q|q\pm 2k_{\mathrm{sp}}(\omega))t^*_{\omega}(\pm k_{\mathrm{sp}}(\omega)|\pm k_{\mathrm{sp}}(\omega))\rangle \\ \times \langle t_{2\omega}(q|q\pm 2k_{\mathrm{sp}}(\omega))t_{\omega}(\pm k_{\mathrm{sp}}(\omega)|\pm k_{\mathrm{sp}}(\omega)) \\ \times \tau_{\omega}(\pm k_{\mathrm{sp}}(\omega)|k)\tau_{\omega}(\pm k_{\mathrm{sp}}(\omega)|k). \qquad (99)\end{aligned}$$

The factor $|G(q \pm 2k_{\rm sp}(\omega), 2\omega)|^2$ displays Lorentzian peaks or dips depending on the effective nonlinearities $\widehat{Q}_{ss}(q, \pm k_{\rm sp}(\omega), \pm k_{\rm sp}(\omega))$, at $q = \pm[k_{\rm sp}(2\omega) - 2k_{\rm sp}(\omega)]$. For the particular values of the dielectric functions $\epsilon(\omega)$ and $\epsilon(2\omega)$ assumed in our calculations, these peaks/dips occur at $\theta_{\rm s} = \pm 1.1^\circ$, that is in the vicinity of the direction normal to the surface. Possibly, the additional weak structure in the vicinity of the normal direction present in Figs. 6–9 can be attributed to this mechanism.

6. Conclusions

In summary, in this paper we have presented results of perturbative calculations of the second harmonic generation of light in reflection from a weakly rough random metal surface. We solved the linear problem of the scattering of light of the fundamental frequency ω, and used its solution to determine the surface nonlinear polarization at the harmonic frequency 2ω. In solving the scattering problem for the harmonic fields we used the nonlinear boundary conditions. The results obtained display a peak or a dip (at small angles of incidence) in the retroreflection direction and in the direction normal to the mean surface in the angular distribution of the intensity of the second harmonic light, depending on the model of the nonlinear response of the surface. The analysis of the linear and nonlinear processes that occur at a rough metal surface presented in this paper allowed us to separate the mechanisms leading to the experimentally observed features. Our analysis provided a complete explanation of all the features present in the angular dependence of the second harmonic generation intensity measured in the experiments of [21, 22].

The comparison with the experimental results of [22], where no distinct dip or peak was observed, except for a shallow minimum in the direction normal to the mean surface, suggests that an important role in the surface nonlinear polarization is played by the nonlinear constant $\chi^{\rm s}_{ztt}$. Thus, on the basis of the results of our analysis we can conclude that the Agranovich and Darmanyan model of the surface nonlinear polarization, rather than the free-electron model, should be used to describe the available experimental results. From our results it also follows that the nonlinear coefficient $\chi^{\rm s}_{zzz}$, which is the largest nonlinear coefficient in the free-electron model, should have a considerably smaller value, while $\chi^{\rm s}_{ztt}$ should have a nonzero value. We believe our analysis can provide a complete explanation of all the features present in the angular dependence of the second harmonic intensity measured in the experiments of [21, 22].

Acknowledgments

Disorder and randomness have played a central role in much of Roger Elliott's research throughout his distinguished career, and this has inevitably influenced the research interests of those of us who have been fortunate enough to work with him. It is therefore a pleasure for us to dedicate to Roger this example of where our interest in random systems has led us in recent years, in commemoration of his 75th birthday. We add our best wishes for many more years of good health, happiness, and creative energy.

The research of A.A. Maradudin and T.A. Leskova was supported in part by Army Research Office Grant No. DAAD 19-02-1-0256. The research of E.R. Méndez and M. Leyva-Lucero was supported by CONACYT Grant No. 3804P-A.

References

[1] Sipe, J.E. and Stegeman, G.I., in *Surface Plasmon Polaritons*, eds. Agranovich, V.M. and Mills, D.L. (North-Holland, Amsterdam, 1982), p. 661.
[2] Heinz, T., in *Nonlinear Surface Electromagnetic Waves*, eds. Ponath, H.-E. and Stegeman, G.I. (North-Holland, Amsterdam, 1991), p. 323.
[3] Farias, G.A. and Maradudin, A.A., *Phys. Rev. B* **30** (1984) 3002.
[4] Coutaz, J.L., Neviere, M., Pic, E. and Reinisch, R., *Phys. Rev. B* **32** (1985) 2227.
[5] Simon, H.J., Huang, C., Quail, J.C. and Chen, Z., *Phys. Rev. B* **38** (1988) 7408.
[6] Chen, C.K., de Castro, A.R.B. and Shen, Y.R., *Opt. Lett.* **4** (1979) 393.
[7] Deck, R.T. and Grygier, R.K., *Appl. Opt.* **23** (1984) 3202.
[8] McGurn, A.R., Maradudin, A.A. and Celli, V., *Phys. Rev. B* **31** (1985) 4866.
[9] West, C.S. and O'Donnell, K.A., *J. Opt. Soc. Am. A* **12** (1995) 390.
[10] Agranovich, V.M. and Grigorishin, K.I., *Nonlinear Optics* **5** (1993) 3.
[11] McGurn, A.R., Leskova, T.A. and Agranovich, V.M., *Phys. Rev. B* **44** (1991) 11441.
[12] Wang, X. and Simon, H.J., *Opt. Lett.* **16** (1991) 1475.
[13] Simon, H.J., Wang, Y., Zhou, L.B. and Chen, Z., *Opt. Lett.* **17** (1992) 1268.
[14] Aktsipetrov, O.A., Golovkina, V.N., Kapusta, O.I., Leskova, T.A. and Novikova, N.N., *Phys. Lett. A* **170** (1992) 231.
[15] Wang, Y. and Simon, H.J., *Phys. Rev. B* **47** (1993) 13695.
[16] Kuang, L. and Simon, H.J., *Phys. Lett. A* **197** (1995) 257.
[17] Bozhevolnyi, S.I. and Pedersen, K., *Surf. Sci.* **377–379** (1997) 384.
[18] Kretschmann, E., *Z. Physik.* **241** (1971) 313.
[19] Leskova, T.A., Leyva-Lucero, M., Mendez, E.R., Maradudin, A.A. and Novikov, I.V., *Opt. Commun.* **183** (2000) 529.
[20] Novikov, I.V., Maradudin, A.A., Leskova, T.A., Méndez, E.R. and Leyva-Lucero, M., *Wave. Random Media* **11** (2001) 1.
[21] O'Donnell, K.A., Torre, R. and West, C.S., *Opt. Lett.* **21** (1996) 1738.

[22] O'Donnell, K.A., Torre, R. and West, C.S., *Phys. Rev. B* **55** (1997) 7985.
[23] O'Donnell, K.A. and Torre, R., *Opt. Commun.* **138** (1997) 341.
[24] Leyva-Lucero, M., Méndez, E.R., Leskova, T.A., Maradudin, A.A. and Lu, J.Q., *Opt. Lett.* **21** (1998) 1809.
[25] Leyva-Lucero, M., Méndez, E.R., Leskova, T.A. and Maradudin, A.A., *Optics Commun.* **161** (1999) 79.
[26] Bloembergen, N., Chang, R.K., Jha, S.S. and Lee, C.H., *Phys. Rev.* **74** (1968) 813.
[27] Maystre, D. and Neviére, M., *Appl. Phys. A* **39** (1986) 115.
[28] Reinisch, R., Neviere, M., Akouayri, H., Coutaz, J.L., *Opt. Eng.* **27** (1988) 961.
[29] Agranovich, V.M. and Darmanyan, S.A., *JETP Lett.* **35** (1982) 80.
[30] Mendoza, B.S. and Mochán, W.L., *Phys. Rev.* **53** (1996) 4999.
[31] Mendoza, B.S. and Mochán, W.L., *Phys. Rev. B* **55** (1997) 2489.
[32] Leskova, T.A., Maradudin, A.A. and Méndez, E.R., in *Optical Properties of Nanostructured Random Media*, ed. Shalaev, V.M. (Springer, New York, 2002), p. 359.
[33] Shchegrov, A.V., Maradudin, A.A. and Méndez, E.R., *Progress in Optics* (to appear).
[34] Maradudin, A.A. and Méndez, E.R., *Appl. Opt.* **32** (1993) 3355.
[35] Lord Rayleigh, *The Theory of Sound*, 2nd edn. Vol. II (MacMillan, London, 1895), pp. 89 and 297.
[36] Petit, R. and Cadilhac, M., *C. R. Acad. Sci. Paris B* **262** (1966) 468.
[37] Hill, N.R. and Celli, V., *Phys. Rev. B* **17** (1978) 2478.
[38] Brown, G.C., Celli, V., Coopersmith, M. and Haller, M., *Surf. Sci.* **129** (1983) 507.
[39] Brown, G.C., Celli, V., Haller, M., Maradudin, A.A. and Marvin, A., *Phys. Rev. B* **31** (1985) 4993.
[40] Kubo, R., *Phys. Soc. Japan* **17** (1962) 1100.
[41] Johnson, P.B. and Christy, R.W., *Phys. Rev. B* **6** (1972) 4370.

CHAPTER 14

THEORY FOR LARGE-SCALE ELECTRONIC STRUCTURE CALCULATIONS: FRACTURE AND SURFACE RECONSTRUCTION OF SILICON

T. Fujiwara*, T. Hoshi* and R. Takayama*,†

*Department of Applied Physics, University of Tokyo, 7-3-1 Hongo, Bunkyo-ku, Tokyo 113-8656, Japan

†Research and Development for Applying Advanced Computational Science and Technology, Japan Science and Technology Agency, 4-1-8 Honcho, Kawaguchi-shi, Saitama 332-0012, Japan

Any physical quantity including the total energy can be rigorously expressed by the density matrix. We propose several novel methods, the Local Orbital-Order N formalism and the Krylov subspace method, to calculate the density matrix without obtaining eigenstates. Practically, these methods are of great use for large-scale electronic structure calculations and molecular dynamics simulation. We then propose the hybrid scheme combining these novel methods for quantum mechanical simulation. The hybrid method is applied to the problem of crack propagation on the (001) atomic plane of crystalline silicon of systems of more than 10^5 atoms. The simulation shows the two-stage process of the dynamical surface formation in fracture: electronic and elastic processes. In subnanoscale systems, steps of atomic planes are created systematically in order to release the atomic elastic energy loss. Systematic step formation can never be observed in an electronic structure molecular dynamics simulation without the present novel computational scheme.

1. Introduction

Accurate large-scale atomistic simulations are very important to investigate and predict various properties of materials. For this purpose, the early principle electronic structure theories have been extended to calculations of the total energy and forces. The first principle molecular dynamics (MD) simulation or the Car–Parrinello method [1] are now used quite widely in condensed matter physics. However, the systems investigated by the first principle MD simulations are practically limited to a much smaller size, at most, of hundreds of atoms and a much shorter time period of a few tens of pico-seconds. The other extreme is classical MD simulations with short-range inter-atomic potentials, which are applied to systems of millions or ten millions atoms with time periods of a few hundreds pico-seconds [2, 3]. Though classical MD simulations are very useful to investigate nanoscale systems with accurate inter-atomic potentials, the applicability of classical MD simulations is limited to phenomena in which the electronic process does not play an essential role.

Modern materials technology is deeply involved in electronic processes. Intense attention has been paid to the order-N method for electronic structure calculations, whose computational cost increases in *linear* proportion to the number of electrons [4, 5]. We have developed a novel order-N method on the basis of the Wannier states [6, 7]. The Wannier states are formally defined with the unitary transformation of the occupied eigen-states. We are also developing a different method based on the Krylov subspace, which provides an efficient way to extract the essential character of the original Hamiltonian within a limited number of basis sets [8].

In order to get physical quantities or execute a molecular dynamics simulation, one should obtain either eigen-states $|\psi_\alpha\rangle$ or the density matrix defined as

$$\rho = \sum_\alpha |\psi_\alpha\rangle\langle\psi_\alpha| f\left(\frac{\varepsilon_\alpha - \mu}{k_{\mathrm{B}}T}\right). \tag{1}$$

Here $f\left(\frac{\varepsilon_\alpha-\mu}{k_{\mathrm{B}}T}\right)$ is the Fermi–Dirac distribution function as a function of the eigen-energy ε_α of the eigen-states $|\psi_\alpha\rangle$, and the chemical potential μ of the system. The expectation value of any physical quantity X can be obtained with the density matrix as

$$\langle X\rangle = \mathrm{Tr}[\rho X] = \sum_{ij} \rho_{ij} X_{ji}. \tag{2}$$

Here i and j are suffices of atom sites and orbitals. The energy and forces acting on an atom are contributed only by elements having non-zero values of the Hamiltonian. Therefore, we only need elements of the short-ranged atomic pairs of the density matrix, and no other restriction such as the localization constraint is required.

In this paper we give a review of our recently proposed novel methods with MD simulation for the electronic structure calculations of large systems [6–8]. Examples will also be shown for the crack propagation and surface formation along the (001) plane of Si.

2. Generalized Wannier State

Pioneering work in the generalized Wannier states was done by Walter Kohn for large-scale electronic structure calculations [9, 10]. The generalized Wannier states $\{\phi_i\}$ are defined as localized wave functions that satisfy the equation

$$H|\phi_i\rangle = \sum_{j=1}^{\mathrm{occ}} \varepsilon_{ij}|\phi_j\rangle \tag{3}$$

with the orthonormalization relation $\langle\phi_i|\phi_j\rangle = \delta_{ij}$. The parameter ε_{ij} is the Lagrange multiplier and satisfies $\varepsilon_{ij} = \langle\phi_j|H|\phi_i\rangle$.

The solutions of Eq. (3) hold a freedom of the unitary transformation

$$|\phi_i'\rangle = \sum_{k}^{\mathrm{occ}} U_{ik}|\phi_k\rangle. \tag{4}$$

The Bloch states or the eigen-states of the Hamiltonian H are obtained once one fixes the freedom of the unitary transformation so that the matrix $\{\varepsilon_{ij}\}$ is diagonal.

The theory of the generalized Wannier states is closely related to the localized-orbital order-N formulation [11], and we derived an eigen-value equation for the generalized Wannier states [6, 7]:

$$H_{\mathrm{WS}}^{(i)}|\phi_i\rangle = \varepsilon_{ii}|\phi_i\rangle, \tag{5}$$

where

$$H_{\mathrm{WS}}^{(i)} \equiv H - \bar{\rho}_i\Omega - \Omega\bar{\rho}_i, \quad \Omega \equiv H - \eta, \tag{6}$$

$$\bar{\rho}_i \equiv \rho - |\phi_i\rangle\langle\phi_i| = \sum_{j(\neq i)}^{N} |\phi_j\rangle\langle\phi_j|, \tag{7}$$

and η is an arbitrary parameter chosen to be sufficiently high. This eigen-value problem corresponds to the variational procedure of a specified Wannier state (ϕ_i), while all other Wannier states ($\{\phi_k\}_{k\neq i}$) are fixed. If

$\bar{\rho}_i|\phi_i\rangle = 0$ is satisfied, Eq. (5) reduces to Eq. (3). Then the band structure energy is calculated as

$$E_{\rm bs} = \sum_i^{\rm occ} \langle\phi_i|H|\phi_i\rangle = \sum_i^{\rm occ} \varepsilon_{ii}, \tag{8}$$

where the summation runs over the occupied Wannier states. The locality of the Wannier states is satisfied by a choice of the starting localized wavefunction in an iterative procedure of solving Eqs. (5)–(7), which we call the variational method.

The Wannier states centered on the j-bond can be expressed as

$$|\phi_j\rangle = C_j^{(0)}|b_j\rangle + \sum_{i(\neq j)} C_j^{(i)}|a_i\rangle + \text{terms of more distant bond sites}, \tag{9}$$

where $C_j^{(0)}$ is the mixing coefficient of the central bonding orbital $|b_j\rangle$ and $C_j^{(i)}$ is that of the anti-bonding orbital $|a_i\rangle$ on the neighboring i-bond [6, 7]. The mixing of the bonding orbitals on the neighboring bonds is negligibly small due to orthogonality and completeness, because they contribute to other Wannier states. By the first-order perturbation theory, the coefficients $C_j^{(i)}$ for the first bond-step can be given by [6, 7]

$$\frac{C_j^{(i)}}{C_j^{(0)}} = \frac{\langle a_i|H|b_j\rangle}{\varepsilon_b - \varepsilon_a}, \tag{10}$$

and $\varepsilon_{b/a}$ is the energy of the bonding/antibonding orbital. We named this procedure the perturbation method. Therefore, there could be several different procedures to solve Eq. (5): (i) the exact diagonalization method, (ii) the variational method (iterative), and (iii) the perturbative method.

3. Krylov Subspace Method

The KS (Krylov subspace) method gives the mathematical foundation of many numerical algorithms such as the conjugate gradient method [12, 13]. Starting from a certain vector $|i\rangle$, a subspace of the original Hilbert space is generated by a set of vectors;

$$|i\rangle,\ \ H|i\rangle,\ \ H^2|i\rangle, \ldots,\ H^{\nu_{\rm K}-1}|i\rangle. \tag{11}$$

The subspace spanned by the basis vectors $\{H^n|i\rangle\}$ in Eq. (11) is generally called the Krylov subspace (KS) in mathematical textbooks [12]. We will denote the orthonormalized basis vectors in the KS as

$$|K_1^{(i)}\rangle (\equiv |i\rangle), \quad |K_2^{(i)}\rangle, \quad |K_3^{(i)}\rangle, \ldots, |K_{\nu_{\mathrm{K}}}^{(i)}\rangle. \tag{12}$$

The dimension of the KS is ν_{K}.

Information about H's external eigen-values tends to emerge long before the procedure is completed. So the procedure can be terminated in a finite number of steps. From the practical viewpoint of calculations, the procedure to create $H^n|i\rangle$ consumes most CPU time, so the number of bases in the KS (ν_{K}) should be chosen to be much smaller than that of the original Hamiltonian matrix. This drastic reduction of the matrix size or the dimension of the KS is a great advantage for practical large-scale calculations. We then denote the reduced Hamiltonian as $H^{K(i)}$ for the KS $\{|K_n^{(i)}\rangle\}$ [8].

In order to extract the desired density matrix, we diagonalize the reduced Hamiltonian matrix $H^{K(i)}$. Once obtaining the eigen-value $\varepsilon_\alpha^{(i)}$ and eigen-vector $|w_\alpha^{(i)}\rangle$ as

$$H^{K(i)}|w_\alpha^{(i)}\rangle = \varepsilon_\alpha^{(i)}|w_\alpha^{(i)}\rangle, \tag{13}$$

we then introduce the density matrix within the KS:

$$\rho^{K(i)} \equiv \sum_{\alpha}^{\nu_{\mathrm{K}}} |w_\alpha^{(i)}\rangle\langle w_\alpha^{(i)}| f\left(\frac{\varepsilon_\alpha^{(i)} - \mu}{k_{\mathrm{B}}T}\right). \tag{14}$$

An approximation is the replacement of the density matrix $\langle i|\rho|j\rangle$ by the KS $\langle i|\rho^{K(i)}|j\rangle$:

$$\langle i|\rho|j\rangle \Rightarrow \langle i|\rho^{K(i)}|j\rangle \quad \text{or} \quad \rho \Rightarrow \rho^{K(i)}. \tag{15}$$

The validity of this procedure can be mathematically proved [8].

While methods of the density matrix may not usually provide information about the energy spectrum of electronic structure, the Krylov subspace (KS) method can at the same time. To discuss the electronic spectra in the framework of the KS method, we introduce the Green's function $G_{ij}(\varepsilon)$:

$$G_{ij}(\varepsilon) = \left[(\varepsilon + \mathrm{i}\delta - H)^{-1}\right]_{ij}, \tag{16}$$

where δ is an infinitesimally small positive number. Since the replacement for the density matrix (15) is guaranteed, a similar replacement for the

Green's function is also allowed:

$$G_{ij}(\varepsilon) \Rightarrow G_{ij}^{K(i)}(\varepsilon). \tag{17}$$

The matrix elements of the Green's function in the KS is defined as

$$G_{in}^{K(i)}(\varepsilon) = \sum_{\alpha}^{\nu_{\mathrm{K}}} \frac{C_{\alpha i}^{*} C_{\alpha n}}{\varepsilon + \mathrm{i}\delta - \varepsilon_{\alpha}^{(i)}}. \tag{18}$$

The coefficient of $C_{\alpha i}$ is the expansion coefficient of the eigen-vector $|w_{\alpha}^{(i)}\rangle$ of $H^{K(i)}$ in terms of the basis $|K_n^{(i)}\rangle$:

$$|w_{\alpha}^{(i)}\rangle = \sum_{n=1}^{\nu_{\mathrm{K}}} C_{\alpha n}^{*} |K_n^{(i)}\rangle. \tag{19}$$

In fact, the Green's function $G_{ij}(\varepsilon)$ can be calculated with the Green's function $G_{in}^{K(i)}(\varepsilon)$ in the KS as

$$G_{ij}^{K(i)}(\varepsilon) = \sum_{n}^{\nu_{\mathrm{K}}} G_{in}^{K(i)}(\varepsilon) \langle K_n^{(i)} | j \rangle. \tag{20}$$

4. Hybrid Scheme

As another fundamental methodology for large-scale calculations, we developed the hybrid scheme within quantum mechanics [14, 15]. The basic idea is the following. The density matrix is decomposed into two partial matrices or "subsystems" each of which is constructed from several occupied wave functions. This decomposition of density matrix corresponds to the decomposition of the Hilbert space of occupied states. The *different* partial density matrices are solved by *different* methods. Each subsystem is obtained with a well-defined mapped Hamiltonian and a well-defined electron number [15]. The hybrid scheme can be applied to several systems with the combinations between (a) the diagonalization method and perturbative Wannier state methods, (b) the variational and perturbative Wannier state methods [14], and (c) the Krylov subspace method and the perturbative Wannier state method [15].

5. Linearity of the CPU Load and Parallelism

Parallel computation is important for large-scale calculations. A test calculation of the perturbative Wannier state method is carried out with up to 10^6 atoms [16] using the Message Passing Interface technique [17] and

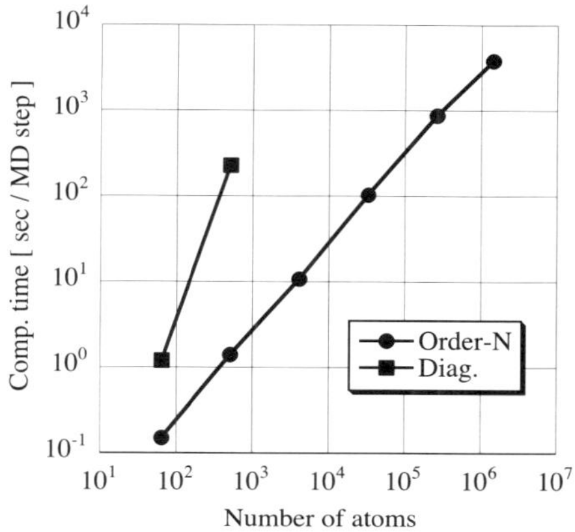

Fig. 1. The computational time for bulk silicon as a function of the number of atoms (N), up to 1,423,909 atoms, using one standard work station with one Pentium 4™ processor and 2 GB of RAM. The CPU time is measured for one time step in the molecular dynamics (MD) simulation. A tight-binding Hamiltonian is solved using the exact diagonalization method and an "order-N" method with the perturbative Wannier state.

with up to 10^7 atoms [15] using the Open-MP technique [18]. We are now developing the parallelization of other methods [8, 15].

Figure 1 shows the CPU time for an MD simulation of bulk silicon as a function of the number of atoms by using one CPU standard workstation for our Wannier state density matrix order-N method together with the standard matrix diagonalization method. The result shows that the computational cost of the order-N method is really linearly proportional to the system size.

6. Applications and Discussions

6.1. *Fracture propagation and reconstruction of the Si (001) surface*

The molecular dynamics simulation was performed for fracture of nanocrystalline silicon as a practical nanoscale application [14]. A standard workstation was used for the simulations with up to 10^5 atoms. Here we used the hybrid scheme of the variational Wannier state method and the perturbative Wannier state method.

In the continuum theory of fracture [19, 20], a critical crack length is defined by a dimensional analysis of the competitive energy terms of the bulk strain (3D) energy and the surface formation (2D) energy, which is independent of the sample size. In smaller systems of nanoscale size, the critical crack length is much larger than the system size. Then we can expect a crossover in fracture phenomena between macroscale and nanoscale samples; the fundamental process is governed by either the elastic energy balance as in the conventional fracture theory or other principles. Our simulation was done with the purpose of the investigation of the above crossover and the points of how and why the fracture path is formed and propagates in a specific crystallographic surface [21].

Dynamical fracture processes are simulated under external loads in the [001] direction. As an elementary process in fracture, we observe a two-stage surface reconstruction process. The process contains the drastic change of the Wannier states from the bulk (sp^3) bonding state to surface ones.

Figure 2 shows several snapshots of our simulation, in which the fracture propagates anisotropically on the (001) plane and reconstructed surfaces appear with asymmetric dimers [14]. An anisotropic bond-breaking propagation is seen in the [110] and [$1\bar{1}0$] directions, especially in the early snapshots. In the [110] direction, the successive bond breakings propagate along the *nearest neighbor* bond sites, which forms a zigzag path. A bond breaking process drastically weakens the *nearest neighbor* bonds, due to the local electronic instability. Therefore, the successive bond breakings propagate easily in the [110] direction. In the [$1\bar{1}0$] direction, on the other hand, the bond-breaking paths are not connected. In this direction, the bond

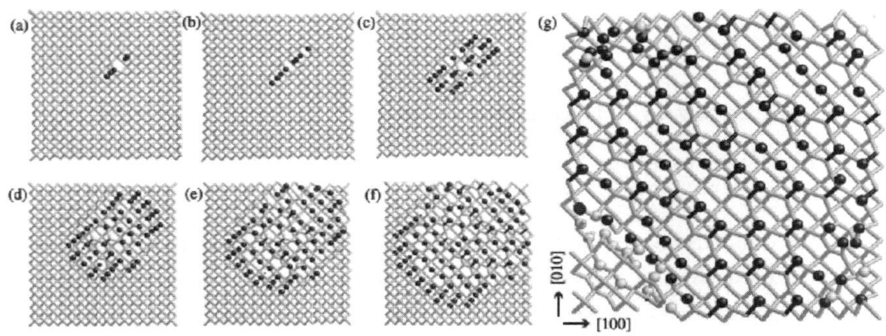

Fig. 2. Snapshots of a fracture process in the (001) plane. The sample contains 4,501 atoms and one initial defect bond as the fracture *seed*. The time interval between two successive snapshots is 0.3 ps, except between (f) and (g) (approximately 1.3 ps). A set of connected black rods and black balls corresponds to an asymmetric dimer.

breakings are propagated through the local strain relaxation, not by the local electronic instability.

In larger systems, step structures are formed so as to reduce the anisotropic surface strain energy within a flat (001) surface [14]. Such a step formation might be a beginning of a crossover between nanoscale and macroscale samples. Further investigation should be done for direct observation of the crossover. This crossover may indicate the crucial role of quantum mechanical freedoms in the fracture phenomena.

Another important issue is the origin why the easiest cleavage plane in macroscale samples of silicon is the (111) plane having a metastable 2×1 structure, instead of the ground-state (7×7) structure.

"Multiscale mechanics" is one of the most important current issues in material technology. The present simulation may be very important as a material simulation method, since the present MD simulation can reproduce phenomenon where quantum and classical mechanical effects appear in different ways in different scale length. The present work gives a guiding principle and a typical example for the concept, which is carried out by simplifying the total energy functional.

The present calculations are carried out using a tight-binding Hamiltonian within s and p orbitals. We should say that the applicability of the sp orbital model is rather limited due to the simplicity of the Hamiltonian. However, this parameter theory reproduces systematically essential results of *ab initio* calculations among different elements or phases, and we believe that the qualitative feature of the results of our simulation is not model-dependent. More sophisticated and practical tight-binding Hamiltonian can be systematically constructed through the *ab initio* theory [22].

6.2. *Energy spectrum of the asymmetric dimer on the Si (001) surface*

In order to single out the physical insight behind the asymmetric dimer of the Si (001) surface, we calculate the local density of states (lDOS) per atom of the system with reconstructed surface with dimer by using the Krylov subspace method and the results are shown in Fig. 3(a). The lDOS can be defined as

$$n_I(\varepsilon) = -\frac{1}{\pi} \sum_{\alpha} \operatorname{Im} G_{I\alpha, I\alpha}(\varepsilon) = \sum_{\alpha,\kappa}^{\nu_{\mathrm{K}}} |\langle I\alpha | w_{\kappa}^{(I\alpha)} \rangle|^2 \delta(\varepsilon - \varepsilon_{\kappa}^{(I\alpha)}), \tag{21}$$

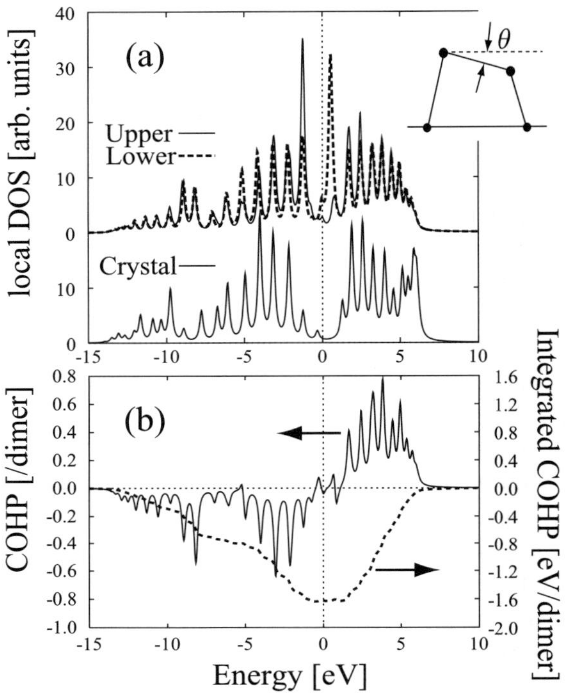

Fig. 3. (a) Local density of states (lDOS) per atom for the system with asymmetric dimer and that for the system of crystal. Solid line (broken line) in upper panel represents an upper (lower) atom of the asymmetric dimer. (b) COHP and integrated COHP for the corresponding dimer. The energy zeroth both in (a) and (b) are common and is set to be the top of the occupied states in the bulk. In order to show the structure we introduce finite imaginary part, $\delta = 0.136\,\text{eV}$, in the energy denominator of the Green function. The size of the reduced matrix is $\nu_{\text{K}} = 30$ and the temperature factor of the system in Eq. (14) is $T = 1580\,\text{K}$ $(= 0.136\,\text{eV})$. The chemical potential is estimated as $\mu = 0.126\,\text{eV}$.

where I and α are the atomic site and orbitals, respectively, and κ is suffix for eigen-states of the KS. First of all, we see the lDOS of crystal. Because of the finite number of computed levels, $\nu_{\text{K}} = 30$, the shown lDOS has thirty spikes with weight factor $|\langle I\alpha|w_\kappa^{(I\alpha)}\rangle|^2$ distributed from bottom to top of the band. Here we have an introduced finite imaginary part, $\delta = 0.136\,\text{eV}$ $(10^{-2}$ Ryd), to smooth out these spiky structures. The calculated lDOS of crystal reproduces the gap that lies within $0 \sim 1\,\text{eV}$ satisfactorily. The lDOS of the deeper layer of the present slab system is similar to this and does not change before and after the surface reconstruction as it should. In the lDOS for dimerized surface atoms, the lDOS of the upper (lower) atom has a peak at -1.25 ($+0.54$) eV in Fig. 3(a). The former (latter) peak corresponds to the occupied (unoccupied) surface state and the difference of the spectra represents the electron charge transfer from the lower atom to the upper atom in the asymmetric dimer.

The choice of $\nu_{\rm K}$ is important to reproduce the asymmetric dimer since the surface dimer reflects the electronic structure close to the chemical potential, in particular the occupied and unoccupied surface states. The size of the KS should be chosen to be large enough that the profile of the surface states is well reproduced. In fact, the calculation with $\nu_{\rm K} < 20$ leads to an unstable value of θ, for example, $\theta = 0.2, 9.8, 14.5, 4.6°$ for $\nu_{\rm K} = 15, 16, 17, 18$, respectively. While those with $\nu_{\rm K} > 25$ gives stable values, $13 \sim 14°$. Here we chose $\nu_{\rm K} = 30$.

To see the chemical bonding in condensed matter, we introduce the following quantity:

$$C_{IJ}(\varepsilon) = -\frac{1}{\pi} \sum_{\alpha,\beta} \mathrm{Im}\, G_{I\alpha,J\beta}(\varepsilon) H_{J\beta,I\alpha}, \tag{22}$$

which is sometimes called the crystal orbital Hamiltonian populations (COHP). The integration of this quantity gives the cohesive energy from a pair of atoms just as the integration of local DOS gives the occupation number. Actually, the total energy is decomposed into contributions of each atom pair as a sum of integration over the energy of C_{IJ}:

$$\mathrm{Tr}(\rho H) = \sum_{I,J} \sum_{\alpha,\beta} \rho_{I\alpha,J\beta} H_{J\beta,I\alpha} = \sum_{I,J} \int_{-\infty}^{\varepsilon_{\rm F}} C_{IJ}(\varepsilon)\, d\varepsilon. \tag{23}$$

The analysis of the COHP and the integrated COHP shows where and how the bond formation energetically stabilizes the system. The COHP for the dangling bond pair (for an ideal surface) is negligible (not zero), because the interaction matrix element $H_{J\beta,I\alpha}$ within the dangling bond pair is very small due to a larger interatomic distance. Once a surface dimer is formed, the interatomic distance is shortened and the COHP gives a finite value (Fig. 3(b)). The integration of the COHP has its minimum almost at the chemical potential.

References

[1] Car, R. and Parrinello, M., *Phys. Rev. Lett.* **55** (1985) 2471.

[2] Broughton, J.Q., Abraham, F.F., Bernstein, N. and Kaxiras, E., *Phys. Rev. B* **60** (1999) 239.

[3] Bachlechner, M.E., Omeltchenko, A., Nakano, A., Kalia, R., Vashishta, P., Ebbsjö, I. and Madhukar, A., *Phys. Rev. Lett.* **84** (2000) 322.

[4] Goedecker, S., *Rev. Mod. Phys.* **71** (1999) 1085.
[5] Ordejón, P., *Comp. Mat. Sci.* **12** (1998) 157.
[6] Hoshi, T. and Fujiwara, T., *J. Phys. Soc. Jpn.* **69** (2000) 3773.
[7] Hoshi, T. and Fujiwara, T., *Surf. Sci.* **493** (2001) 659.
[8] Takayama, R., Hoshi, T. and Fujiwara, T., *J. Phys. Soc. Jpn.* **73** (2004) 1519.
[9] Kohn, W., *Phys. Rev. B* **7** (1973) 4388.
[10] Kohn, W., *Chem. Phys. Lett.* **208** (1993) 167.
[11] Mauri, F., Galli, G. and Car, R., *Phys. Rev. B* **47** (1993) 9973; Ordejón, P., Drabold, D.A., Grumbach, M.P. and Martin, R.M., *Phys. Rev. B* **48** (1993) 14646.
[12] Golub, G.H. and Van Loan, C.F., in *Matrix Computations*, 2nd edn. (The Johns Hopkins University Press, Baltimore and London, 1989).
[13] Henk A. van der Vorst, in *Iterative Krylov Methods for Large Linear Systems* (Cambridge University Press, Cambridge, 2003).
[14] Hoshi, T. and Fujiwara, T., *J. Phys. Soc. Jpn.* **72** (2003) 2429.
[15] Hoshi, T., Takayama, R. and Fujiwara, T., in preparation.
[16] Geshi, M., Hoshi, T. and Fujiwara, T., *J. Phys. Soc. Jpn.* **72** (2003) 2880.
[17] http://www.mpi-forum.org/.
[18] http://www.openmp.org/.
[19] Griffith, A.A., *Philos. Trans. R. Soc. London Ser. A* **221** (1920) 163.
[20] As a textbook on fractures, see Lawn, B., *Fracture of Brittle Solids*, 2nd edn. (Cambridge University Press, Cambridge, 1993).
[21] See Ref. 14 and references therein.
[22] Andersen, O.K., Jepsen, O. and Glötzel, D., in *Highlights of Condensed Matter Theory* (North Holland, 1985).

CHAPTER 15

SYMMETRIC MAGNETIC CLUSTERS

J.B. Parkinson
Mathematics Department, UMIST, P.O. Box 88, Sackville St., Manchester M60 1QD, UK

R.J. Elliott
Theoretical Physics, 1 Keble Rd., Oxford OX1 3NP, UK

J. Timonen
Department of Physics, University of Jyväskylä, P.O. Box 35, FIN-40351 Jyväskylä, Finland

We study small clusters of magnetic atoms (spins) with an antiferromagnetic exchange interaction. If there is no frustration (paths of odd length) then the classical ground state is simple although the quantum ground state is not trivial. In the case of frustration the classical ground state is much more complicated and is well known only for very small clusters such as the triangle. In either case, if a magnetic field B is applied then the atoms gradually align with the field and become parallel to it at some critical field. We have found that in the frustrated case different "phases" can occur as B changes, characterized by different symmetry. The symmetry can be described in terms of the irreducible representations of the space group of the cluster. If the exchange interaction is XXZ-like with anisotropy $J_z/J_x \equiv q$, rather than Heisenberg, then a very rich "phase diagram" can occur for the frustrated systems, even for quite small clusters in terms of B and q.

1. Two Spins in a Magnetic Field

As an introduction let us consider the smallest possible cluster with just two spins, $\mathbf{s}_1$ and $\mathbf{s}_2$. We take these to have equal length S and assume that the interaction between them is isotropic Heisenberg exchange, described by the Hamiltonian

$$\mathcal{H} = J\mathbf{s}_1 \cdot \mathbf{s}_2.$$

The magnetic field $\mathbf{B}$ is in the z-direction, $\mathbf{B} = (0, 0, B)$. The energy E is the same as the Hamiltonian since we only consider static properties, so that

$$E = \mathcal{H} = J\mathbf{s}_1 \cdot \mathbf{s}_2 - B\left(s_1^z + s_2^z\right).$$

The exchange interaction is antiferromagnetic and we put $J = +1$. We now find the lowest energy (ground state in the quantum case) for a given J, B which is the actual state at $T = 0$.

Classically, we find the spins align at equal and opposite angles to the field, as shown in Fig. 1.

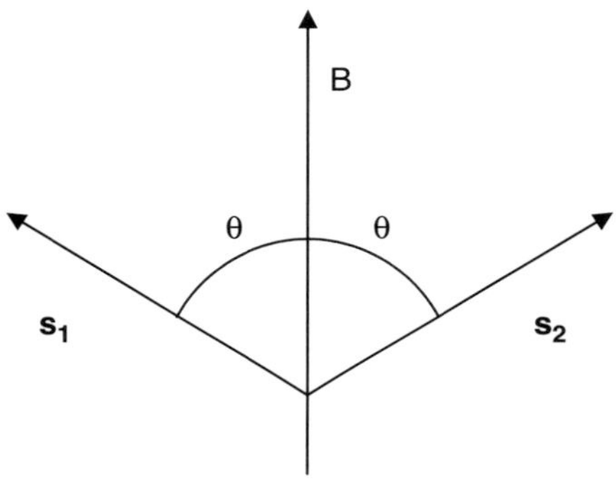

Fig. 1. Classical pair configuration in the presence of a magnetic field.

Clearly, the energy is $E = S^2 \cos(2\theta) - BS \cos\theta$, which is a minimum for $\cos\theta = B/S$. The magnetization $M \equiv s_1^z + s_2^z$ is $M = 2S\cos\theta = 2B$ with a maximum value $M_{\max} = 2S$ when $\theta = 0$. The fractional magnetization $m \equiv M/M_{\max} = B/S$. The magnetization curve is shown in Fig. 2.

Quantum mechanically, we first diagonalize $\mathcal{H}$. The magnetization M is a good quantum number with allowed values $-2S, -2S+1, \ldots, 2S$. We only need to consider the lowest state with a given M. As B varies these lowest states can cross so that the ground state becomes a state with a different magnetization, as shown in Fig. 3.

The quantum magnetization curve thus consists of a series of steps, shown in Fig. 2 for $S = 2$. As $S \to \infty$ the steps get more numerous

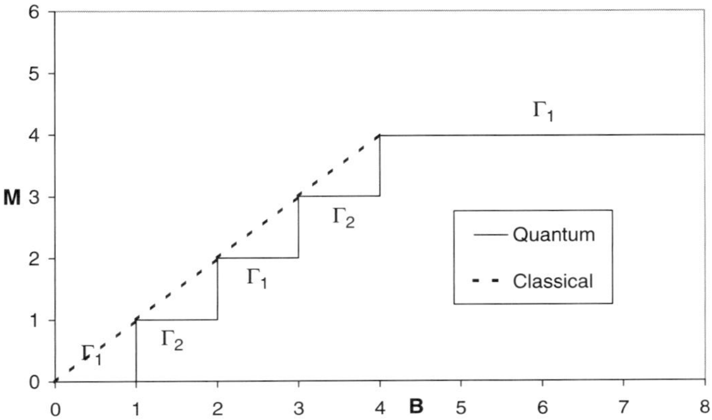

Fig. 2. Magnetization curve of a quantum pair of $S = 2$ atoms with isotropic exchange. Also shown is the corresponding classical curve. The Γ_i are the symmetries of the lowest states for each M.

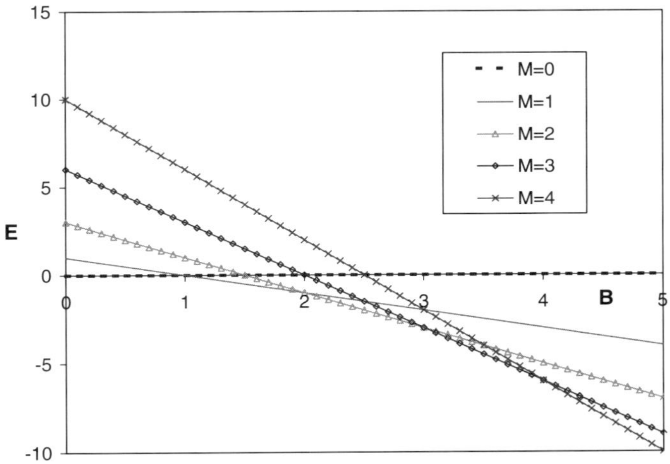

Fig. 3. Crossing of levels as a function of magnetic field in the quantum case (schematic).

and smaller, leading to the same smooth magnetization curve as the classical case.

We can introduce the symmetry in terms of the irreducible representations of the space group. Note that the operations of the space group only affect the positions of the spins, not their orientations. For the pair, the space group is C_{2v} with the following character table:

		Group operations	
	C_{2v}	E	C_2
Irreducible representations	Γ_1	1	1
	Γ_2	1	-1

In the classical case, if $\theta \neq 0$, the spins have different orientations so only the identity operation (E in the table) leaves the pair invariant. This leads to the regular representation $\Gamma_1 + \Gamma_2$ with characters:

	E	C_2
$\Gamma_1 + \Gamma_2$	2	0

For $\theta = 0$, the fully aligned state, the representation is Γ_1.

For the quantum case the representations of each step of the magnetization curve alternate between Γ_1 and Γ_2, as shown in Fig. 2, starting with Γ_1 for the fully aligned state. This indicates that in the limit $S \to \infty$ the representation is $\Gamma_1 + \Gamma_2$ as expected.

2. Triangle

For three spins of length S with isotropic Heisenberg exchange and a magnetic field B in the z-direction, the Hamiltonian is

$$\mathcal{H} = J(\mathbf{s}_1 \cdot \mathbf{s}_2 + \mathbf{s}_2 \cdot \mathbf{s}_3 + \mathbf{s}_3 \cdot \mathbf{s}_1) - B\big(s_1^z + s_2^z + s_3^z\big).$$

If $B = 0$ then the minimum energy of a given pair is when they are antiparallel. However, this is not possible for all three bonds and this situation is known as frustration, illustrated in Fig. 4.

For this isotropic Hamiltonian a factorization is possible:

$$\mathcal{H} = \frac{1}{2}J(\mathbf{s}_1 + \mathbf{s}_2 + \mathbf{s}_3)^2 - \frac{3}{2}JS^2 - B\big(s_1^z + s_2^z + s_3^z\big).$$

So if $B = 0$, any configuration with $\mathbf{s}_1 + \mathbf{s}_2 + \mathbf{s}_3 = 0$ has lowest energy. One possibility is shown in Fig. 5.

When the magnetic field is non-zero we find that the classical magnetization curve is a straight line given by $M = B$ for $0 \leq B \leq 3S$. The saturation magnetization $3S$ occurs for $B \geq 3S$ when all spins are pointing in the z-direction. The classical curve is shown in Fig. 6.

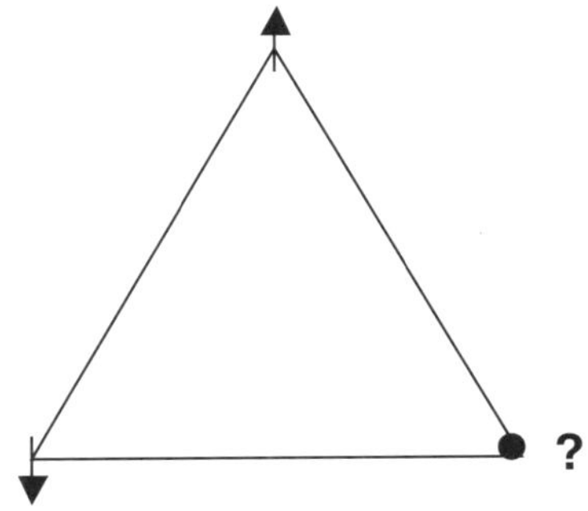

Fig. 4. Frustration in an antiferromagnetically coupled triangle.

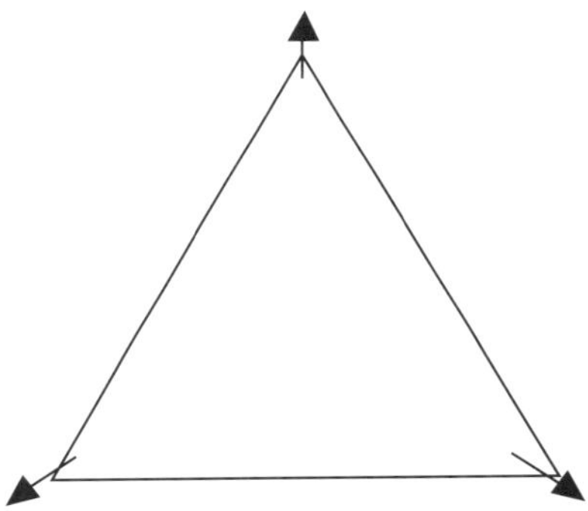

Fig. 5. One possible classical arrangement for an antiferromagnetic triangle.

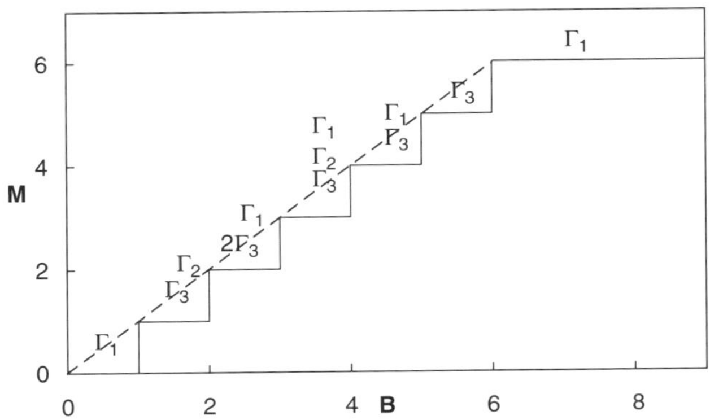

Fig. 6. Magnetization curves of quantum ($S = 2$) and classical triangle, showing symmetries of the quantum states.

The space group[a] is C_{3v} with character table:

C_{3v}	E	$2C_3$	$3\sigma_v$
Γ_1	1	1	1
Γ_2	1	1	-1
Γ_3	2	-1	0

The regular representation is $\Gamma_1 + \Gamma_2 + 2\Gamma_3$, which has characters:

	E	$2C_3$	$3\sigma_v$
$\Gamma_1 + \Gamma_2 + 2\Gamma_3$	6	0	0

Clearly, this is the representation for the case in which all spins are pointing in different directions so that only the identity E of the group operations leaves the configuration unchanged.

The factorization of the Hamiltonian is associated with a *hidden symmetry* such as is found in the Hamiltonians of the H-atom or the 3D harmonic oscillator. It results in increased degeneracy in the quantum states. In this case the degeneracy becomes very large as the length of the spin S increases. Nevertheless, we have been able to show that the dominant representation as $S \to \infty$ is the regular representation for $0 \leq B \leq 3S$.

The magnetization curve and the symmetries for $S = 2$ are shown in Fig. 6.

3. Four-Atom Ring (Square)

Here there is no frustration. In the absence of the field the atoms can arrange themselves so that each atom is antiparallel to its neighbors as shown in Fig. 7.

The Hamiltonian is

$$\mathcal{H} = J \sum_{i=1}^{4} \mathbf{s}_i \cdot \mathbf{s}_{i+1} - B \sum_{i=1}^{4} s_i^z$$

with $i + 4 \equiv i$.

[a]The atoms do not need to be equidistant as long as the interactions are equal nor does the plane of the triangle need to be at any particular angle to the direction of the magnetic field, the z-direction. Strictly speaking the group should be the permutation group rather than C_{3v} but it is convenient to use this to visualize the effects of the group operations.

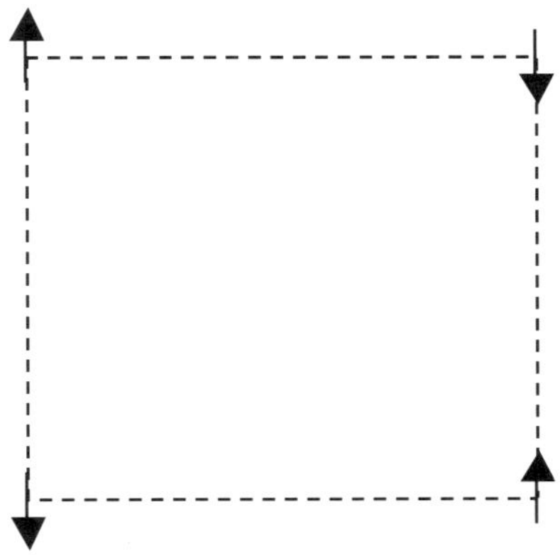

Fig. 7. Unfrustrated ring of four spins.

Again, factorization is possible:

$$\mathcal{H} = J\mathbf{t}_1 \cdot \mathbf{t}_2 - B\big(t_1^z + t_2^z\big),$$

where $\mathbf{t}_1 = \mathbf{s}_1 + \mathbf{s}_3$, $\mathbf{t}_2 = \mathbf{s}_2 + \mathbf{s}_4$. The hidden symmetry does result in extra degeneracy in the quantum states in this case but not in the ground state. This is presumably because there is no frustration.

Classically, we find that for all fields the atoms at opposite corners remain parallel and the system behaves exactly as the two-spin system discussed earlier.

The space group is C_{4v} with the following character table:

C_{4v}	E	C_2	$2C_4$	$2\sigma_v$	$2\sigma_d$
Γ_1	1	1	1	1	1
Γ_2	1	1	1	-1	-1
Γ_3	1	1	-1	1	-1
Γ_4	1	1	-1	-1	1
Γ_5	2	-2	0	0	0

With opposite spins parallel group operations E, C_2 and σ_d leave the configuration invariant, leading to the representation $\Gamma_1 + \Gamma_4$. In the quantum case, we find that the steps alternate between these Γ_1 and Γ_4, consistent with the classical picture.

Nevertheless, this four-atom ring is typical of the non-frustrated clusters in that the lowest state always has half the spins parallel in one direction and the other half parallel in another direction, the two directions corresponding to the directions of the spins in the pair.

4. Anisotropic Exchange

All the systems discussed so far show a very simple classical magnetization curve consisting of a straight line from $(B, M) = (0, 0)$ to $(B, M) = (B_c, M_{\max})$ where $B_c = NSz/2$ and z is the number of nearest neighbors of a given atom. However, in earlier work on larger clusters [1], we had observed that sometimes the curve was more complicated. In particular, for an icosahedral cluster with 12 atoms, there are two distinct parts to the curve, corresponding to configurations with different symmetries. We also found similar behavior for a 19-atom cluster with the FCC structure. On the other hand for some larger clusters such as the octahedron and 12-atom clusters with FCC and HCP structures there is just a single smooth curve.

To try to understand these structures we examined the symmetries of the low-lying eigen-states which give the stepped magnetization curve in the quantum case. This gives some information, but in order to study the transition to classical behavior one would like to examine quantum systems with large S. For a cluster of 12 or more atoms it is not feasible to obtain all the eigenstates for $S > \frac{3}{2}$ so there are not sufficient steps to fully categorize a section of the curve corresponding to a particular classical symmetry.

We have found, however, that much smaller clusters can show magnetization curves with regions of different symmetry, provided that the exchange interaction is allowed to be anisotropic. We have chosen an XXZ type of interaction, i.e., $J_x = J_y \neq J_z$ with the z-axis the direction of the magnetic field. Clearly, other forms of interaction would be possible and the direction of the anisotropy could be chosen to be in a different direction to the magnetic field, but we have not investigated this at present.

We now find that the behavior of the non-frustrated systems is still very similar to the two-spin system, the magnetization curve consisting of a straight line as before but with one additional step at $M = 0$ for $J_z > J_x$.

The frustrated systems, however, show quite interesting behavior. For example we show in Fig. 8 the magnetization curve for the triangle in the case $J_x = 0.7, J_z = 1.0$. There are clearly three different regions as well as the fully aligned state for large B. An additional region is obtained

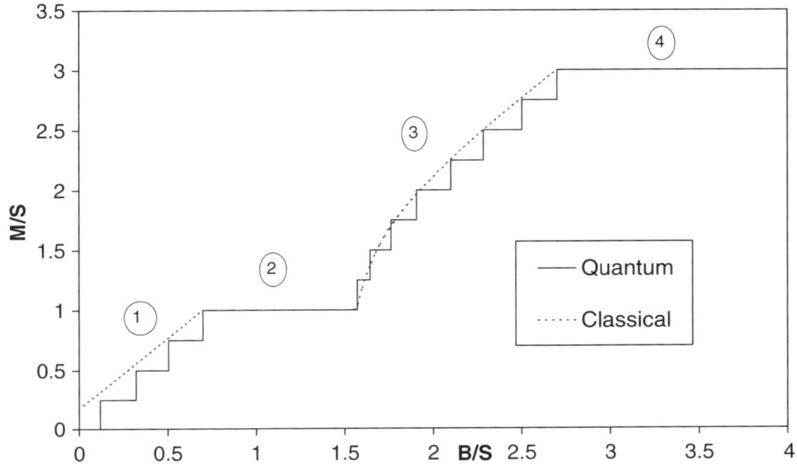

Fig. 8. Magnetization curve of the triangle for $J_z = 1, J_x = 0.7$ as a function of B, indicating the different regions. The stepped curve is the corresponding quantum curve for $S = 4$.

for $J_z < J_x$. Note that we refer to these regions as *phases* although of course these are not thermodynamic phases but rather regions of different symmetry at $T = 0$. By *phase change* we mean a transition between regions as the applied magnetic field B is varied.

The configurations in the different regions are given in the following table:

	Region 1		Region 2		Region 3		Region 4		Region 5	
Atom	θ	ϕ	θ	ϕ	θ	ϕ	θ	ϕ	θ	ϕ
1	θ_1	0	0	—	θ_1	0	0	—	θ_1	0
2	θ_1	π	0	—	θ_1	0	0	—	θ_1	$2\pi/3$
3	π	—	π	—	θ_2	π	0	—	θ_1	$4\pi/3$

The phase diagram is shown in Fig. 9.

Some of the phase boundaries can be calculated analytically, others are only known numerically. Note the vertical boundary at $J_x = J_z$. This is associated with the factorization of the Hamiltonian and the massive degeneracy in the quantum case.

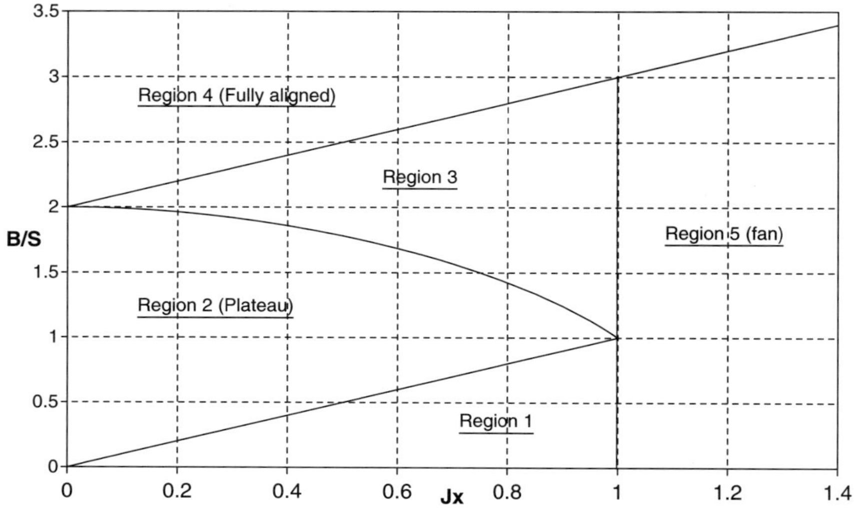

Fig. 9. The $T = 0$ phase diagram of the classical triangle with $J_z = 1$ as a function of B and J_x.

The classical symmetries can now be described in terms of the irreducible representations of the space group C_{3v}, given in the following table:

Region	Symmetry
1	$\Gamma_1 + \Gamma_2 + 2\Gamma_3$
2	$\Gamma_1 + \Gamma_3$
3	$\Gamma_1 + \Gamma_3$
4	Γ_1
5	$\Gamma_1 + \Gamma_2 + 2\Gamma_3$

These are consistent with the symmetries of the steps in the quantum curve for corresponding regions.

5. Five-Atom Ring

As a final example we show results for a five-atom ring with anisotropic exchange. The Hamiltonian does not factorize even for isotropic exchange so there are no massive degeneracies for this system. Classically, we find five distinct regions with orientation angles θ, ϕ shown in the following table:

	Region 1		Region 2		Region 3		Region 4		Region 5	
Atom	θ	ϕ	θ	ϕ	θ	ϕ	θ	ϕ	θ	ϕ
1	θ_1	0	0	—	θ_1	0	θ_1	0	0	—
2	θ_1	π	0	—	θ_1	0	θ_1	$4\pi/5$	0	—
3	θ_2	0	π	—	θ_2	π	θ_1	$8\pi/5$	0	—
4	0	—	0	—	θ_3	0	θ_1	$12\pi/5$	0	—
5	θ_2	π	π	—	θ_2	π	θ_1	$16\pi/5$	0	—

The classical phase diagram is shown in Fig. 10. Region 3 is a narrow strip between regions 2 and 4 from $J_x = 0$ to the point P_1 in the figure.

The space group is C_{5v} and the classical symmetries are summarized in the following table:

Region	Symmetry
1	$\Gamma_1 + \Gamma_2 + 2\Gamma_3 + 2\Gamma_4$
2	$\Gamma_1 + \Gamma_3 + \Gamma_4$
3	$\Gamma_1 + \Gamma_3 + \Gamma_4$
4	$\Gamma_1 + \Gamma_2 + 2\Gamma_3 + 2\Gamma_4$
5	Γ_1

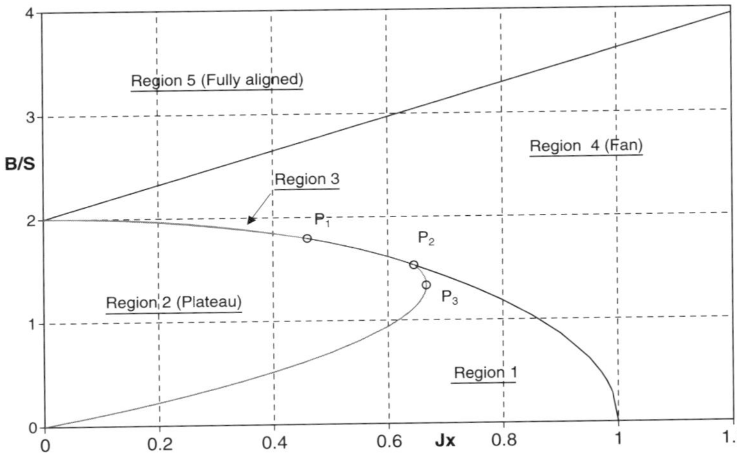

Fig. 10. The $T = 0$ phase diagram of the classical five-atom ring with $J_z = 1$ as a function of B and J_x.

6. Conclusion

We have shown that small clusters with antiferromagnetic nearest neighbor exchange and frustration can have interesting magnetization curves showing different phases with different symmetry at $T = 0$ as B is varied. If the exchange is anisotropic, phase changes can occur in clusters as small as the triangle. These phase changes occur in the classical systems and we observe the transition from quantum to classical behavior as $S \to \infty$.

The symmetry of the different phases is characterized by different representations of the space group of the cluster, i.e., different linear combinations of the irreducible representations. Two special cases are (a) if all spins point in the same direction we obtain the Γ_1 representation and (b) if all spins point in different directions we obtain the regular representation.

The use of the space group means that the description of the symmetry is not complete. For example, if a phase is such that the spins are all in different directions but with a uniform rotation from one to the next there is clearly additional symmetry which would need a spin-space group to describe correctly, and we have not yet been able to do this.

An additional feature is the hidden symmetry noted for the triangle and the four-atom ring (and also the tetrahedron, not discussed here [3]) in the case that the exchange is isotropic. We can see that this symmetry exists from the fact that the Hamiltonian in these cases factorizes in a simple way, resulting in large degeneracies. It is destroyed by making the exchange anisotropic and presumably by other changes to the Hamiltonian, but we still do not have a good understanding of this extra symmetry.

Finally, we note that the earlier work on somewhat larger clusters [1] can now be reassessed. We obtained the classical configurations for many of these including the octahedron, icosahedra with 12 and 13 atoms, FCC clusters with 12, 13 and 19 atoms, and HCP clusters with 12 and 13 atoms. We expect to be able to characterize the classical symmetries of these in terms of the irreducible representations of the space group, even though quantum results are only available for small values of S. In addition new phases may be expected for anisotropic exchange.

References

[1] Parkinson, J.B. and Timonen, J., *J. Phys.: CM* **12** (2000) 8669–8682.
[2] Parkinson, J.B., Elliott, R.J., Timonen, J. and Viitala, E., *J. Phys.: CM* **14** (2002) 45–58.
[3] Parkinson, J.B., Elliott, R.J. and Timonen, J., *J. Phys.: CM* **16** (2004) 2407–2419.

Final Note by J.B. Parkinson

Jussi Timonen and I had been working on the larger clusters for some time and we had noticed some phase changes. However, we did not know how to describe these in terms of symmetry changes. In April 1998 I was visiting Mike Thorpe at Michigan State University and Roger Elliott also visited. I described our work to him and he expressed some interest, but I thought no more about it. To my astonishment, in August I received a letter from him with all the details of the symmetry of the $S = \frac{1}{2}$ octahedron calculated by hand. I quote from his letter:

'With nothing better to do in the dog days of August I decided to work through the octahedral case. I was partly motivated by the fact that I have agreed to give my Group Theory lectures in the autumn to cover for someone's leave and I thought an "exercise for the student" would be good for me.'

This "exercise" showed how to include the symmetry for the first time and led to the detailed symmetry analysis presented above.

CHAPTER 16

OPTICAL AND FERMI-EDGE SINGULARITIES IN ONE-DIMENSIONAL SEMICONDUCTOR QUANTUM WIRES

K.P. Jain

Nuclear Science Centre, New Delhi — 110067, India

We present a simple many-body treatment of Fermi-edge singularities (FES) in a quasi one-dimensional quantum wire associated with the absorption and photoemission spectrum using the Fermi golden rule for the transition probabilities with approximate many-body wave functions. The functions are expressed in terms of Hartree–Fock determinants so that the problem of computing the transition probabilities reduces to calculating these determinants. The edge singularity exponents are related to the phase shifts of the scattering states at the Fermi surface, which depend substantially on the electron density. The essential result of this work is that it is possible to infer the edge singularity exponents of the infinite system from the size dependence of the many-body determinants.

1. Introduction

The fabrication of semiconductor nanostructures in one or two dimensions has led to the observation of new phenomena both in optics and in transport. The properties of doped semiconductor nanostructures, where electrons are confined in reduced dimensions, are radically different from those in the bulk. This is because elementary excitations, both single particle as well as collective and phase space filling, exhibit profound changes on account of reduced dimensionality. For instance, the dynamical behavior of an electron gas constrained to one-dimensional (1D) motion exhibits novel properties since the electron or hole can only scatter forwards or backwards. Of special significance is the appearance of sharp peaks at the

Fermi level in the optical spectra associated with FES and the behavior of 1D plasmons.

FES have recently been observed at low temperatures in photoluminescence excitation (PLE) experiments on doped GaAs and InGaAs quantum wires [1–3], with electron densities in the 1D quantum limit, where only the lowest 1D conduction subband is occupied. The modulation doped quantum well wires are difficult to produce due to patterning fluctuations in the 1D potential. These effects are also seen in 2D GaAs/AlGaAs and the modulation doped single quantum well [4]. This requires high electron density and mobility and gives a FES power law exponent for PLE ranging from 0.3 to 0.4. Another aspect of this is the non-Coulombian intersubband scattering, which can induce FES [5]. Here, the coupling of the electrons at the Fermi level to empty conduction subbands leads to the tunability of FES: the Fano effect enhances singularity effects. FES is the combined effect of both Coulombian many-body scattering processes and extrinsic effects, which include intersubband coupling and alloy disorder. More recently, FES effects have been studied in remotely doped AlAs/GaAs coupled quantum wire arrays [6]. Here again the role of extrinsic scattering, i.e., non-Coulomb FES effects in the optical spectrum of degenerate electron systems have been emphasized.

These singularities have also been observed in X-ray absorption in metals and in two-dimensional semiconductor quantum-well structures [7, 8]. However, these effects are expected to be much more pronounced in 1D as compared to 2D since for a 2D electron gas, the hole recoil process has a much greater phase space for the indirect transition, which smears out the Fermi distribution with sufficiently high electron concentration. In 1D, on the other hand, no indirect transitions are possible since there are only two electronic states $+k$ for each $E(k)$ and the hole recoil which cannot compensate the electron momentum is likely to be much less significant. The suppression of valence hole recoil processes in 1D is expected to enhance FES. With these qualitative remarks it is nevertheless clear that the effect of a finite mass hole is an important factor and must be taken into account in a quantitative description of the FES. Another factor is the role played by disorder, impurity scattering and hole localization in the robustness of the FES against low-energy single-particle or collective excitations of the Fermi sea [9, 10].

The FES, originally proposed by Mahan [11] for X-ray absorption in metals and later refined by Nozières and Dominicis [12] is due to many-body absorption transitions in which the Fermi sea responds collectively and dynamically to the sudden switching on of the localized hole potential. The response of the Fermi sea depends upon two competing effects: Anderson's orthogonality catastrophe [13] and final state electron–hole interaction (vertex corrections or excitonic effects). The former effect, which corresponds to a direct transition of the electron above the Fermi sea, and the concomitant re-arrangement of the Fermi sea electrons to new states, tends to suppress the absorption near the edge. The latter is due to transitions mediated by electrons inside the Fermi sea which enhance it, so that the power-law divergence is a result of the two. In the rigid Fermi sea approximation where the dynamical response of the Fermi sea electrons is neglected, the interaction between the hole and rigid Fermi sea electrons leads to a bound state (Mahan exciton) below the Fermi energy. The dynamic response of the Fermi sea to the hole potential softens the Mahan exciton, leading to a power-law singularity.

Here we present a simple many-body treatment of the edge singularity in a quasi-dimensional quantum wire associated with the absorption and photoemission spectra using the Fermi golden rule for the transition probabilities with appropriate many-body wave functions. These functions are expressed in terms of Hartree–Fock determinants so that the problem of calculating the transition probabilities reduces to numerically calculating a sufficient number of determinants. A random-phase approximation (RPA) screened Coulomb interaction is used for the electron–hole potential, which determines the single particle properties such as the bound state and the phase shifts of the scattering states. Both these depend substantially on the electron density. The hole is taken to be localized at a point and has an infinite mass. The spectra can now be calculated as indicated above, where the electron–hole interaction plays the crucial role, as expected. In the presence of a bound state, a second threshold corresponding to the final state in which bound state is not occupied is resolved, as originally surmised by Nozières and Combescot.

One of the most important aspects of this work is that it is possible to explicitly determine the exponent of the singularities for $N \to \infty$ from a finite N calculation by studying the N dependence of the projection determinants. We therefore proceed in the same way as Anderson in his classic paper [13], who predicted the orthogonality catastrophe by investigating N dependence of a very similar determinant. The singularity exponents have been determined as a function of the electron density [14].

Clearly, it is of some importance to understand the dependence of the singularity exponents for primary and secondary thresholds on the electron density. Our results are in quantitative agreement with Hopfield's thumb rule [15]: the exponents are given by $1 - n_f^2$, where n_f is the excess localized charge near the excited core in the ground state and is related to the scattering phase shifts by the Friedel sum rule. The primary threshold of the FES becomes sharper as the electron density decreases, finally going over into the exciton profile in the insulating limit. The secondary threshold, on the other hand, also evolves continuously towards the continuum absorption as the electron density is reduced.

2. Theoretical Background

2.1. *Many-body aspect*

The first task is to define many-electron wave functions for processes in which there is a transition of an electron from the valence band of a semiconductor in the presence of a Fermi sea in the conduction band. The many-body wavefunction of the ground state $|\Phi_I\rangle$ is constructed from the N_f free single-particle states with a k-vector lower than the Fermi wave vector k_f so that

$$|\Phi_I\rangle = \prod_{k<k_\mathrm{f}} c_k^+ |0\rangle, \tag{1}$$

where $c_k^+(c_k)$ denotes the creation (annihilation) operator of a conduction electron with energy $(m/m^*)(ka_0)^2 R_y$ (with $Ry = e^2/2a_0$ and m^* being the effective mass of the conduction electrons) and where, furthermore, $|0\rangle$ denotes the vacuum state consisting of a filled valence band and an empty conduction band. The final state may be written in terms of a new set of creation operators, labelled by the quantum number $|\lambda\rangle$:

$$\mathrm{c}_\lambda^+ = \sum_{k=1}^{N} \psi_{k,\lambda} c_k^+, \tag{2}$$

which takes into account the effect of the hole in the valence band on the conduction electrons. Here N is the number of states. Assuming that the hole is fixed at the lattice site 0, taken as the origin, we can calculate the coefficients $\Psi_{k,\lambda}$ by introducing a perturbing electron–hole potential $V(k)$ into the single particle Schrödinger equation:

$$(ka_0)^2 \psi_{k,\lambda} + \frac{1}{Na_l} \sum_{k'} \frac{m^*}{m} V(k-k') \psi_{kl,\lambda} = \frac{m^*}{m} E_\lambda \psi_{k,\lambda}. \tag{3}$$

Here a_l is the lattice constant. Thus the wave function of the final state reads

$$|\Phi_F'\rangle = c_v(0)|\Phi_F\rangle = c_v(0) \prod_{\lambda(F)} c_\lambda^+ |0\rangle, \tag{4}$$

where $c_v(0)$ is the annhilation operator of an electron in the valence band, at the lattice site 0. The absorption (ABS) and photoemission (PES) spectrum can be calculated by Fermi's golden rule, so that for PES one gets

$$\rho(\omega)\,\Delta\omega = 2\pi\,|\mu|^2 \sum_{F(\Delta\omega)} |\langle\Phi'_F|c_v(0)|\Phi_I\rangle|^2 \tag{5}$$

and for ABS

$$\alpha(\omega)\,\Delta\omega = 2\pi\,|\mu|^2 \sum_{F(\Delta\omega)} \left|\langle\Phi'_F|c_c^+(0)\,c_v(0)|\Phi_I\rangle\right|^2, \tag{6}$$

where μ is the dipole transition matrix element. The sum runs over all final states with energy $E_F - E_I$ lying in the interval ω and $\omega + \Delta\omega$.

Our approach is based on rewriting the matrix elements between the many-body states in Eq. (5) and Eq. (6) as determinantal quantities, which can be done easily by using Eqs. (1) and (4) and standard commutation relations between the creation and destruction operators. Thus we obtain for the PES

$$\langle\Phi'_F|c_v(0)|\Phi_I\rangle = \langle\Phi_F|\,\Phi_I\rangle = \mathrm{Det}\begin{Vmatrix} \psi_{k_1,\lambda_1^F} & \cdots & \cdots & \cdots & \psi_{k_1,\lambda_{N_f}^F} \\ \cdots & \cdots & \cdots & \cdots & \cdots \\ \cdots & \cdots & \cdots & \cdots & \cdots \\ \cdots & \cdots & \cdots & \cdots & \cdots \\ \psi_{k_f,\lambda_1^F} & \cdots & \cdots & \cdots & \psi_{k_f,\lambda_{N_f}^F} \end{Vmatrix} \tag{7}$$

and for ABS

$$\begin{aligned}\langle\Phi'_F|\,c_c^+(0)\,c_v(0)\,|\Phi_I\rangle &= \langle\Phi_F|\,c_c^+(0)\,|\Phi_I\rangle \\ &= \mathrm{Det}\begin{Vmatrix} \psi_{\lambda_1^F}(0) & \cdots & \cdots & \psi_{\lambda_{N_{f+1}}^F}(0) \\ \psi_{k_1,\lambda_1^F}(0) & \cdots & \cdots & \psi_{k_1,\lambda_{N_{f+1}}^F} \\ \cdots & \cdots & \cdots & \cdots \\ \cdots & \cdots & \cdots & \cdots \\ \psi_{k_f,\lambda_1^F}(0) & \cdots & \cdots & \psi_{k_f,\lambda_{N_{f+1}}^F} \end{Vmatrix},\end{aligned} \tag{8}$$

where

$$\psi_\lambda(0) = \frac{1}{\sqrt{N}} \sum_{k=1}^{N} \psi_{k,\lambda} \tag{9}$$

Note that the determinant in Eq. (7) is a minor of the determinant in Eq. (8). The latter possesses one more column than the former due to the additional electron from the valence band and one more row which is furnished by the electron probability amplitude $\psi_\lambda(0)$ at the hole site. The

ABS and PES can now be determined from Eqs. (5) and (6) by numerically calculating a sufficient number of determinants, Eqs. (7) and (8) whose elements $\psi_{k,\lambda}$ are provided by the numerical solution of Eq. (3). The latter equation marks the point where the basic assumptions of our physical model enter, and shall be discussed in more detail now.

2.2. *Single-particle aspects*

It is now necessary to choose an appropriate electron–hole interaction potential that describes the essential physics of the ID quantum wire, which we assumed to have a finite width. A harmonic confinement model [16, 17] is used for the Q1D wire where the electrons in a zero thickness xy-plane are confined in the y-direction by a harmonic potential while being free to move in the x-direction. The actual 1D electron–hole potential is then obtained by calculating the matrix elements of the 2D potential between the wave functions of the harmonic confinement potential. For the one subband system considered here, only the ground state wave functions $\sim \exp(-y^2/2b^2)$, with b being the wire width, are necessary. Hence, the unscreened potential $V^{\mathrm{us}}(q)$ of a finite width wire is

$$V^{\mathrm{us}}(q) = -\frac{e^2}{\varepsilon_{\mathrm{s}}} B(bq), \tag{10}$$

where

$$B(bq) = e^{b^2q^2/4} K_0(b^2q^2/4). \tag{11}$$

K_0 is the modified Bessel function of the second kind and ε_{s} the effective dielectric constant of the system.

We now introduce screening by taking the electronic polarizability $\chi(q)$ in the random-phase approximation (RPA) so that

$$\chi(q) = \frac{m^*}{\pi \hbar^2 q} L(q) = \frac{2m^*}{\pi \hbar^2 q} \log \left| \frac{q + 2k_{\mathrm{f}}}{q - 2k_{\mathrm{f}}} \right| . \tag{12}$$

We are thus led to the dielectric screening function [17]:

$$\varepsilon(q) = 1 + \chi(q) V^{\mathrm{us}}(q) = 1 + \frac{1}{a_{\mathrm{B}} q \pi} B(bq) L(q) , \tag{13}$$

where $a_{\mathrm{B}} = \varepsilon_{\mathrm{s}} \hbar^2 / m^* e^2$ is the radius of the first Wannier exciton and the electron–electron interaction is assumed to be equal to $-V^{\mathrm{us}}$. The screened electron–hole potential in Eq. (3) becomes $V(q) = V^{\mathrm{us}}(q)/\varepsilon(q)$. The sum over k in Eq. (3) is carried out by introducing a cut-off so that $\langle 0|k| \leq K(a_l = \pi/K)$ and a step width in k-space given by $\Delta k = 2K/N$.

To actually calculate the single particle energy spectrum from Eq. (3) a choice for the parameters a_{B} and b has to be made. Bearing in mind GaAs quantum wires we take $b = 800a_0$ and $a_{\mathrm{B}} = 270a_0$.

We have found that a cut-off of k at $K = 0.03\pi/a_0$ and a number of $N = 1000$ yielded a good convergence for the lowest eigenvalues and eigenstates of Eq. (3). There are two types of solutions of Eq. (3) corresponding to both bound and scattering states which will be considered in turn. Figure 1 shows the bound state energy as a function of the Fermi energy obtained by numerically solving Eq. (3). The range considered is typical for doped quantum wires in the extreme quantum limit in which the only lowest 1D subband is occupied by electrons ($k_{\mathrm{f}}a_{\mathrm{B}} \sim 0.2$ in [1]). The energies $m/m^* E_\lambda$, are given in units of $m^*/mE_{\mathrm{ex}} = (a_0/a_{\mathrm{B}})^2 Ry$, the binding energy of the first Wannier exciton of the system.

In the 1D case there always exists a bound state, irrespective of the form of the potential, as opposed to the 3D systems where its existence depends on the strength of the potential. We see from Fig. 1 that it shows a pronounced dependence on the carrier density. In the region $k_{\mathrm{f}}a_{\mathrm{B}} < 0.2$, it rapidly approaches the bottom of the conduction band. The efficiency of Fermi-sea electrons in screening the electron–hole interaction can once

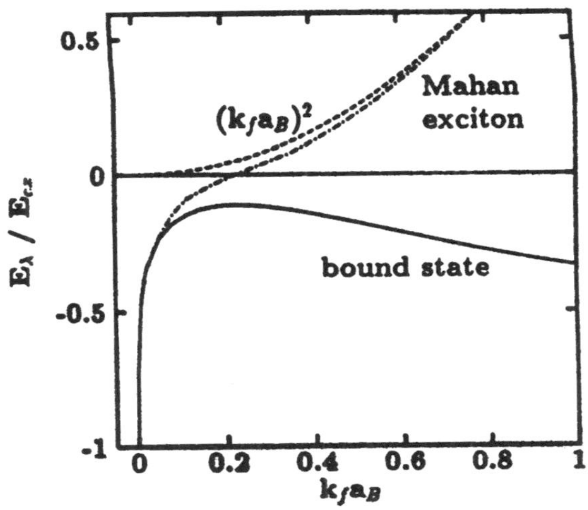

Fig. 1. Mahan exciton and bound state of the single-particle energy spectrum, Eq. (3), as a function of the Fermi energy in a range typical for doped 1D GaAs quantum wires. Energies and k_{f} are given in units of the binding energy E_{ex} and the Bohr radius a_{B}, respectively, of the first Wannier exciton.

more be seen from the fact that the higher Wannier states are completely screened out in the range considered and can only be observed in the very low-density regime $k_f a_B < 0.001$.

The binding energy of the bound state reveals somewhat unexpected behavior for $k_f a_B > 0.4$ where it increases again when one would expect a monotonic decrease (or at least saturation behavior) with increasing k_f because it is not obvious why a higher carrier density should lead to less efficient screening. However, this effect can be traced to the complicated behavior of the polarization function $\chi(q)$ in Eq. (12) as k_f is varied. We presume that the increase of the binding energy as k_f increases as shown in Fig. 1 is an artefact due to breakdown of the RPA.

Besides the bound state, the energy spectrum of Eq. (3) possesses $N-1$ scattering states with $E_\lambda = 0$. In Fig. 2 we have plotted the phase shifts δ_λ as a function of ka_B for four different Fermi energies. From these four curves, the phase shift δ_f at $k = k_f$ is of particular interest for the following. We therefore added another graph which connects these points and shows δ_λ as a function of $k_f a_B$. To provide a reference curve for the latter graph we

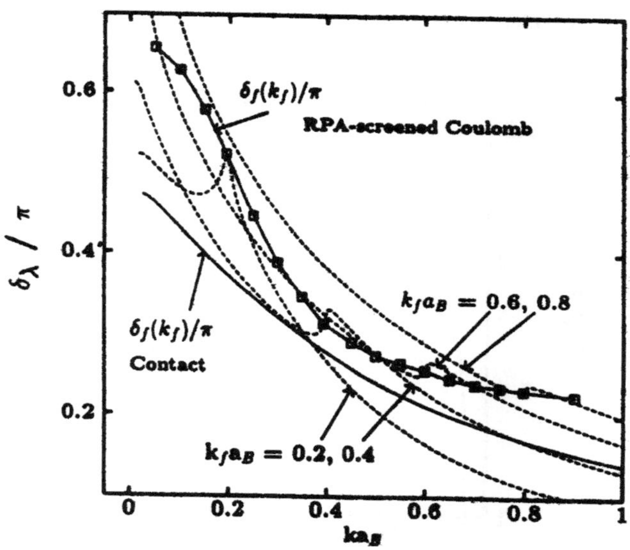

Fig. 2. (i) Phase shifts for the scattering states of Eq. (3) at four electron densities corresponding to $k_f a_B = 0.2, 0.4, 0.6, 0.8$ (dashed curves). (ii) Phase shifts δ_f the Fermi surface as a function of k_f (calculated values marked by empty boxes and connected with solid line). Each dashed curve intersects the δ_f-*curve* at its respective k_f value. (iii) Phase shifts δ_f for an on-site potential (simple solid line).

have performed another calculation using a simple contact potential (setting $m^*/mV(q) = \text{const.} = U$) instead of the screened Coulomb potential. The potential parameter U was chosen to give a comparative bound state energy of $E_\lambda/E_{\text{ex}} = -0.2$.

Comparing the different $\delta_{\lambda(k)}$ curves, one observes that, for a fixed $k, \delta_{\lambda(k)}$ tends to increase with growing k_{f}. This is due to the fact that the screening of the electron–hole potential for very low k values becomes less efficient. That the δ_{f} curve is falling in spite of this tendency originates in the simple fact that with greater k_{f} the distance to the low k values, being mostly affected by the perturbation of the hole, increases. It is interesting to note that although the curves of the phase shifts for a particular k_{f} have some structure, we obtain a smooth and structureless δ_{f}-curve being not very different from that obtained for the simple contact potential, though we failed to find a value for the parameter U resulting in better agreement of both curves than that of Fig. 2.

3. Results

3.1. *Fermi-edge singularity exponent*

We are now in a position to calculate the ABS and PES from Eqs. (5) and (6) with the determinants (7) and (8). The results are given in Fig. 3, which displays both spectra for $k_{\text{f}}a_{\text{B}} = 0.6$ and $N = 500$ plotting against $k \sim \omega^{1/2}$ rather than ω to reveal FES with clarity.

Consider the non-interacting limit when the electron–hole interaction is switched off. We get a series of equidistant peaks of the height $1/N$ for the ABS starting at $ka_{\text{B}} = 0.6$, reminiscent of a step function in the $N \to \infty$ limit. For PES one gets a single peak of height one at $ka_{\text{B}} = 0$, since the only non-vanishing determinant describes the transition to the state $|\Phi_F\rangle = |\Phi_I\rangle$.

Switching on the electron–hole interaction in Eq. (3) has a dramatic effect on the spectra. In absorption, one observes a considerable increase of oscillator strength at k_{f}. Obviously, transitions to the states near the Fermi surface become more favorable, at the expense of transitions to the states $ka_{\text{B}} > 0.8$. For the limit $N \to \infty$ this edge structure evolves into the FES. In systems having a bound state, there exists a second threshold corresponding to final states in which the bound state is not occupied. This was first pointed out by Combescot and Nozières [7]. For absorption this second threshold has a very low transition probability and is consequently not resolved in Fig. 3. However, it can be clearly seen in the PES. The

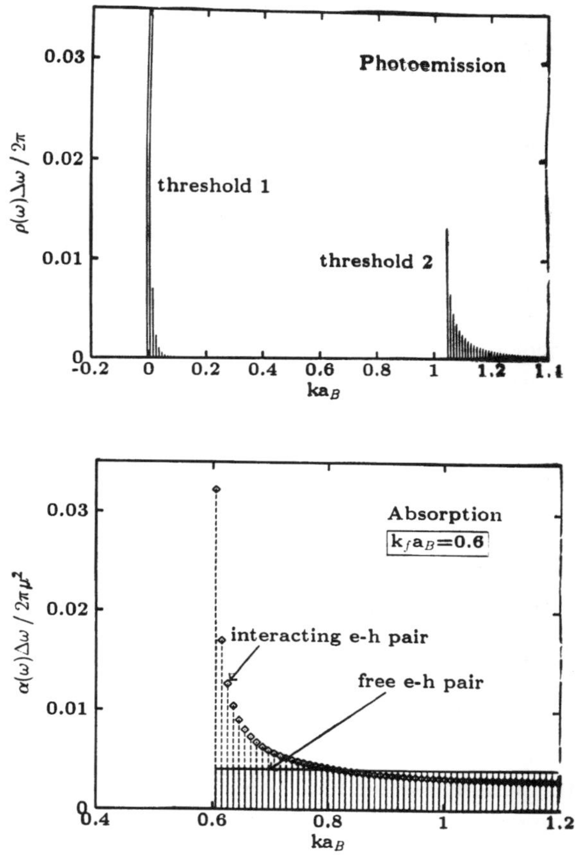

Fig. 3. Photoemission and absorption spectrum for a QID GaAs-like system with an electron density corresponding to $k_{\rm f}a_{\rm B} = 0.6$, calculated by evaluating determinants of the form of Eqs. (7) and (8), for a system of $N = 500$ single-particle states with properties depicted in Figs. 1 and 2. Peak structure near the Fermi edge (at $k_{\rm f}a_{\rm B} = 0.6$ in absorption) evolves into FES for $N \to \infty$.

energy difference between both thresholds is the sum of Fermi energy and the binding energy of the bound state, i.e., just the energy needed to take one electron from the bound state to the first level above the Fermi surface.

To satisfy the sum rules it is necessary to calculate the determinants corresponding to all different final states, clearly an impossible task. Fortunately, not every excitation is equally important. There are three sorts of excitations to be distinguished. In absorption these are, first, states of the form $|\Phi_F\rangle = c^+_{\lambda\nu}\,|\Phi^0_F\rangle$, where only the valence electron is excited and the remaining N_f electrons are in the ground state $\left|\Phi_F^0\right\rangle$ of the system with hole. Second, excitations leaving just one hole in the Fermi sea (Auger-like process), and, finally, all the rest, i.e., excitations with more than one hole in the Fermi sea (which we call simply multiple particle excitations). For the spectra of Fig. 3, we only took the valence electron and the Auger-like processes into account. The higher excitations die out very near the edge so that the spectrum is exclusively governed by valence electron excitations. That means that although the overall accuracy of the spectrum certainly suffers from neglect of the higher excitations, the very first part of the spectrum near the edge is exact even if one considers valence electron excitations only.

At this point we focus on the main question addressed in this paper: a calculation of the FES exponents. The idea is that if for any given N the first transition in a spectrum is accurately calculated as far as the higher excitations of the Fermi sea are concerned, it should also provide a reasonable approximation for the $N \to \infty$ spectrum near k_f, which consequently should be obtainable from the first transitions plotted as a function of N. This is done in Fig. 4 for different N between $N = 300$ and $N = 1,200$. Since $\Delta\omega \sim \omega^{1/2}\ \Delta k \sim \omega^{1/2}/N$ have made the spectra comparable to each other by multiplying Eq. (6) by N. We see that due to the decrease of the energy quantum with expanding system size, the onset of the spectra

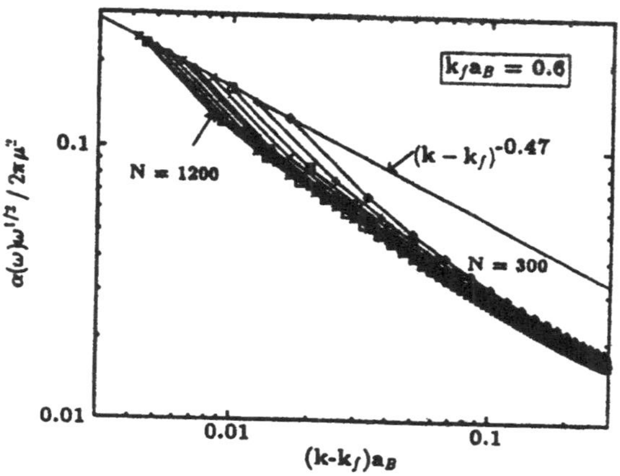

Fig. 4. Study of the N-dependence of finite size absorption spectra ($N = 300, 400, 1,200$), for determining singularity exponent of the FES of the infinite system. N-dependence of the first transition of each spectrum results in a FES exponent of -0.47 for $k_\mathrm{f}a_\mathrm{B} = 0.6$.

is continuously shifted towards the edge. Yet, on approaching the edge, the probability of the first transition of each spectrum is growing and lies on a common straight line in a log–log plot. In other words, $\alpha(\omega)$ displays singular behavior for $k \to k_{\mathrm{f}}$. The exponent of this singularity can be determined from the slope of this line and is 0.47 for $k_{\mathrm{f}} a_{\mathrm{B}} = 0.6$. We thus see that by studying the N-dependence of the determinant one can infer the FES exponent for the infinite system from a finite size calculation.

3.2. *Electron-density dependence of the singularity exponents*

In Fig. 5 the first and second threshold exponents as a function of the electron density are given. In his classic paper [13], Anderson showed that the overlap of the ground states $|\Phi_F^0\rangle$ and $|\Phi_I\rangle$ of the system with and without hole, respectively, vanishes as $N \to \infty$, leading to the orthogonality catastrophe. This result emerged from investigating a determinant of the form (7) furnished with the coefficients of the N_{f} lowest single-particle states. The N-dependence of this determinant was shown to be given by $N_{\mathrm{f}} \exp(-\delta_{\mathrm{f}}^2/\pi^2)$, where δ_{f} is the phase shift at the Fermi surface. The exponent of the first singularity of the PES in Fig. 5 has been calculated by

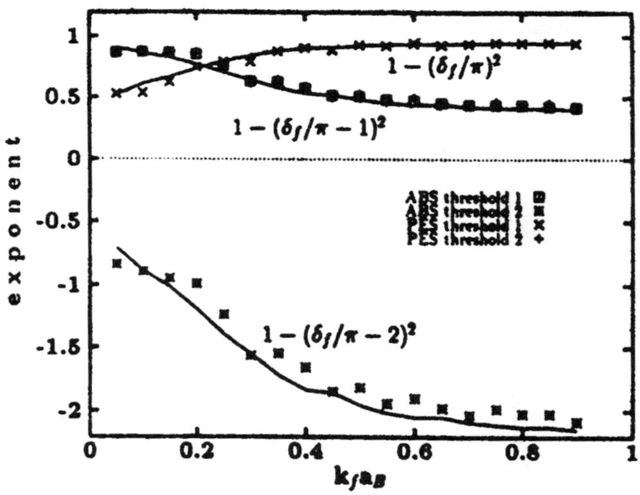

Fig. 5. Exponents for the FES and the second threshold edge features in absorption and photoemission as a function of k_{f}, determined from finite size spectra in the way depicted in Fig. 5 (symbols) and compared with the exponents derived from Hopfield's rule with the $\delta_{\mathrm{f}}(k_{\mathrm{f}})$ from Fig. 2 (continuous lines). (ABS = absorption spectrum, PES = photoemission spectrum.)

means of the determinant $\langle \Phi_F^0 | \Phi_I \rangle$, which is just the determinant that Anderson studied. And indeed, as is evident from Fig. 5 the calculated exponents agree well with the Anderson result $1 - \delta_{\rm f}^2/\pi^2$, where the phase shifts are those of Fig. 2.

With respect to the other cases of Fig. 5, our approach can be viewed as a natural extension of Anderson's approach, in the sense that the edge features of both ABS and PES and their thresholds can be described by a class of determinants similar to that of Anderson's. According to Hopfield's rule [15], the exponents are given by $1 - n_{\rm f}^2$ where $n_{\rm f}$ is the excess localized charge near the excited core in the ground state $\langle \Phi_F^0 |$. This rule was found by Combescot and Nozières [7] to be generally correct. According to the Friedel sum rule [18], the hole potential gives rise to an excess charge $n_{\rm f}$ of $\delta_{\rm f}/\pi$ so that with Hopfield the exponent for the first threshold of the PES is just $1 - \delta_{\rm f}^2/\pi^2$. However, for the second PES threshold the bound state electron is absent, and so the net charge $n_{\rm f}$ becomes $\delta_{\rm f}/\pi - 1$ and consequently the exponent is $1 - (\delta_{\rm f}/\pi - 1)^2$. In the absorption process on the other hand, the electron from the valence band has to be taken into account so that the exponents are $1 - (\delta_{\rm f}/\pi - 1)^2$ and $1 - (\delta_{\rm f}/\pi - 2)^2$ for the first and the second threshold, respectively. All of these exponents are plotted in Fig. 5 and show good agreement with our numerically determined exponents. The phase shifts $\delta_{\rm f}$ are those due to the screened Coulomb potential.

4. Conclusions

We have studied FES in quasi-one dimensional quantum wires within a simple framework in which the transition probabilities are calculated by considering Hartree–Fock determinants appropriate to many-body transitions. A screened Coulomb interaction in the RPA has been used for the electron–hole interaction to determine the single-particle properties both with respect to the bound and scattering states of the system, which constitutes the starting-off point of our many-body calculations.

The essential feature of this contribution is that it is possible to infer the edge singularity exponents of the finite system from the size dependence of the many-body determinants. This procedure is in the spirit of the method used by Anderson to predict the orthogonality catastrophe. These exponents depend on the scattering phase shifts at the Fermi surface and, via the $k_{\rm f}$ dependence of these phase shifts, on the electron density. We have been able to check the connection for 1D wires by calculating both the phase shifts as well as exponents. To our knowledge, this is the first explicit calculation of the singularity exponents of the FES in quantum wires. What is more interesting is that this result emerges from an extrapolation of a finite size calculation. The FES intensities and exponents are sensitive to electron correlation and density.

At this point we inject a caveat and ask a crucial question: what is the effect of the hole recoil on the singularities and how are they modified by this? Our treatment has been limited to an infinite hole mass case. Drawing attention to his earlier work [19], Nozières [20] has recently focussed attention on the effect of recoil on edge singularities. He showed that for dimensions $d \geq 2$ the singularities vanish if recoil allows the hole to diffuse to infinity. However, when $d = 1$, or for $d > 1$ if localization is included, the FES exponents are reduced due to recoil, by a factor that involves an angular average over the Fermi surface. For 1D, the FES persists but the exponent is half as a result of recoil. Obviously more work is needed to elucidate the effect of hole recoil on FES.

Another question of interest is the robustness of FES against low-energy collective excitations of the Fermi-sea in quasi-1D systems, since one knows that these excitations are more significant here than single particle electron–hole transitions.

It should be added parenthetically that the original deviation of orthogonality exponents by Anderson made a variety of approximations including Taylor expansion of certain quantities in powers of the phase shift. The FES and othogonality exponents were studied in great generality by Zagoskin and Affleck [21] using conformal field theory, an exact sum rule and numerics on a tractable 1D tight binding model of spinless electrons, where the core potential produces a bound state. They conclude that the Anderson and FES exponents are determined by δ_{F} for both primary and secondary thresholds. Also the behavior of the FES exponent α is determined as the electron density $\to 0$: $\alpha_{\mathrm{f}} \to 0$ and $\alpha_{\mathrm{e}} \to 1$ since the phase shift δ_{k}, from Levinson's theorem, approaches π at the bottom of the band where there is a bound state. Here $\alpha_{\mathrm{f}} = 1 - 2x_{\mathrm{f}}$, where $x_{\mathrm{f}} = 1/2(\delta E_{\mathrm{F}}/\pi)^2$. They also calculate the electron density dependence of the FES amplitude over a wide range.

Acknowledgments

The work presented here was done together with Prof. Sir Roger Elliott, F.R.S. and H.H. von Gruenberg in the Department of Theoretical Physics, University of Oxford some years ago. It is, therefore, gratifying that this should find a place in the Proceedings of a Symposium organized by the University of Mexico to honor Prof. Elliott, particularly in view of my long association with him stretching over several decades. I wish to thank Professors Rafael Barrio and Kimmo Kaski for their hospitality there. I also wish to thank Prof. Philippe Nozières for a discussion on the nature of FES in reduced dimensions.

References

[1] Calleja, J.M. *et al.*, *Surf. Sci.* **263** (1992) 346.
[2] Fritze, M., Nurmikko, A.V. and Hawrylak, P., *Surf. Sci.* **305** (1994) 580.
[3] van der Meulen, H.P., *Phys. Rev. B* **58** (1998) 10705.
[4] van der Meulen, H.P., *Phys. Rev. B* **60** (1999) 897.
[5] Melin, T. and Laruelle, F., *Phys. Rev. Lett.* **85** (2000) 852.
[6] Melin, T. and Laruelle, F., *Phys. Rev. B* **65** (2002) 195303.
[7] Combescot, M. and Nozières, P., *J. Phys.* (Paris) **32** (1971) 913.
[8] Schmitt-Rink, S., Chemla, D.S. and Miller, D.A.B., *Adv. Phys.* **38** (1989) 89.
[9] Mueller, J.F., Ruckenstein, A.E. and Schmitt-Rink, S., *Phys. Rev. B* **45** (1992) 8902.
[10] Hawrylak, P., *Solid State Commun.* **81** (1992) 525.
[11] Mahan, G.D., *Many-Particle Systems*, 2nd edn. (Plenum, New York, 1981).
[12] Nozières, P. and De Dominicis, C.T., *Phys. Rev.* **178** (1969) 1097.
[13] Anderson, P.W., *Phys. Rev. Lett.* **18** (1967) 1049.
[14] von Gruenberg, H.H., Jain, K.P. and Elliott, R.J., *Phys. Rev. B* **54** (1996) 1987.
[15] Hopfield, J.J., *Comment. Solid State Phys.* **2** (1969) 40.
[16] Hu, G.Y. and O'Connell, R.F., *Phys. Rev. B* **42** (1990) 1290.
[17] Hu, G.Y. and O'Connell, R.F., *J. Phys. Cond. Matt.* **2** (1990) 9381.
[18] Friedel, J., *Comment. Solid State Phys.* **2** (1969) 21.
[19] Gavoret, J., Nozières, P., Roulet, B. and Combescot, M., *J. Physique* **30** (1969) 987.
[20] Nozières, P., *J. Phys. I* (France) **4** (1994) 1275.
[21] Zagoskin, A.M. and Affleck, I., preprint, cond-mat/97-04248 (1997).

CHAPTER 17

PROBING THE MAGNETIC COUPLING IN MULTILAYERS USING DOMAIN WALL EXCITATIONS

A.S. Carriço
Departamento de Física Teórica e Experimental, Universidade Federal do Rio Grande do Norte, 59072-970 — Natal, RN, Brazil
acarrico@dfte.ufrn.br

Ana L. Dantas
Departamento de Física, Universidade do Estado do Rio Grande do Norte, 59610-210 — Mossoró, RN, Brazil
ana@uern.br

New magnetic phases and phenomena have been observed in artificial multilayered systems consisting of thin films of magnetic or non-magnetic materials. A key parameter of these systems is the interaction between the magnetic layers and a continuing experimental problem is the measurement of the inter-layer magnetic coupling. The sign and average strength of the coupling are usually found through magnetization measurements, ferromagnetic resonance, light scattering and magnetoresistance measurements. These techniques sample large areas and thus average out the microscopic details. The frequency of domain wall excitations is due to restoring forces localized to the region of the domain wall. Thus, the excitations of domain walls stabilized by the interlayer coupling may be used to study the magnetic coupling in length scales of the order of the width of the domain wall. We discuss the use of domain wall excitations as a means of probing the magnetic coupling. We consider the case of a Nèel-like domain wall pinned by a one-dimensional defect and show that the frequency of rigid displacement domain wall oscillations results from energy fluctuations within the domain wall width. We apply the results to the study of the interface exchange coupling in a ferromagnet–antiferromagnet bilayer, consisting of a thin uniaxial ferromagnetic film on a two-sublattice antiferromagnetic substrate. We consider the limits of weak and strong interface exchange coupling, and show that in both cases the spectrum of domain wall excitations is in the frequency range of the uniform excitations (FMR) of the ferromagnetic film.

1. Introduction

Exchange coupling between magnetic layers in multilayer geometries can strongly influence magnetization behavior and spin-wave energies by correlating the motion of spins in one layer with the motion of spins in adjacent layers.

The most commonly used techniques for magnetization measurements sample large areas of the multilayer. Therefore, these measurements do not reveal the microscopic magnetic structure. Instead, they may only be used to investigate the average magnetic coupling between films. The large amount of experimental data produced within this strategy has lead to a corresponding theoretical effort. Key properties of artificial magnetic multilayers have so far been investigated using theoretical models that rely on the average exchange coupling between the magnetic layers. For instance, the influence of surfaces on the critical field for producing an instability in the antiferromagnetic order of transition metal multilayers, has been investigated using models that are based on average parameters [1–3]. However, there are also phenomena of current interest which require further understanding of the magnetic structure. This is the case of phenomena of technical relevance, such as the exchange bias observed in ferromagnet–antiferromagnet bilayers.

The exchange coupling of ferromagnetic (F) and antiferromagnetic (AF) films across their common interface may modify some of their properties. One of the leading effects is a shift in the hysteresis loop of the F film, called exchange bias. There has been considerable interest in exchange-biased F films because the shift can be useful in controlling the magnetization in spin valve devices used for read heads based on the giant magnetoresistance effect [4].

A possible spin valve system consists of a pair ferromagnetic layers, separated by a non-magnetic spacer, grown on an antiferromagnetic substrate. The ferromagnetic layer exchange coupled to the AF substrate requires a large value of the external field strength to switch the magnetization. Thus, it serves as a reference layer, and the magnetoresistance results from the field induced misalignment of the magnetization in the free layer relative to the magnetization of the pinned layer.

Another interesting physical phenomena, the current driven switching of the magnetic layers in magnetic tunnel junctions, is controlled by the relative orientation of the magnetization of the electrodes [5–7]. For this reason, it is useful to have a reference layer, either much thicker or stabilized by an AF substrate. Also, the junction magnetoresistance has been found to increase for exchange-biased tunnel junctions, using an AF substrate [8].

Noteworthy theoretical models have been proposed to describe key features of the exchange bias phenomena [9]. However, the exchange bias is still a subject of discussion [10], which requires detailed a understanding of the microscopic nature of the F/AF interface.

Techniques that probe small areas of the system would be helpful to investigate the magnetic coupling between layers on a microscopic scale. Domain wall excitations may be used to study the magnetic structure in a microscopic scale, if the domain walls are stabilized by the inter-layer exchange coupling. Such is the case of correlated domain wall pairs in neighboring films in metallic magnetic multilayers. In these systems made of transition metal layers separated by non-magnetic spacer layers, significant changes in the effective exchange coupling between magnetic layers may be produced by fluctuations in the spacer thickness. It has been shown that the measurement of domain wall resonance in magnetic trilayers might be a promising means to probe the local value of the inter-layer exchange coupling. The restoring force of an optical mode of domain wall pairs was shown to be proportional to the exchange coupling energy between the magnetic layers integrated within the domain wall width [11].

Domain wall excitations may also be used to investigate the interface exchange coupling in F/AF bilayers. The current view about the magnetic coupling in these systems is based on a contact interaction, representing the exchange coupling between interface spins, which tends to align the spins of the ferromagnetic film with the AF spins at the interface [10]. Interface roughness is a key aspect of F/AF bilayers and may affect significantly the effective magnetic coupling. The correlation between the interface pattern and the magnetic structure of the F layer depends on the strength of the interface exchange coupling and the average size of regions in the AF interface plane with a single sublattice. If the dimensions of the interface AF domains are much larger than the domain wall width of the F layer, one finds a one-to-one correspondence between the pattern of the AF interface domains and the domains in the F layer [12].

The AF substrate used in F/AF bilayers is commonly a two sublattice material and the interface magnetic structure depends on the crystalographic orientation and on the degree of interface roughness. The F/AF interfaces are not perfectly smooth and the presence of atomic steps at the interface may favor the nucleation of domain walls in the F film. A long interface step divides the AF interface plane into two regions, each one with magnetic moments from a single AF sublattice, if the step height corresponds to an odd number of lattice parameters. Furthermore, if the distance between the interface step defects is much larger than the domain wall width of the ferromagnetic film, one may find domains in the ferromagnetic film in a one-to-one correspondence with the AF substrate magnetic pattern [13]. For ultra-thin ferromagnetic films, the magnetic pattern does not change in the direction perpendicular to the interface. In this case the domain wall profile can be represented by Nèel-like, one-dimensional models and the effect of external fields on the domain wall excitations may be used to estimate the strength of the interface exchange [14–17].

The starting point to investigate domain wall excitations is the static configuration which minimizes the magnetic energy. The domain wall excitations are small fluctuations around the equilibrium structure. Most studies of domain walls in artificial multilayers focus on their static properties. The interaction between the ferromagnetic thin layers separated by the nonmagnetic intermediate layers modifies the domain and domain wall structures. The coupling between magnetic layers may lead to changes in essential features of the domain wall structure of thin films. For instance, while thin Permalloy films tend to exhibit cross-tie walls, superimposed Nèel walls and quasi-Nèel walls are found in trilayers consisting of Permalloy thin films separated by a carbon spacer [18]. The essential difference between the trilayers and the single Permalloy layers of the same thickness (300 Å) is that when a second film is deposited on top of the first one, a new way of reducing the magnetostatic energy of the domain walls becomes possible. The flux of the stray field can be closed through the superimposed domain walls, provided they have opposite polarity.

One-dimensional walls with the magnetization in-plane have also been observed in trilayers used in spin valve devices [19]. The main interest in this study has been the origin of coercivity in giant magnetoresistance NiO-Co-Cu-Co spin valves. It has been found that domain walls are nucleated at defects of the NiO-Co interface. Since the ferromagnetic film coupled to the AF substrate is harder (the coercivity of Co on CoO is five times the coercivity of a Co free film), the stray field of the domain wall pinned at this layer induces the nucleation of a domain wall of opposite chirality in the free layer. The magnetic coupling of these superimposed domain walls produces an undesirable increase in the coercivity of the free layer, which makes the switching field larger. The measured value of the escape field has been reproduced using the stray field of Nèel walls in the pinned layer.

In this paper we discuss recent work on domain wall excitations in F/AF bilayers. We show that the excitations of domain walls, pinned by local defects, are controlled by the magnetic structure in regions of microscopic dimensions. We study rigid domain wall displacement modes (RDWDM) and we show that, provided the pinning energy is of the same order of magnitude as the anisotropy energy of the ferromagnet, these domain wall excitations can be accessed by resonance experiments in experimental set-ups designed for ferromagnetic resonance (FMR).

In the second section, we obtain the field dependence of the frequency of RDWDM for a general model of a Nèel-like domain wall. We keep the energy density of the wall in general form and obtain the frequency of excitations by examining energy fluctuations around the equilibrium state. At this stage, except for being a line defect, no assumption is made of the nature of the defect which pins the domain wall. We allow the field to displace the wall from the pinning center and calculate the restoring force constant and the Döring mass in terms of the equilibrium profile functions. In the following section, we apply the results to the study of excitations of domain walls pinned by a step defect in an F/AF bilayer, considering the limits of weak and strong interface exchange coupling. The final section is devoted to discussing the applications to systems of current interest.

2. Domain Wall Pinning at a One-Dimensional Defect

We consider a π-wall of a uniaxial ferromagnet, pinned by a line-defect running along the z-axis at $y = 0$. The magnetization is in the yz-plane and its orientation with respect to the uniaxial z-axis, in the plane, is given by the function $\theta(y)$.

In the absence of external magnetic field, the domain wall center position is at $y = 0$. In this case, we have $\theta(0) = \pi/2$ and at the domains we have $\theta = 0$ and $\theta = \pi$. In Fig. 1, we show an schematic representation of the domain wall, for the particular case of an interface step defect. This structure will be investigated in the following sections. In this section we

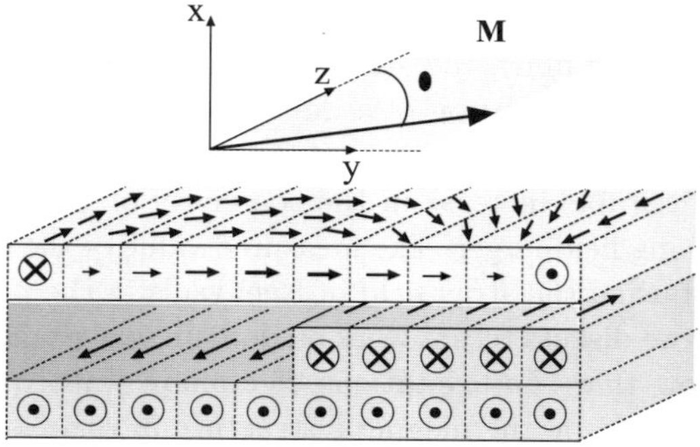

Fig. 1. Schematic representation of a Nèel wall pinned at a step defect on an antiferromagnetic substrate.

discuss general properties of excitations of domain walls pinned by interface defects. However, Fig. 1 might help to discuss a few general features. For instance, while $y = 0$ is the position of the domain wall center in the absence of external magnetic field, if an external field H is applied parallel to one of the domains, it will displace the domain wall, favoring the increase of the domain parallel to H. The position of the domain wall center is shifted to $y = q_H$, and q_H is an increasing function of the strength of the external field. As the external field is increased, the distance of the domain wall center to the defect increases. Since the restoring force which stabilizes the fluctuations of the domain wall around the position $y = q_H$ is due to the pinning potential of the defect, which is localized at $y = 0$, one might expect that the frequency of oscillations must decrease with the strength of the external field. There is a critical value of the external field strength H^* at which the domain wall is liberated from the defect. At this value of H, the frequency of the domain wall oscillations must be zero. In the following, we describe these features for a general model of a domain wall pinned at a line defect. For a given value of the external field strength, H, the equilibrium profile, represented by $\theta^0(y)$, includes the field-induced displacement of the domain wall center.

We start from an equilibrium profile $\theta^0(y)$ which minimizes the magnetic energy:

$$E_{\text{eq}} = \int_{-L}^{L} dy\, f(\theta, \theta_y), \tag{1}$$

where L is the width of the domains at each side of the domain wall and $f(\theta, \theta_y)$ is the magnetic energy density, including intrinsic exchange and anisotropy energies of the ferromagnet as well as Zeeman energy and the domain wall pinning energy. $\theta^0(y)$ is a solution of the Euler–Lagrange equation:

$$\frac{\partial f}{\partial \theta} - \frac{\partial}{\partial y}\frac{\partial f}{\partial \theta_y} = 0. \tag{2}$$

Rigid displacement domain wall excitations are characterized by a rigid displacement of the angular profile of the domain wall. We consider the variations induced in the energy by small amplitude displacements around the equilibrium pattern, using the function $\theta(y-q)$, with $q = q_0 e^{i\Omega t}$. We also introduce an extra term in the energy corresponding to a small out-of-plane angle $\psi = \psi_0 e^{i\Omega t}$. The out-of-plane oscillations induce surface charges and the demagnetizing energy is approximated by:

$$E_M = \int_{-L}^{L} 2\pi M^2 \sin^2\psi \sin^2\theta \, dy. \tag{3}$$

The total energy is the sum of Eqs. (1) and (3). We calculate the variations in E_{eq}, when $\theta(y) = \theta^0(y-q)$ is used in Eq. (1) in the place of $\theta^0(y)$ and add to it the demagnetizing energy, given by Eq. (3). The variations in θ and θ_y are given by $\delta\theta = -q\theta^0_y(y)$ and $\delta\theta_y = -q\theta^0_{yy}(y)$, where $\theta^0_y(y)$ is the y-derivative of $\theta^0(y)$. In order to calculate the leading term of the excitation energy, we expand the function $f(\theta, \theta_y)$ up to second order of the displacement variable q. Considering that the function $f(\theta^0, \theta^0_y)$ is a solution of the Euler–Lagrange equations, we find that

$$E_{\mathrm{eq}}(\{\theta, \theta_y\}) = E_{\mathrm{eq}}(\{\theta^0, \theta^0_y\}) + \delta E, \tag{4}$$

where

$$\delta E = \frac{q^2}{2}\int_{-L}^{L} dy \left\{ \frac{\partial^2 f}{\partial \theta^2}\left(\theta^0_y\right)^2 + \frac{\partial^2 f}{\partial \theta_y^2}\left(\theta^0_{yy}\right)^2 + \frac{\partial^2 f}{\partial\theta\partial\theta_y}\theta^0_y\theta^0_{yy} \right\}. \tag{5}$$

For a good number of magnetic systems of current interest there is no cross-derivative of the energy density ($\partial^2 f/\partial\theta\partial\theta_y = 0$). Furthermore, for a rigid displacement, the intrinsic exchange energy does not change. Thus, we do not have a term involving the θ_y-derivative of the energy density. We then find

$$\delta E = \frac{q^2}{2}\int_{-L}^{L} dy \, \frac{\partial^2 f}{\partial\theta^2}\left(\theta^0_y\right)^2. \tag{6}$$

The factor $(\theta^0_y)^2$ in the integrand of Eq. (6) restricts the contribution to the excitation energy δE to the region of the domain wall. Notice also, from Eq. (3), that the main contribution to the magnetostatic energy comes from the domain wall region, since the function $\sin^2\theta$ is zero in the domains. We assume that both q/Δ_0 and ψ are small quantities. Furthermore, we assume that the out-of-plane fluctuation ψ is uniform within the domain wall. Since the equilibrium value of ψ is zero, to calculate the excitation energy up to second order in the variables q and ψ, we use the equilibrium function $\theta^0(y)$ in Eq. (3).

The leading contributions to the excitation energy for small amplitude rigid displacement oscillations are given by Eqs. (3) and (6). The total energy, $E = E_{\rm eq} + E_M$, is of the form

$$E = E^0 + \frac{1}{2}\,k\,q^2 + \frac{1}{2}\,b\,\psi^2, \tag{7}$$

where E^0 is the equilibrium value of the energy, as given by Eq. (1), using the profile $\theta^0(y)$. The constants k and b are given by integrals in which the leading contribution comes from the domain wall region. As seen in Eqs. (6) and (3), these integrals involve the second derivative of the magnetic energy density $(\partial^2 f/\partial\theta^2)$, calculated using $\theta^0(y)$, and the functions $(\theta^0_y)^2$ and $\sin^2\theta^0_y$. These three functions are defined by the equilibrium magnetic profile, which in turn is determined by the domain wall pinning energy and the intrinsic properties of the ferromagnetic film, such as the intrinsic exchange energy and the anisotropy as well as the saturation magnetization.

In order to obtain the equations of motion for the domain wall variables q and ψ, we integrate the Landau–Lifshitz torque equations throughout the domain wall [20] and obtain a pair of coupled equations given by

$$\frac{dq}{dt} = \frac{\gamma}{2M}\frac{\partial E}{\partial\psi}, \tag{8}$$

$$\frac{d\psi}{dt} = -\frac{\gamma}{2M}\frac{\partial E}{\partial q}, \tag{9}$$

where γ is the gyromagnetic factor.

From Eqs. (8) and (9), we obtain the frequency of domain wall oscillations as

$$\Omega = \frac{\gamma}{2M}\sqrt{k\,b}. \tag{10}$$

The restoring force constant k is a decreasing function of the external field strength. When the external field approaches the threshold value, H^*, which makes the domain wall free from the defect, the center of the domain wall is far from the defect line at $y = 0$. Assuming the defect contribution to the magnetic energy to be of finite range, centered at $y = 0$, when

$H \approx H^*$ the function $\partial^2 f/\partial\theta^2$ is practically zero, since in the defect range, the magnetization is uniform. Thus, the fluctuations in the domain wall position produce no extra energy, and $k = 0$.

Notice that the results, so far, are valid for any kind of one-dimensional magnetic domain wall structure, provided that the equilibrium structure corresponds to having the magnetization in a plane. This covers Nèel walls as well as Bloch walls. Furthermore, the domain wall pinning mechanism, as well as the internal structure of the ferromagnet have not been specified. Thus, the results apply equally well for a variety of systems [21–26].

3. Interface Step Defect in an F/AF Bilayer

A self-consistent local field algorithm has been used to investigate the nucleation and pinning of Nèel walls at an interface step defect of an uncompensated F/AF interface [13]. In this study, only the defect induced modification of the domain wall profile was explored. It has been shown that ferromagnetic narrow domain walls are nucleated at interface step defects. Furthermore, even for moderately large values of the interface exchange coupling, the magnetic structure corresponds to a Nèel wall. It has been found that the domain wall profile is of the form $\tan\frac{\theta}{2} = e^{\frac{y}{\Delta}}$. For small values of the interface exchange field, the domain wall parameter approaches the free Nèel wall value $\Delta = \sqrt{A/K}$, where A and K are the exchange stiffness parameter and the uniaxial anisotropy energy parameter of the F layer, respectively. It was also found that the domain wall width parameter Δ may be significantly reduced by the interface exchange field if the anisotropy energy is small. Furthermore, these numerical results also indicated that in the presence of an external field parallel to one of the domains, the domain wall center is displaced from the step defect according to $\tan\frac{\theta}{2} = e^{\frac{y-q_H}{\Delta}}$. These results suggest that in order to study domain wall excitations, one may consider a Nèel-like wall profile with adjustable parameters. In the weak exchange coupling case, an interface modified Nèel wall profile is used, where the domain wall width parameter and the position of the domain wall center are found by minimizing the magnetic energy.

Domain wall excitations were calculated for a system consisting of a Nèel wall pinned at an step defect in an F/AF interface [14, 15]. The system consists of a thin ferromagnetic film with in-plane magnetization on a two-sublattice uniaxial antiferromagnetic substrate as shown in Fig. 1. The anisotropy axis of the antiferromagnet is parallel to the easy direction of the ferromagnet (the z-axis). The substrate step edge runs along the z-axis and divides the interface into two regions, each one containing spins from a sublattice of the antiferromagnet. In our model, no relaxation is allowed for the substrate spins, which are held fixed along the anisotropy direction.

If, as appropriate for thin F layers, we do not consider any variation of the magnetization along the z- or x-axis directions, the magnetic pattern is described by the y-dependence of θ, the angle between the magnetization and the easy axis ($\hat{z}$). The nucleation of a Nèel wall in the ferromagnetic film follows from the discontinuous change of direction of the interface exchange field at the step edge. The magnetic energy density is given by

$$f(\theta, \theta_y) = A(\theta_y)^2 - (HM + J(y))\cos\theta - K\cos^2\theta, \tag{11}$$

where

$$J(y) = \begin{cases} J; & -L < y < 0, \\ -J; & 0 < y < L. \end{cases} \tag{12}$$

The first term in the Eq. (11) is the intrinsic exchange energy density, the second term is the Zeeman energy density for an external field of strength H applied along the direction $\hat{z}$, the third term is the interface coupling energy density and the last term is the uniaxial anisotropy energy.

Compared to the uniform state, magnetized along the easy axis of the uniaxial anisotropy, a Nèel wall has a positive energy of $4\sqrt{AK}$. Thus, there is an energy barrier for wall nucleation. The interface energy depends on the strength of the interface exchange coupling J as well as on the width L of the regions on each side of the defect. We assume that L is much larger then the intrinsic domain wall width of the ferromagnet ($\sqrt{A/K}$).

For large values of L, the interface exchange energy may overcome the energy barrier for wall nucleation even if the interface exchange coupling J is not large. We may estimate the threshold value of J for wall nucleation by comparing the interface energy in the domains (of width L) with the intrinsic energy of the wall. Compared to the uniform state the wall has an energy of $\delta\varepsilon = 4\sqrt{AK} - 2JL$. Thus, the threshold value of J for wall nucleation is determined by the energy density of the wall ($2\sqrt{AK}/L$), and may become vanishingly small if L is large. This is a question of interest for the study of F/AF interfaces, since the value of the interface field is not well known for most systems of interest. Thus, although it is intuitive that wall nucleation should occur for large values of J, it is important to know that even if J is small, domain wall nucleation occurs provided that the distance between step defects L is sufficiently large.

Furthermore the interface coupling tends to reduce the domain wall width, so as to favor larger areas of the interface complying with the trends imposed by the interface field [13]. Thus, even if L is not much larger than the intrinsic domain wall width, the reduction of the domain wall width imposed by the interface coupling may favor the domain wall nucleation.

A few features of the domain wall excitations may be anticipated. For rigid displacement domain wall excitations, the intrinsic exchange and the anisotropy energies make no contribution to the restoring force. The fluctuations around the equilibrium position of the domain wall center induce no change in the relative orientation of neighboring spins or in the overall orientation of the spins with respect to the anisotropy axis, since the magnetic structure moves rigidly. As the wall moves rigidly out of the equilibrium position by a small displacement, it induces a change in the Zeeman energy due to the modification in the sizes of the domains. The interface exchange energy is also changed since the displacement of the wall induces changes in the orientation of the magnetization, within the domain wall, with respect to the interface field. Thus the restoring force is given by $k = \frac{1}{q}\frac{\partial E_{J,H}}{\partial q}$, where $E_{J,H}$ is the sum of the Zeeman energy and the interface coupling energy.

3.1. *Weak interface coupling case*

In order to study rigid domain wall displacement oscillations around the equilibrium position, we use a Nèel-like domain wall profile in which both the domain wall width (Δ) and the position of the domain wall center (q_H) are determined as functions of the interface exchange field and the external field strengths. We use

$$\tan\left(\frac{\theta(y,t)}{2}\right) = \exp\left(\frac{y - q_H - \eta(t)}{\Delta}\right) \tag{13}$$

and

$$\Psi = \psi(t), \tag{14}$$

where $\eta(t)$ is the dynamical variable which describes the oscillations of the domain wall center around the equilibrium position q_H, and $\psi(t)$, the angle between the projection of the magnetization in the yx-plane and the y-axis, describes the out-of-plane component of the magnetization.

q_H and Δ are the equilibrium values of the position of the domain wall center and the domain wall width, respectively. They are obtained from the minimization of the energy and are given by

$$q_H = \Delta \tanh^{-1}\left(\frac{H}{H_J}\right) \tag{15}$$

and

$$\frac{\Delta}{\Delta_0} = \left[1 + 2\frac{H_J}{H_A}\ln 2 - \frac{H+H_J}{H_A}\ln\left(1+\frac{H}{H_J}\right) + \frac{H-H_J}{H_A}\ln\left(1-\frac{H}{H_J}\right)\right]^{-1/2}, \tag{16}$$

where $\Delta_0 = \sqrt{A/K}$, $H_J = J/M$ and $H_A = 2K/M$.

Using the magnetic profile defined by Eqs. (13) and (14), we get

$$E(\eta,\psi) = E(q_H,0) + 4\pi M^2 \Delta \psi^2 + \frac{J}{\Delta\cosh^2\left(\frac{q_H}{\Delta}\right)}\eta^2. \tag{17}$$

In Eq. (17), $E(q_H,0)$ is the equilibrium value of the energy ($\eta = 0, \psi = 0$). Note that both q_H and Δ are functions of the strength of the external field.

From Eqs. (10) and (17), we obtain the frequency of the domain wall oscillations:

$$\left(\frac{\Omega}{\Omega_0}\right)^2 = \frac{4\pi M H_J}{H_A(H_A+4\pi M)\cosh^2\left[\tanh^{-1}\left(\frac{H}{H_J}\right)\right]}, \tag{18}$$

where $\Omega_0 = \gamma\sqrt{H_A(H_A+4\pi M)}$ is the frequency of the uniform mode of the domains in the absence of interface effects and external field.

In Fig. 2, we show $\Omega(H)/\Omega_0$. We selected a few values of the interface exchange field for an anisotropy field of $H_A = 0.55\,\mathrm{kOe}$. $\Omega(H)$ is a monotonically decreasing function of H with an upper limit of the order of Ω_0.

The upper limit of the excitation frequency, $\Omega(H)$, is for $H = 0$. As seen in Eq. (18), $\Omega(0)/\Omega_0$ is proportional to the square root of H_J/H_A. Thus, a large increase in H_J/H_A does not lead to a correspondingly large increase in $\Omega(0)$.

The restoring force constant k is a decreasing function of H and becomes zero for $H = H_J$. For $H = 0$, the energy fluctuations include in full the oscillations of the domain wall around the step edge. The equilibrium position of the wall center moves away from the step defect when H increases. For $H \cong H_J$ the step defect is at the tail of the domain wall. Thus, there is no variation in the angular profile near the step edge for small displacement oscillations ($\theta(y) \cong 0$ and $\theta_y(y) \cong 0$), and there is no variation of the interface energy due to small oscillations of the domain wall position.

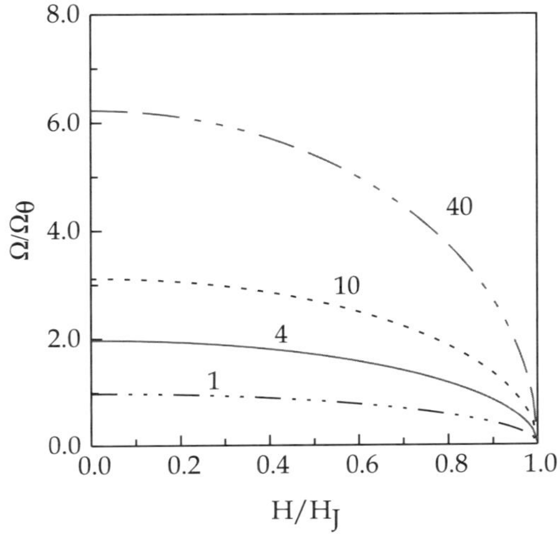

Fig. 2. Frequency of rigid displacement domain wall oscillations. The numbers by the curves indicate the values of H_J/H_A.

3.2. *Strong interface coupling case*

There is a controversy regarding the strength of the interface exchange field in F/AF bilayers. The effective interface exchange field estimated from the shift of the hysteresis loop is of the order of the coercive field of the F layer. On the other hand, one expects the interface field to be three orders of magnitude larger than the anisotropy field since it originates in the exchange coupling of interface spins [10]. In this section, we investigate the domain wall excitations, starting with solutions of the Euler–Lagrange equations, which are valid in the limit of strong interface exchange coupling. In this case, the variational approach used in the previous section is not valid since the interface defect may modify significantly the standard Nèel wall profile.

The magnetic energy density is the same as in Eq. (11), and in order to account for the variation of the interface field, we use two independent functions $\theta(y)$, one for each side of the step defect. The equilibrium profile is found by matching these function at $y = 0$.

The Euler–Lagrange equations corresponding to independent functional variations of the integrand for $y > 0$ and $y < 0$ lead to the pair of second order equations

$$-\frac{A}{K}\frac{\partial^2\theta}{\partial y^2} + (h + j(y))\sin\theta + \sin\theta\cos\theta = 0, \tag{19}$$

where $h = HM/2K$, $j(y) = J(y)/2K$, and there is one equation for each sign of $J(y)$. Throughout the text, we also make reference to the interface exchange field, given by $H_J = J/M$, and to the anisotropy field $H_A = 2K/M$.

$\theta(y)$ is obtained by integrating Eq. (19), with the integration constants found by using the boundary conditions $\theta(-L) = 0, \theta(L) = \pi$, and $\left.\frac{d\theta}{dy}\right|_{y=\pm L} = 0$. Also, the continuity of $\theta(y)$ at $y = 0$ is required. Thus, when integrating the Euler–Lagrange equations, we impose this condition. Furthermore, it is necessary that the derivative of $\theta(y)$ be continuous. It can be shown that the continuity of the derivative of $\theta(y)$ at $y = 0$ corresponds to choosing the value of $\theta(0) = \theta_0$ that minimizes the magnetic energy.

The first integrals of the Euler–Lagrange equations are

$$\frac{d}{dy'}\left[\theta_{y'}^2 + 2\left(h + j(y')\right)\cos\theta + \cos^2\theta\right] = 0, \tag{20}$$

where $y' = y/\Delta_0$ and $\theta_{y'} = d\theta/dy'$. Thus, we have

$$\theta_{y'}^2 + 2\left(h + j(y')\right)\cos\theta + \cos^2\theta = C_\pm, \tag{21}$$

where the integration constants $C_\pm$ are obtained from the boundary conditions at $y' = \pm L'$, with $L' = L/\Delta_0$. Imposing the boundary conditions, we get $C_\pm = 1 + 2\left(h \mp j\right)$. Integrating the pair of equations in (21), we get

$$\tan\frac{\theta}{2} = \alpha_-\mathrm{cosech}\left(\beta_- - \frac{y'}{\Delta_-}\right), \quad y' < 0, \tag{22}$$

$$\tan\frac{\theta}{2} = \alpha_+\sinh\left(\beta_+ + \frac{y'}{\Delta_+}\right), \quad y' > 0, \tag{23}$$

where

$$\alpha_- = \sqrt{\frac{1 + j + h}{j + h}}, \tag{24}$$

$$\Delta_- = \frac{\Delta_0}{\sqrt{1 + j + h}}, \tag{25}$$

$$\beta_- = \ln\left[\alpha_-\cot\frac{\theta_0}{2} + \sqrt{1 + \alpha_-^2\cot^2\frac{\theta_0}{2}}\right], \tag{26}$$

and

$$\alpha_+ = \sqrt{\frac{j-h}{1+j-h}}, \tag{27}$$

$$\Delta_+ = \frac{\Delta_0}{\sqrt{1+j-h}}, \tag{28}$$

$$\beta_+ = \ln\left[\frac{\tan\frac{\theta_0}{2}}{\alpha_+} + \sqrt{1+\frac{\tan^2\frac{\theta_0}{2}}{\alpha_+^2}}\right], \tag{29}$$

where θ_0 is found from energy minimization. It turns out that θ_0 is given by

$$\theta_0 = \arccos\left(\frac{h}{j}\right). \tag{30}$$

These results are valid for any value of j and $h < j$.

Note that when $h = j$, we have $\cos\theta(0) = 1$. Thus, the domain wall is liberated from the step defect. Also, in the limit of very small values of J, for $h = 0$, we should recover the usual π-wall profile ($\tan\frac{\theta}{2} = \exp(\frac{y}{\Delta_0})$). This is indeed what is found from Eqs. (22) and (23), when the limit of small values of J is used.

For any value of the applied field strength ($h < j$), the center of the domain wall y_0 is found by imposing the condition: $\theta(y_0) = \frac{\pi}{2}$. For $H \neq 0$, the domain wall center is in the $y > 0$ side of the step defect. Thus, we use Eq. (23) to derive the position of the domain wall center. We have

$$y_0 = \frac{\Delta_0}{\sqrt{1+j-h}}\left[\operatorname{arcsinh}\left(\sqrt{\frac{1+j-h}{j-h}}\right) - \ln\left(\frac{\sqrt{1+j-h}+\sqrt{1+2j}}{\sqrt{j+h}}\right)\right]. \tag{31}$$

Note that $y_0 = 0$ for $H = 0$ and that y_0 diverges when $h \cong j$ ($H \cong H_J$). Thus, for $h \cong j$, the wall is detached and moves away from the interface defect.

We obtain the domain wall width $\Delta(H)$ from the inverse of the derivative of $\theta(y)$, calculated at the position of the domain wall center. It is given by

$$\Delta = \frac{\pi}{\left.\frac{d\theta}{dy}\right|_{y=y_0}}. \tag{32}$$

Using Eq. (22), we have

$$\Delta = \frac{\Delta_0}{\sqrt{1+2\,(j-h)}}. \tag{33}$$

Note that the squeezing of the domain wall width is maximum for $h = 0$. The reduction of domain wall width is controlled by the ratio between the interface exchange energy and the uniaxial anisotropy energy ($j = J/2K$). Thus, large reductions may occur for low anisotropy materials.

For very small values of the applied field strength, compared to H_J, there is practically no variation of the domain wall width. The wall is displaced rigidly from the step defect and there is no variation of either the intrinsic exchange energy or the anisotropy energy. The position of the wall center is determined by the balance between the interface and Zeeman energies. Thus, large displacements of the domain wall center are produced if the interface exchange coupling is weak.

The frequency of domain rigid displacement oscillations is given by Eq. (10). As in the preceding section, the contributions to k come from the terms in the energy that are changed by a rigid displacement of the wall from the equilibrium position. Therefore, the intrinsic exchange or anisotropy energies make no contribution and $k = \frac{1}{q}\frac{\partial E_{J,H}}{\partial q}$, where $E_{J,H}$ is the sum of the Zeeman energy and the interface coupling energy. Thus, we use Eq. (6), restricting the energy density f to the Zeeman and interface coupling contributions, in order to calculate the restoring force constant k.

We have

$$k = (H_J + H)M \int_0^{\theta(0)} \cos\theta\,(y)\frac{d\theta}{dy}\,d\theta - (H_J - H)M \int_{\theta(0)}^{\pi} \cos\theta\,(y)\frac{d\theta}{dy}\,d\theta, \tag{34}$$

where terms of higher order in the q and ψ have been neglected. Using the equilibrium profile $\theta(y)$ given by Eqs. (22) and (23), we find

$$k = \frac{2\,(H_J + H)M}{\Delta_0} f(\chi,\eta) + \frac{2\,(H_J - H)M}{\Delta_0} f(\zeta,\xi), \tag{35}$$

where

$$\begin{aligned} f(\chi,\eta) &= \chi\sqrt{\chi^2+\eta^2}\left(1-\chi^2-\frac{\eta^2}{2}\right) \\ &\quad + \frac{\eta^2}{2}\sqrt{1+\eta^2} + \frac{\eta^2\left(2+\eta^2\right)}{2}\ln\frac{\chi+\sqrt{\chi^2+\eta^2}}{1+\sqrt{1+\eta^2}} \end{aligned} \tag{36}$$

and $\eta = \sqrt{\frac{H_J+H}{H_a}}, \xi = \sqrt{\frac{H_J-H}{H_a}}, \chi = \cos(\theta(0)/2)$ and $\zeta = \sin(\theta(0)/2)$.

The restoring force constant k is a decreasing function of h and becomes zero for $h = j$. In fact, for $h = j$, we have $\theta(0) = 0$. Thus, $\chi = 1$ and

$f(\chi, \eta) = 0$. The equilibrium position of the domain wall center for $h \cong j$ is much larger than the domain wall width ($\Delta \cong \Delta_0$ for $h \cong j$), thus there is no variation of the interface energy due to small oscillations of the domain wall position.

b is calculated from the magnetostatic energy variations. We have $b = \frac{1}{\psi} \frac{\partial E_M}{\partial \psi}$. Using the equilibrium profile θ given by Eqs. (22) and (23) and assuming that the amplitude of the out-of-plane angle ψ_0 is small and constant within the domain wall, we find

$$b = 8\pi M^2 \Delta_0 \left\{ \sqrt{1+\eta^2} - \chi\sqrt{\chi^2+\eta^2} + \eta^2 \ln \frac{\chi + \sqrt{\chi^2+\eta^2}}{1+\sqrt{1+\eta^2}} \right.$$
$$\left. + \sqrt{1+\xi^2} - \zeta\sqrt{\zeta^2+\xi^2} + \xi^2 \ln \frac{\zeta + \sqrt{\zeta^2+\xi^2}}{1+\sqrt{1+\xi^2}} \right\}. \qquad (37)$$

The magnetostatic energy is proportional to the domain wall width and increases as the applied field increases, as seen from Eq. (33). Thus, $\frac{db}{dH} > 0$. However, the restoring force constant is a strongly decreasing function of the applied field strength, and the frequency of the domain wall rigid displacement oscillations Ω is a monotonically decreasing function of H, and $\Omega = 0$ for $h \cong j$.

The restoring force originates in the energy variations when the wall moves out of the equilibrium position. For $H = 0$, the energy variations includes in full the oscillation of the wall around the step edge, where the interface energy changes sign. However, as the external field increases, the equilibrium position of the wall center moves away from the line (at the step edge) where the interface field has a discontinuity. For large values of the external field, the restoring force is strongly reduced. It is due to small interface energy variations in the region of the step defect. For large values of H, the step defect is at the tail of the domain wall. Since in the tail of the domain wall there is little variation in the angular profile for small displacement oscillations ($\theta(y) \cong 0$ and $\theta_y(y) \cong 0$), the restoring force is rather small.

$\Omega(H)$ is a monotonically decreasing function of the external field. Thus, the spectrum of domain wall oscillations is to a large extent controlled by the value of $\Omega(H)$ for $H = 0$. In Fig. 3, we show $\Omega(0)/\Omega_0$ as function of H_J/H_A. $\Omega(0)$ is the value of $\Omega(H)$ for $H = 0$ and $\Omega_0 = \gamma\sqrt{4\pi M\, H_A}$. Note that $\Omega(0)/\Omega_0$ is a rapidly increasing function for small values of H_J/H_A. However, only modest changes in $\Omega(0)/\Omega_0$ occur when H_J/H_A increases by three orders of magnitude.

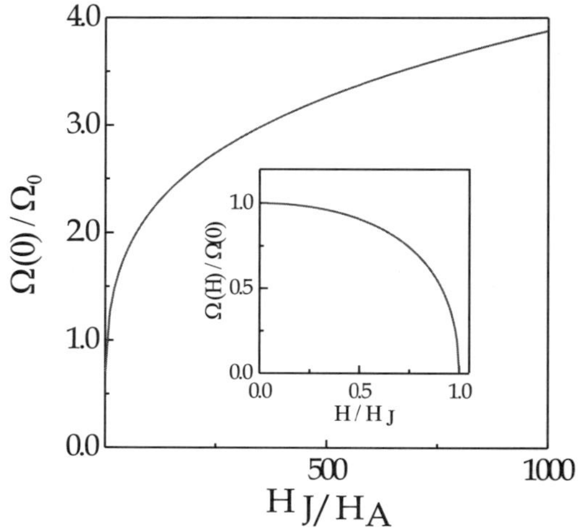

Fig. 3. Frequency of domain wall rigid displacement oscillations. The upper frequency limit $\Omega(0)$ is a function of the ratio between the interface effective field and the anisotropy field (H_J/H_A). In the inset, we show the field dependence of the frequency of domain wall oscillations. $\Omega(H)$ is shown in units of $\Omega(0)$, and the field is shown in units of the interface effective field (H_J).

One might expect $\Omega(0)/\Omega_0$ to increase rapidly with H_J/H_A since, as seen in Eq. (30), the restoring force constant seems to be proportional to H_J. However, the energy fluctuations imposed by the rigid domain wall displacement are due to the magnetostatic and interface coupling energies integrated within the domain wall width. From Eq. (33), we see that the domain wall width is strongly reduced for large values of H_J/H_A. Thus, a large increase in H_J/H_A does not lead to a correspondingly large increase in $\Omega(0)$.

In the inset of Fig. 3, we show $\Omega(H)/\Omega(0)$. $\Omega(H)$ is a monotonically decreasing function of H with an upper limit of the order of Ω_0 for a wide range of values of H_J/H_A. This is a point of interest for the possible use of domain wall excitations to investigate the interface exchange coupling, as we discuss below.

4. Final Remarks

The shift of the hysteresis in F/AF bilayers is commonly found to be of the order of the anisotropy field of the F-film [10]. Attributing this shift to the interface exchange coupling, one would expect $H_J/H_A \cong 1$. However, H_J/H_A may be much larger than the average effect shown in magnetization measurements. Assuming H_J to be of the same order of magnitude of the intrinsic exchange field of the thin ferromagnetic film, one may have H_J larger than H_A by two to three orders of magnitude [10]. As seen in Fig. 3, for $1 < H_J/H_A < 10^3$ the response of an interface defect pinned domain wall is at the same frequency range of the uniform mode of oscillations of the magnetization in the domains (which is of the order of Ω_0). Thus, except for the large fields that might be required to reach the $\Omega(H) = 0$ condition, it should be possible to observe interface pinned domain wall modes in experimental set-ups designed for ferromagnetic resonance measurements.

We have studied the equilibrium and excitations of a Nèel wall for a single interface step defect pinning center. Our results for the domain wall excitations might be helpful to estimate the interface contact interaction in vicinal interfaces formed on Fe/Cr/Fe wedge samples [23] or in Fe/Cr bilayers, where the Cr substrate surface is oriented along a vicinal direction [22].

In these systems the density of domain wall pinning centers may be controlled by the vicinal angle of the antiferromagnetic spacer wedge [23] or the antiferromagnetic substrate [22]. By choosing small vicinal angles, the terraces formed are wide and the excitations of the domain walls pinned at each step edge should not be significantly affected by the existence, and excitations, of the other domain walls of the periodic stripe domain pattern. Furthermore, by changing the vicinal angle, one also changes the step defect density. Thus, the intensity of the response of the domain wall excitations localized at the step edges can be controlled. This should help to identify the domain wall excitations, contribution to the absorption spectrum.

Our results are valid for any value of the interface field strength. Provided the reduction in the domain wall width is not so strong as to render the continuous medium calculation invalid, the expressions derived in the present work can be used to interpret the absorption lines in FMR experiments due to the excitations of Nèel walls pinned at step defects in F/AF bilayers. In these systems, the detection of domain wall excitations might turn out to be a delicate process. The domains are likely to occupy a large fraction of the interface area. Therefore, the ratio between the interface domain area and the total interface area occupied by domain walls are likely to be large numbers. Thus, the FMR lines due to domain wall absorption might present difficulties to be seen. Although the domain walls might be of microscopic size and constitute a minor fraction of the whole sample, the measurement of the field effects on the frequency of the domain wall excitations provides a promising means for accessing the magnetic structure in a local manner. We showed that, contrary to the long wave-length domain excitations, measured by FMR, the frequency of RDWDM is a decreasing function of the external field and turns zero at the value of the external field, which depins the domain wall from the local pinning center. Furthermore, by choosing the polarization of the excitation field in the FMR experiment parallel to the easy axis, the absorption of energy will occur only in the domain wall region. This kind of technique has been used recently in a ferromagnetic resonance study of two-dimensional wall structures in Co films [27].

In a rough F/AF interface, the borders of the interface regions, separating areas where different sublattices of the substrate adjoin the interface, may act as domain wall nucleation centers. The degree of interface magnetic roughness may thus be estimated from the intensity of the response of narrow domain walls, pinned at the borders of the interface step defects. Thus, domain wall resonance measurements might be helpful to study the scale of interface roughness in F/AF bilayers. This might be particularly helpful in the cases where the interface roughness leads to an almost compensated interface, as recently observed [12]. In this case, the effective field acting on the F layer, representing the exchange interaction with the AF substrate, may be vanishingly small. This might mislead the interpretation of the magnetic coupling, if the exchange bias from hysteresis loops is used to investigate the interface structure.

Acknowledgments

This research was partially supported by the CNPq.

References

[1] Wang, R.W., Mills, D.L., Fullerton, E.E., Mattson, J.E. and Bader, S.D., *Phys. Rev. Lett.* **72** (1994) 920.
[2] Dantas, A.L. and Carriço, A.S., *Phys. Rev. B* **59** (1999) 1223.
[3] Dantas, A.L., Vieira, S.R. and Carriço, A.S., *Phys. Rev. B* **65** (2002) 172414.
[4] Dieny, B., Speriosu, V.S., Parkin, S.S.P., Gurney, B.A., Wilhoit, D.R. and Mauri, D., *Phys. Rev. B* **43** (1991) 1297.
[5] Heide, C., Ziberman, P.E. and Elliott, R.J., *Phys. Rev. B* **63** (2001) 6424.
[6] Elliott, R.J., Chmil, A.I., Epshtein, E.M., Guylaev, Y.G., Krikunov, A.I., Ogrin, Y.F. and Ziberman, P.E., *IEEE Trans. Magn.* **38** (2002) 2869.
[7] Heide, C., Krukinov, A.I., Ogrin, Y.F., Ziberman, P.E. and Elliott, R.J., *J. Appl. Phys.* **87** (2000) 5221.
[8] Chang, C.H., Berera, G.P. and Moodera, J.S., *Appl. Phys. Lett.* **72** (1998) 605.

[9] Mauri, D., Siegman, H.C., Bagus, P.S. and Kay, E., *J. Appl. Phys.* **62** (1987) 3047; Malozemoff, A.P., *J. Appl. Phys.* **63** (1988) 3874; Koon, N.C., *Phys. Rev. Lett.* **78** (1997) 4865; Stiles, M. and McMichael, R.D., *Phys. Rev. B* **59** (1999) 3722.

[10] Nogués, J. and Schuller, I.K., *J. Mag. Mag. Mater.* **192** (1999) 203.

[11] Stamps, R.L., Carriço, A.S. and Wigen, R.P., *Phys. Rev. B* **55** (1997) 6473.

[12] Nolting, F., Scholl, A., Stohr, J., Seo, J.W., Fompeyrine, J., Siegwart, H., Locquet, J.-P., Anders, S., Lunning, J., Fullerton, E.E., Toney, M.F., Scheinfein, M.R. and Padmore, H.A., *Nature* **405** (2000) 767.

[13] Dantas, A.L. and Carriço, A.S., *J. Phys.: Condens. Matter* **11** (1999) 2707.

[14] Dantas, A.L., Carriço, A.S. and Stamps, R.L., *Phys. Rev. B* **62** (2000) 8650.

[15] Dantas, A.L. and Carriço, A.S., *IEEE Trans. Magn.* **36** (2000) 3053.

[16] Dantas, A.L. and Carriço, A.S., *Mater. Sci. Forum* **302** (1999) 101.

[17] Dantas, A.L., Vasconcelos, M.S. and Carriço, A.S., *J. Mag. Mag. Mater.* **226** (2001) 1604.

[18] Niedoba, H., Huberta, A., Mirecki, B. and Puchalska, I.B., *J. Mag. Mag. Mater.* **80** (1989) 379.

[19] Chopra, H.D., Yang, D.X., Chen, P.J., Parks, D.C. and Egelhoff, W.F. Jr, *Phys. Rev. B* **61** (2000) 9642.

[20] Malozemoff, A.P. and Slonckzewsky, J.C., *Magnetic Domain Walls in Bubble Materials* (Academic Press, New York, 1979).

[21] Berger, A. and Hopster, H., *Phys. Rev. Lett.* **73** (1994) 193.

[22] Escorcia-Aparicio, E.J., Choi, H.J., Ling, W.L., Kawakami, R.K. and Qiu, Z.Q., *Phys. Rev. Lett.* **81** (1998) 2144; Escorcia-Aparicio, E.J., Wolfe, J.H., Choi, H.J., Ling, W.L., Kawakami, R.K. and Qiu, Z.Q., *Phys. Rev. B* **59** (1999) 11892.

[23] Unguris, J., Celotta, R.J. and Pierce, D.T., *Phys. Rev. Lett.* **67** (1991) 140; Unguris, J., Celotta, R.J. and Pierce, D.T., *Phys. Rev. Lett.* **69** (1992) 1125.

[24] Machado, F.L.A. and Rezende, S.M., *J. Appl. Phys.* **79** (1996) 6558.

[25] Carara, M., Baibich, M.N., Gundel, A. and Sommer, R.L., *J. Appl. Phys.* **84** (1998) 3792.

[26] Gregg, J.F., Allen, W., Ounadjela, K., Viret, M., Hehn, M., Thompson, S.M. and Coey, J.M.D., *Phys. Rev. Lett.* **77** (1996) 1580.

[27] Ebels, U., Buda, L., Ounadjelaand, K. and Wigen, E., *Phys. Rev. B* **63** (2001) 174437.

CHAPTER 18

DENSITY OF ELECTRONIC STATES IN THE QUANTUM PERCOLATION PROBLEM

Gerardo G. Naumis* and Rafael A. Barrio

Instituto de Fisica, Universidad Nacional Autónoma de México (UNAM), Apdo. Postal 20-364, 01000, Mexico D.F., Mexico

**naumis@fisica.unam.mx*

We present numerical calculations for the density of states in the quantum percolation problem. A pseudogap is found at the center of the lower subband. This behavior is studied in two ways: first, by analyzing the first spectral moments, and second, by making a renormalization of the lattice. This shows that the pseudogap appears as a consequence of frustration effects in the renormalized sublattice.

1. Introduction

Quantum percolation in two dimensions (2D) is a problem with a long history and has been the focus of intensive research, since it is believed that is not possible to have metallic conductivity due to Anderson localization. However, there is controversy in the literature about the corresponding spectral structure and the localization properties. In the literature, the transfer matrix formalism has been used [4–6], and the results seem to agree with the scaling approach [7] in the sense that all states are localized even above the geometrical percolation threshold. However, Meir *et al.* [8] found a localization transition for finite concentration near the geometrical percolation threshold. Later on, it was argued that the previous results were inexact [9] and that there is no transition in the localization type. The study of the quantum percolation problem is also relevant for the random binary alloy, since in the split-band limit, geometrical percolation has an impact not only on the band structure but also on the different degrees of localization of electronic states. Although in this work we do not address the problem of metallic conductivity, we shall focus our attention on the pseudogap that appears in the density of states, which certainly has a direct impact on the optical and other physical properties of the alloy. One of the interesting features is that there is a tail of localized states at the center of the spectrum, and not at the edges, as in the usual Anderson localization. Kirkpatrick and Eggarter (KE) [2] investigated numerically a random binary alloy of 1,500 sites, and they found that in the split-band limit, many degenerate localized states appear exactly at the center of the lower subband, and a pseudogap that starts building up around these localized states. The number of these states at the center and the depth of the pseudogap increases as the concentration of atoms with larger site energy increases. Even more, this pseudogap appears before the concentration of forbidden sites (with infinite site energies) attains percolation. The numerical work by KE has been very useful to show the failure of the coherent potential approximation (CPA) in the split-band limit [3], but not much theoretical work has been done in order to explain their interesting results. It is worthwhile mentioning that the spectrum of the vertex problem in a quasiperiodic Penrose lattice shares many features with the random binary alloy [10–12]. In fact, in the Penrose lattice there are confined states [13, 14] that appear precisely at the center of the band, separated from the rest of the states by a gap [13, 15].

Here, we will concentrate on a disordered two-dimensional square lattice, to show the existence of a pseudogap at the center of the spectrum. This problem is addressed by studying the frustration of the wavefunction in a renormalized sublattice, which is obtained from the bipartite property of the lattice. A lattice is bipartite if it can be subdivided into two alternating sublattices, say α and β, and an electron can only hop from an α site onto a β site or back. Then, the Hamiltonian can be renormalized in such a way that the center of the spectrum is mapped into a band edge [15]. We shall realize that the common features between the Penrose tiling and the binary alloy are due to the bipartite character of their lattices. To analyze the opening of a pseudogap, we start by calculating the first spectral moments of the spectrum, using the Cyrot–Lackmann theorem [16], which relates the local density of states (LDOS) to the topology of the local atomic environment.

The structure of this work is as follows. In Sec. 2, the model is described and some numerical results in large lattices are shown. Then, the first spectral moments of the binary alloy are evaluated, and the tendency for a pseudogap to open is obtained by looking at the normalized fourth moment. In

Sec. 3, we show that the bipartite character allows a renormalization of the Hamiltonian, which leads to the appearance of frustration at the center of the lower subband. Finally, we conclude with some discussion about the relevant features of this model in Sec. 4.

2. Hamiltonian and Spectral Moments

We consider an alloy A_xB_{1-x}, in which the two types of atoms, A and B, are distributed randomly on a square lattice, with concentrations x and $1-x$, respectively. Within the single-band tight-binding approximation, the Hamiltonian with diagonal disorder can be written as

$$H = \sum |i\rangle \epsilon_i \langle i| + \sum_{\langle i,j\rangle} |i\rangle V \langle j|, \tag{1}$$

where $|i\rangle$ is the orbital at site i, V is a constant hopping integral between nearest-neighbor sites, and the diagonal elements are $\epsilon_i = 0(\delta)$ on $A(B)$ sites.

When $\delta \gg ZV$, where $Z = 4$ is the coordination number, the spectrum of Eq. (1) splits into two subbands, one centered at $E = 0$ and the other at $E = \delta$. This is the so-called split-band limit. The states in the subband around zero energy, which we call the A subband, are strongly confined on A atoms. In the limit $\delta \to \infty$, it has been shown [2] that the B atoms can be formally removed from the problem and that the A subband can be studied by using a Hamiltonian restricted to A sites only:

$$H_{AA} = \sum_{i,j\in A} |i\rangle V \langle j|. \tag{2}$$

Thus, the problem for the A subband is similar to a square-lattice percolation problem, because B atoms act as perfect barriers. This problem differs from the geometrical percolation, since the quantum wavefunction could loose its coherency, even beyond the percolation threshold (which is $x_c^{(s)} = 0.59$ for the site problem, and $x_c^{(b)} = 0.50$ for the bond problem in the square lattice). The lack of coherency is partly due to the frustration of the wavefunction, as discussed in the next section. We have verified the results given by KE but for larger lattices. Figure 1(a) shows the A subband for $x = 0.65$, obtained from an average of ten randomly chosen configurations of a 3,969-site square lattice with periodic boundary conditions and $V = 1$. Three main features are visible in the DOS: (i) the spectrum is practically symmetric around $E = 0$, since $\delta = 1000V$; (ii) there is a pseudogap around the center of the spectrum; (iii) there are many degenerate states

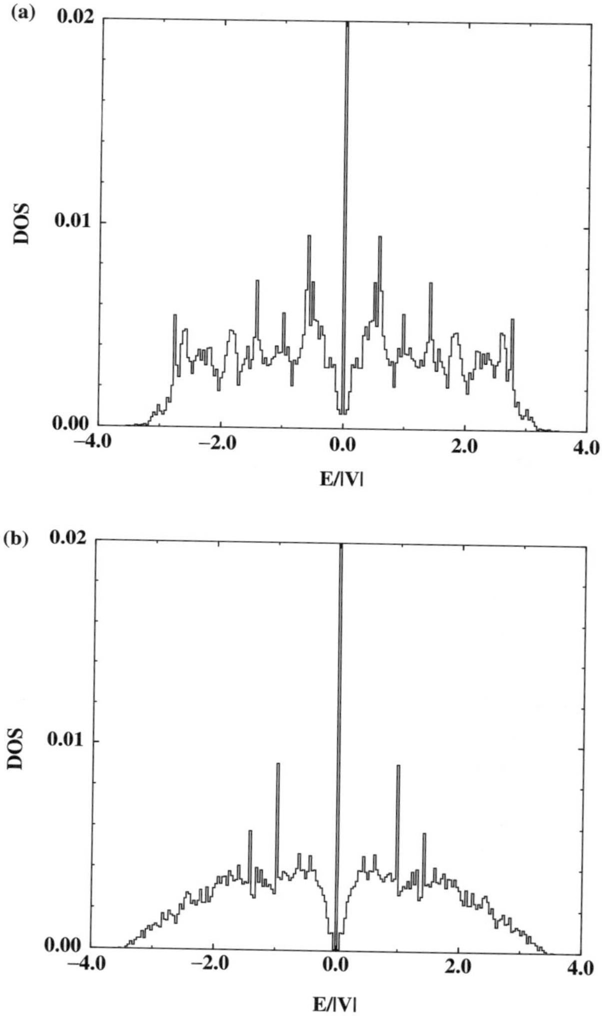

Fig. 1. DOS for ten lattices with 3,969 sites. $V = 1$ and (a) $x = 0.65$, and (b) $x = 0.60$.

at the center. These latter states are strictly confined, even if they can exist in non-isolated clusters [2]. It is worth mentioning that configurations with true-gaps and non-gaps are always statistically present. Therefore, in a strict sense, only a pseudogap should be observed, due to statistical fluctuations. The pseudogap deepens as one approaches the percolation limit, as shown clearly in Fig. 1(b) for $x = 0.60$.

The tendency for a pseudogap to open and the symmetry around $E = 0$ can be obtained from an analysis of the spectral moments. We start by defining the LDOS at site i as $\rho_i(E)$, then the nth moment is [17]

$$\mu_i^{(n)} \equiv \int_{-\infty}^{\infty} (E - H_{ii})^n \rho_i(E)\, dE = \langle i|(H - H_{ii})^n|i\rangle. \tag{3}$$

The last equality is known as the Cyrot–Lackmann theorem [16], from which one can obtain the nth moment by counting all possible closed paths with n steps, starting at site i. In the split-band limit, we can consider the Hamiltonian (2), and site i should be occupied by an A atom.

The moment $\mu_i^{(0)}$ is always unity, because of the normalization condition of the basis ($\langle i|i\rangle = 1$). The first moment, $\mu_i^{(1)}$ is the center of gravity of the LDOS, which is $E = 0$ in this case ($H_{ii} = 0$). The next moment, $\mu_i^{(2)}$, is a measure of the "moment of inertia" of the LDOS with respect to the center of gravity. The third moment, $\mu_i^{(3)}$ measures the skewness about the center of gravity. The fourth moment measures the tendency for a *pseudogap* to form at the middle of the spectrum. A useful criterion to discern this tendency is the dimensionless parameter s_i, defined as [17]:

$$s_i = \frac{\mu_i^{(4)}\mu_i^{(2)} - \left(\mu_i^{(2)}\right)^3 - \left(\mu_i^{(3)}\right)^2}{\left(\mu_i^{(2)}\right)^3}. \tag{4}$$

If $s \geq 1$, the LDOS is unimodal, while for $s < 1$ it is bimodal, which corresponds to two separated peaks in the LDOS [17]. For example, the LDOS of a square lattice is unimodal with a Van Hove singularity at $E = 0$, and $s = 1.25$. A honeycomb lattice has a vanishing LDOS at $E = 0$, and one obtains $s = 0.67$.

In Appendix A, the first four moments of the random binary alloy are calculated in an analytical way, by considering the statistical distribution of paths. In Fig. 2, the full line shows the averaged $\langle s\rangle$ over all sites as a function of x. Notice that $\langle s\rangle < 1$ for $x < 0.55$. This number is very close to the geometrical site percolation threshold.

It is important to notice that confined states at $E = 0$ always give a contribution to the unimodal appearance of the LDOS. In order to examine more exactly the behavior of band states, we should exclude the δ-states at the center. If the fraction of states at $E = 0$ is $f_0(x)$, the band states follow a renormalized LDOS ($\rho_i^*(E)$), related to the complete LDOS by $\rho_i^*(E) = \lambda(x)\rho_i(E)$, where $\lambda(x) = (1 - f_0(x))^{-1}$, due to the normalization condition.

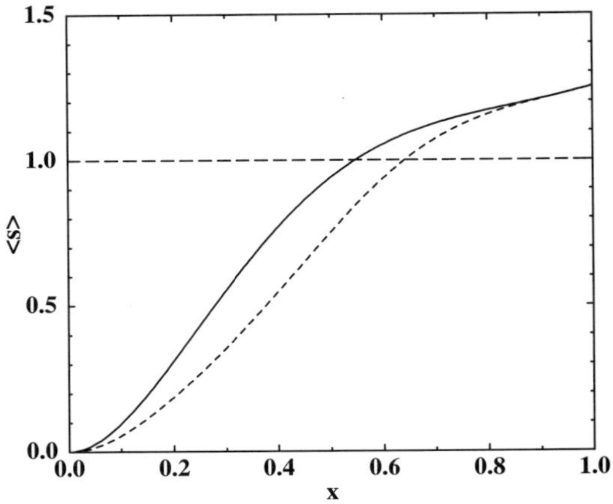

Fig. 2. The full line corresponds to the parameter s. The dashed line is the calculation without considering the δ-states at $E = 0$.

The moments of $\rho_i^*(E)$ should be scaled in the same fashion, that is $\mu_i^{*(n)} = \lambda(x)\mu_i^{(n)}$. The corresponding parameter s^* of $\rho_i^*(E)$ is given by

$$s^* = \frac{s+1}{\lambda(x)} - 1 = s\,(1 - f_0(x)) + f_0(x). \tag{5}$$

The quantity $f_0(x)$ is a function of the concentration, and can be taken from KE, where they used a local counting in finite clusters, and excluding the contribution due to isolated A atoms ($Z = 0$). In Fig. 2 we show the scaled version $\langle s^* \rangle$ as a dashed line. It is interesting to notice that the critical concentration is now $x = 0.64$, which is well beyond the site percolation threshold ($x_c^{(s)} = 0.59$). This fact is consistent with the computational results of KE and ours, where a deep pseudogap in the center of the subband appears even for concentrations higher than $x_c^{(s)}$. We point out that $s = 1$ does not necessarily coincide with the exact percolation limit, since it is only a measure of the mean value of $\rho(E^2)$ in comparison with its average half-width. In the next section, the pseudogap will be analyzed using frustration arguments in a renormalized Hamiltonian.

3. Renormalization of the Hamiltonian

The introduction of B atoms produces a tendency for the spectrum to become bimodal. In order to study this, it is convenient to focus on the renormalized Hamiltonian H_{AA}, which takes advantage of the bipartite nature of the A lattice, once the B atoms are removed. The bipartite character of the A lattice means that it can be separated in two inter-penetrating sublattices, α and β. It is useful to define two orthogonal operators that project each state into one of the sublattices:

$$\begin{aligned} P_\alpha &= \sum_{i\in\alpha} |i\rangle\langle i|, \\ P_\beta &= \sum_{j\in\beta} |j\rangle\langle j|. \end{aligned} \tag{6}$$

Therefore, any eigenvector $|\phi\rangle$ of H_{AA} can be written in terms of these projectors:

$$H_{AA}(P_\alpha + P_\beta)|\phi\rangle = E(P_\alpha + P_\beta)|\phi\rangle. \tag{7}$$

Since H_{AA} produces a hopping in the wave-function between the α and β sublattices, it is clear that

$$H_{AA}P_\alpha|\phi\rangle = EP_\beta|\phi\rangle, \tag{8}$$

$$H_{AA}P_\beta|\phi\rangle = EP_\alpha|\phi\rangle. \tag{9}$$

From these equations, one can see that the spectrum is symmetric around $E = 0$, since if $(P_\alpha + P_\beta)|\phi\rangle$ is an eigenvector with eigenvalue E, $(P_\alpha - P_\beta)|\phi\rangle$ is also an eigenvector with eigenvalue $-E$.

We can decouple the sublattices by further applying H_{AA} to Eqs. (8) and (9):

$$H_{AA}(H_{AA}(P_i|\phi\rangle)) = H_{AA}^2(P_i|\phi\rangle) = E^2(P_i|\phi\rangle), \tag{10}$$

where $i = \alpha, \beta$. Thus, the projection of an eigenvector in each sublattice is a solution of the squared Hamiltonian. Observe that the eigenvalues of H_{AA}^2 are positive definite, and their eigenstates are, at least, doubly degenerate. This spectrum can be regarded as the folding of the original spectrum of H_{AA} around $E = 0$, in such a way that the two band edges of H_{AA}, are mapped into the highest eigenvalue of H_{AA}^2, while the states at the center of the original band are now at the minimum eigenvalue of the squared Hamiltonian (E^2).

When $x < x_c^{(s)}$, all A clusters are finite. Therefore, confinement effects are expected, in particular, the band width of H_{AA}^2 is reduced. This helps to explain the appearance of a gap at the center of the A subband of H_{AA} when $x < x_c^{(s)}$, but it does not predict a pseudogap when $x > x_c^{(s)}$.

The important property of the renormalized Hamiltonian H^2_{AA} is that the states near $E = 0$ have an antibonding nature (the phase between neighbors is π), and since H^2_{AA} contains odd rings, we expect that frustration of the wavefunction can prevent the spectrum from reaching its minimum eigenvalue in a continuous form [15]. Furthermore, since there is a cost in energy due to frustration, wave-functions tend to avoid regions of higher frustration, and the states begin to localize in regions of lower frustration [15]. The amount of frustration can be estimated from the numerical results and using statistics. One can show that this frustration augments with disorder. To see this, it is convenient to separate the contribution for each eigenenergy in three parts: one is due to the self-energy, and the other two are given by the bonds with positive (bonding) and negative (antibonding) contribution to the energy. This separation goes as follows. First, we write the equation of motion for H^2_{AA},

$$(E^2 - Z_i V^2) c_i(E) = \sum_{j \neq i} \left(H^2_{AA}\right)_{ij} c_j(E), \tag{11}$$

where $c_i(E)$ is the amplitude of the wave-function at site i for an eigenenergy-energy E. After summing over all sites i and using the normalization condition of the wave-function, Eq. (11) becomes

$$E^2 = \sum_i Z_i V^2 |c_i(E)|^2 + \sum_{j \neq i} \left(H^2_{AA}\right)_{ij} c_j(E) c_i^*(E) \tag{12}$$

$$\equiv C_1(E^2) - C_2(E^2) + C_3(E^2), \tag{13}$$

where $C_1(E^2) = \sum_i Z_i V^2 |c_i(E)|^2$ is the contribution of the self-energies, which depends on the local coordination. $C_2(E^2) = |\sum'_{i,j} (H^2_{AA})_{ij} c_j(E) \times c_i^*(E)|$, where the prime means that one considers only those bonds whose product $c_j(E)c_i^*(E)$ is negative. This is an antibonding contribution. Finally, $C_3(E^2)$ is similar to $C_2(E^2)$, except that the summation is over bonds with positive $c_j(E)c_i^*(E)$. This equation is valid for all E in the spectrum. At the upper band edge, $C_2(E^2)$ is zero because in a perfect bonding state all the site amplitudes have the same sign. The state $E^2 = 0$ corresponds to a configuration where the sign of the wave amplitude alternates between nearest neighbors, and the bond contribution $(C_3(E^2) - C_2(E^2))$ is equal to the self-energy. $C_3(E^2)$ is a measure of the contribution of bonds that are frustrated, while $C_3(E^2) - C_2(E^2)$ gives the amount of frustration compared with the antibonding term.

These three contributions for the same lattices as in Fig. 1(a) are shown in Fig. 3. Notice that $C_1(E^2)$ (crosses) and $C_3(E^2)$ (circles) decrease

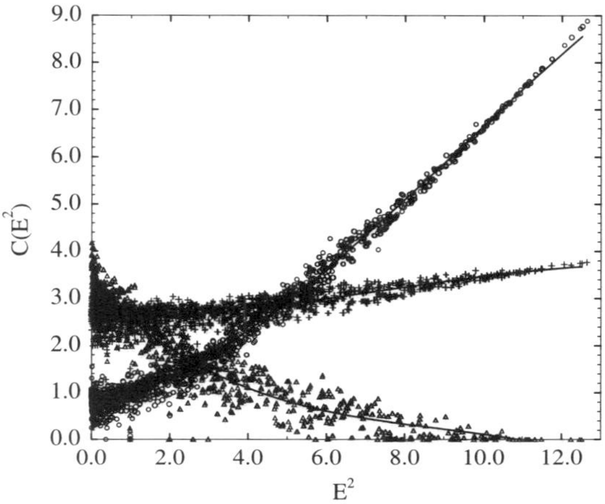

Fig. 3. Contributions to E^2 from $C_2(E^2)$ (circles), $C_3(E^2)$ (triangles), and $C_1(E^2)$ (crosses). These data were obtained from the calculation shown in Fig. 1(a).

towards $E = 0$. The contribution $C_2(E^2)$ (triangles) rises from zero at the band edge, to a maximum value near $E = 0$, except at some energies where $C_2(E^2)$ is zero. A detailed analysis reveals that each of them is a degenerate state, producing high peaks in the DOS. These states correspond to isolated clusters and produce sharp peaks in the DOS (Fig. 1(b)). For example, the state at $E = 1$ corresponds to a doublet of A sites, surrounded by B atoms.

3.1. *Frustration in the lower band edge*

To estimate the effects of frustration as a function of the concentration of impurities, we need $C_3(E^2) - C_2(E^2)$. This can be done, if first we find bounds for $C_1(E^2)$. Writing Z_i as an average $\langle Z \rangle$ plus a fluctuation part δZ_i in the expression for $C_1(E^2)$, one gets

$$C_1(E^2) = \langle Z \rangle V^2 + V^2 \sum_{i=1}^{N} \delta Z_i |c_i(E)|^2. \tag{14}$$

The amplitude $c_i^2(E)$ can be written as an average plus a fluctuation, $\langle c^2(E) \rangle + \delta c_i^2(E)$, and Eq. (14) becomes

$$C_1(E^2) = \langle Z \rangle V^2 + V^2 \sum_i \delta Z_i \delta c_i^2(E), \tag{15}$$

where we have used the fact that the sum over all sites of the coordination fluctuations is zero. The last term in Eq. (15) is not zero, and corresponds to a correlation between amplitude and coordination fluctuations. This contribution is bounded in a statistical sense, since it attains a maximum value when in all sites the sign of the amplitude fluctuation is the same as those fluctuations of the coordination. In a similar way, a minimum is obtained when the fluctuations have opposite signs:

$$-\sum_{i=1}^{N} |\delta Z_i||\delta c_i^2(E)| \leq \sum_{i=1}^{N} \delta Z_i \delta c_i^2(E) \leq \sum_{i=1}^{N} |\delta Z_i||\delta c_i^2(E)|. \tag{16}$$

The size of the fluctuations in the coordination number is estimated by using the standard deviation of the distribution function of the coordination ($P(Z)$), which is a binomial distribution (see Appendix A):

$$\sum_{i=1}^{N} |\delta Z_i||\delta c_i^2(E)| \approx \sqrt{4x(1-x)} \sum_{i=1}^{N} |\delta c_i^2(E)| \leq \sqrt{4x(1-x)}. \tag{17}$$

Finally, we get the statistical bounds for $C_1(E^2)$:

$$V^2\left(4x - \sqrt{4x(1-x)}\right) \leq C_1(E^2) \leq V^2\left(4x + \sqrt{4x(1-x)}\right). \tag{18}$$

This equation can be compared against the results shown in Fig. 3, for $x = 0.65$. Equation (18) gives the maximum value of $C_1(E^2)$ as 3.56, in close agreement with 3.58 observed in the upper band edge of Fig. 3. The calculated lower bound is 1.61, in close agreement with the numerical calculations. Notice that these bounds are not strict, due to their statistical nature.

Now, a lower bound for $C_3(E^2) - C_2(E^2)$ can be obtained from the condition $E^2 \geq 0$. Using this condition, Eqs. (13) and (18), we get

$$C_3(E^2) - C_2(E^2) \geq -V^2\left(4x + \sqrt{4x(1-x)}\right). \tag{19}$$

From this last result, one can say that the frustration increases with the concentration of impurities. If there is no correlation between fluctuations on amplitude and coordination, the lower bound is $-4x$, but if we allow correlation, a lower energy can be reached by reducing the frustration.

We can go further, and obtain a bound for $C_3(E^2)$ alone. The key idea is to get a new equation to separate $C_3(E^2)$ from $C_2(E^2)$. This equation is obtained by observing that in the bonding limit (E_+^2), all the bonds are frustrated, and from the expected value of the energy calculated for a bonding state, we obtain

$$E_+^2 = C_1\left(E_+^2\right) + C_3\left(E_+^2\right). \tag{20}$$

$C_3(E_+^2)$ can be related with $C_3(0)$ and $C_2(0)$, since if we neglect amplitude variations, the main difference between the bonding and antibonding limit is the sign of the amplitude of the wave function between neighbors. In other words, the total number of bonds must remain constant, and if we change the sign of the contribution from bonds with an antibonding nature in the lowest eigenvalue, we obtain a maximum value to the energy. Amplitude variations can only reduce the frustration; this leads to the following inequality:

$$C_3\left(E_+^2\right) \geq C_2(0) + C_3(0). \tag{21}$$

In the perfect square lattice, $C_3(E_+^2) = C_2(0) + C_3(0)$, since each site in H^2 is connected with eight sites: four first neighbors by bonds with hooping integrals $2V^2$, and four second neighbors with hoopings V^2. Thus, when $E = 0$, $C_2(0) = 8V^2$ and $C_3(0) = 4V^2$, since the sign of the wave function alternates between nearest neighbors. In the bonding limit, all the amplitudes have the same sign, and $C_3(E_+^2) = 12V^2$. Using Eq. (12) and that $C_1(E^2) = 4$, one can verify that these values produce the right band edges (0 and 16). As x goes to zero, the difference between $C_3(E_+^2)$ and $C_2(0) + C_3(0)$ grows.

Eliminating $C_2(0)$, using Eq. (13), and the condition that $E^2 \geq 0$, we get

$$C_3(0) \leq \frac{E_+^2 - C_1\left(E_+^2\right) - C_1(0)}{2} = \frac{E_+^2}{2} - 4x. \tag{22}$$

E_+^2 is the band width in H_{AA}^2, and can be calculated using the method of fluctuations, as shown in Appendix B. The statistical bound for the frustration is

$$C_3(0) \leq 6x^2 + 2x\left(\sqrt{3x(1-x)} - 1\right). \tag{23}$$

3.2. *Estimation of the pseudogap position*

In the last subsection, we obtained the frustration that allows to reach the minimum eigenvalue $E = 0$, attained when the correlations in the fluctuations play an important role. The energy where the pseudogap begins (Δ), we need the frustration $C_3(\Delta^2)$ when we do not allow these correlations in the fluctuations near $E = 0$. This could be calculated using a variational procedure similar to that made for Penrose tiling [15]. However, due to the statistical nature of this system, such a calculation is extremely difficult. An easier approach takes advantage of the following observation. $C_3(E^2)$ is two times the number of frustrated bonds (since each bond is shared by two sites), and the number of frustrated bonds is proportional to the number of triangles that appear in H_{AA}^2. This number is proportional to the number of paths with three hops that start and end at the same site, which is the third moment of H_{AA}^2 ($\mu_{H_{AA}^2}^3$). Then, we have for the value of $C_3(E^2)$ near $E = 0$:

$$C_3(E^2) = K\mu_{H_{AA}^2}^{(3)} = K\mu_{H_{AA}}^{(6)}, \tag{24}$$

where K is a constant ($= 3$ in the perfect square lattice) for a given concentration, and we used the fact that the third moment of H_{AA}^2 is the sixth moment of H_{AA}. But from the second section, we know that the states at $E = 0$ produce a weight at $E = 0$ that affects the moments, and we also showed that this effect can be avoided by defining a renormalized set of moments $\mu^{*(n)}$. In a similar way, we can obtain a renormalized value of $C_3(E^2)$, which does not give weight to the states at $E = 0$, and can be associated with the value of the frustration without the fluctuations at a higher energy Δ. Therefore,

$$C_3(\Delta^2) = K\mu_H^{*(6)} = \frac{1}{1 - f_0(x)} C_3(0) \approx (1 + f_0(x))C_3(0). \tag{25}$$

In Eq. (12) we can substitute this result:

$$\Delta^2 = C_1(\Delta^2) - C_2(\Delta^2) + C_3(\Delta^2) \geq \tag{26}$$

$$C_1(0) - C_2(0) + C_3(\Delta^2) \approx f_0(x)C_3(0), \tag{27}$$

where $C_2(0) \geq C_2(\Delta^2)$, and $C_1(\Delta^2) \approx C_1(0)$. Finally, using Eq. (23) we obtain

$$\Delta \geq \sqrt{f_0(x)\left(6x^2 + 2x\left(\sqrt{3x(1-x)} - 1\right)\right)}. \tag{28}$$

In Fig. 4, we show a plot of this equation, giving $\Delta = 0.3$ for $x = 0.65$. This formula is only valid for $x > x_c$, since for lower x, the quantum confinement begins to play an important role, and localization does not allow the reduction of the energy, because there is a competition between frustration and quantum confinement effects, which in fact turns the pseudogap into a real gap.

The frustration of the wave-function also leads to localization, since the energy is reduced by localizing the amplitude in zones of less frustration. This tendency is clear in the inverse participation ratio ($I(E_n)$), defined as

$$I(E_n) = \sum_{i \in A}^{N_A} \|\langle n|i\rangle\|^4,$$

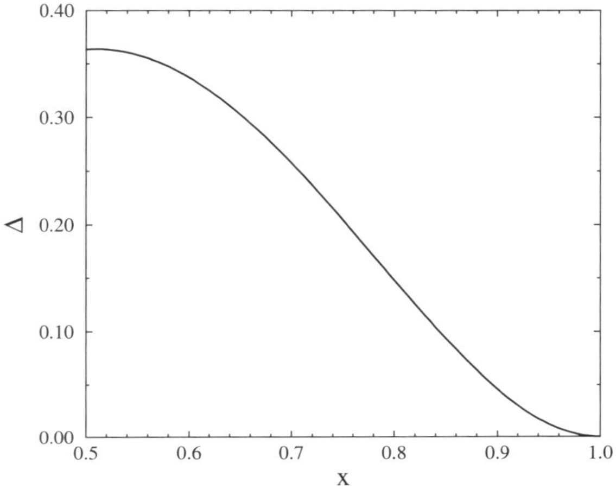

Fig. 4. Pseudogap position as a function of x.

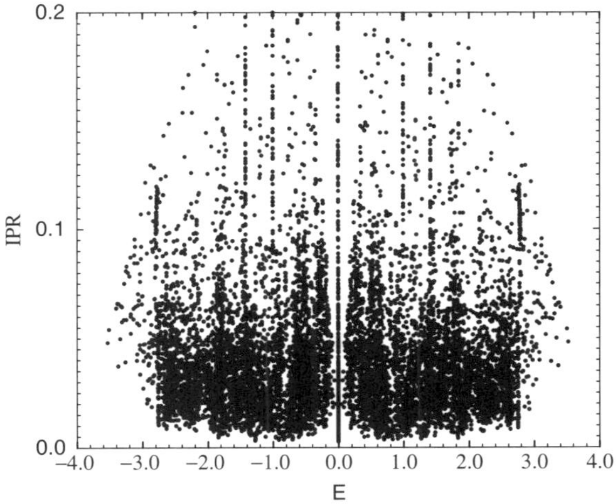

Fig. 5. Inverse participation ratio.

where $\langle n|i\rangle$ is the value of the nth eigenfunction at site i. For extended states, $I(E_n) \sim 1/N$, while for localized states it has a value that does not depend on N. Figure 5 shows that the minimum localization length is increased at the center of the spectrum, and the position where the pseudogap starts is clearly defined.

4. Conclusions

We calculated the first moments of a random binary alloy in a square lattice by using the Cyrot–Lackmann theorem. The results show that there is a transition of the spectrum from unimodal to bimodal behavior as a function of the concentration of impurities. This transition occurs near the geometrical percolation threshold. These ideas are made clear by using the bipartite symmetry of the square lattice once the impurity atoms are removed. This allows one to focus the attention on only one sublattice, which defines a "squared" Hamiltonian that contains odd member rings in the disordered alloy. In this picture, the states near the center of the spectrum are mapped to the lower band edge, and require a large number of nodes, and thus frustration effects are responsible for the depletion of the LDOS near the minimum eigenvalue of H^2.

Acknowledgments

We would like to thank Prof. R.J. Elliott for useful discussions. This work was supported by DGAPA-UNAM Project No. IN108502.

Appendix A: The first spectral moments

To calculate all the required moments in the split-band limit, we need to count all the possible paths that visit A sites that start and return to the same site. One must take into account all possible local configurations of disorder. Thus, Eq. (3) must be considered in a statistical way, by including the probability of a path connecting A sites with n hops. We can define the configurational averaged spectral moments $\langle \mu_i^{(n)} \rangle$ as

$$\langle \mu_i^{(n)} \rangle = \sum_{j_1,\ldots,j_{n-1} \in A} P(i, j_1, \ldots, j_{n-1}) H_{ij_1} H_{j_1 j_2} \cdots H_{j_{n-1} i}, \tag{A.1}$$

where $P(i, j_1, j_2, \ldots, j_{n-1})$ is the probability of a given path.

All the odd moments are zero, because there is no possibility of returning to the starting point with an odd number of steps in the square lattice. If the B sites are forbidden, the clusters of A sites retain this property, while if δ is smaller, the odd moments are not zero, and then the subband A is not symmetric around $E = 0$ any longer.

The second moment is always equal to the local coordination on A sites:

$$\mu_i^{(2)} = Z_i V, \tag{A.2}$$

where Z_i is the number of A type atoms that are first neighbors of i. There are only five different local configurations, with coordination $0, 1, 2, 3$ and 4, respectively. The probability of each coordination $(P(Z))$ around a given site is given by a binomial distribution:

$$P(Z) = C_Z^4 x^Z (1-x)^{4-Z}, \tag{A.3}$$

where C_Z^4 are the combinations of four in Z. This factor takes care of the different geometrical possibilities in which each configuration can occur. The second moment of the DOS corresponds to the sum of the LDOS at all sites. This sum over sites can be performed:

$$\langle \mu^2 \rangle = \sum_i \mu_i^2 = V \sum_{Z=0}^{4} C_Z^4 x^Z (1-x)^{4-Z} Z = 4Vx. \tag{A.4}$$

This number gives an estimation of the band width (W), which for the present case is $W = 2\mu^2 = 8Vx$.

The fourth moment calculation requires counting many different configurations and paths. However, the calculus is simple since we only need to take into account how many different sites are visited on each path, and each different A site has a probability x. Thus, by counting paths in the square lattice, we get

$$\langle \mu^4 \rangle = 4x + 24x^2 + 8x^3.$$

Appendix B: Upper band edge ($\mathbf{E_+^2}$)

The bonding limit of the energy spectrum corresponds to a maximum value of E_+^2, attained when $C_3(E^2) - C_2(E^2)$ and $C_1(E^2)$ are maxima. From Eq. (18), the maximum value of $C_1(E^2)$ is $4x + \sqrt{4x(1-x)}$. The maximum value of $C_3(E^2) - C_2(E^2)$ is obtained from observing that if all the amplitudes have the same sign,

$$\sum_{i,j} \left(H_{AA}^2\right)_{ij} c_i^*(E) c_j(E) \leq \left\langle \left(H_{AA}^2\right)_{ij} \right\rangle + F, \tag{B.1}$$

where F are the fluctuations in the distribution of the squared Hamiltonian.

It is easy to see that $\langle (H_{AA}^2)_{ij} \rangle$ in H_{AA}^2 is exactly the number $Z\langle \sum_j (Z_j - 1) \rangle$. Averaging over Z, one gets

$$\left\langle \left(H_{AA}^2\right)_{ij} \right\rangle = 3xV^2 \sum_{Z=0}^{Z=4} P(Z) Z = 12x^2 V^2. \tag{B.2}$$

The size of the fluctuations is evaluated by an average of the fluctuations for each coordination number:

$$F \approx \sqrt{3x(1-x)}V^2 \sum_{Z=0}^{Z=4} P(Z)Z = 4xV^2\sqrt{3x(1-x)}. \tag{B.3}$$

The band edge of H_{AA}^2 is given by the sum both contributions, (B.2) and (B.3):

$$E_+^2 = \pm V^2\Big(12x^2 + 4x(1+\sqrt{3x(1-x)}) + \sqrt{4x(1-x)}\Big). \tag{B.4}$$

This method gives a much better estimation for the upper band edge, which is usually approximated [17] by $\langle Z\rangle V$. The square root of E_+^2 gives an estimation of the band edges in H_{AA}. For example, if $x = 0.65$, this formula gives $E_+ = 3.3$. This approximation can be compared with Fig. 1, where the band edge is near 3.4. The usual estimation $4x = 2.6$ is not as good as Eq. (B.4). Equation (B.4) gives a better estimation because it includes information about H_{AA}^2 and the size of the fluctuations, which is related with the size of the exponential Lifshitz tails that appear in the band edges.

References

[1] Mott, N.F. and Davis, E.A., *Electronic Processes in Non-Crystalline Materials* (Oxford University Press, Oxford, 1979).
[2] Kirkpatrick, S. and Eggarter, T.P., *Phys. Rev. B* **6** (1972) 3598.
[3] Ziman, J.M., *Models of Disorder* (Cambdrige University Press, Cambridge, 1979).
[4] Soukoulis, C.M., Economou, E.N. and Grest, G.S., *Phys. Rev. B* **36** (1987) 8649.
[5] Berkovits, R. and Avishai, Y., *Phys. Rev. B* (1996) R16125.
[6] Soukoulis, C.M., Li, Q. and Grest, G.S., *Phys. Rev. B* **45** (1992) 8649.
[7] Abrahams, E.N., Anderson, P.W., Licciaredello, D.C. and Ramakrishnan, T.V., *Phys. Rev. Lett.* **42** (1979) 673.
[8] Meir, Y., Aharony, A. and Harris, B., *Europhys. Lett.* **10** (1989) 275.
[9] Soukoulis, C.M. and Grest, G.S., *Phys. Rev.* **44** (1991) 4685.
[10] Naumis, G.G., Barrio, R.A. and Wang, C. in *Proc. 5th Int. Conf. Quasicrystals*, eds. Janot, Ch. and Mosseri, R. (World Scientific, Singapore, 1995), p. 514.
[11] Barrio, R.A., Naumis, G.G. and Wang, Ch., in *Current Problems in Condensed Matter*, ed. Morán López, J.L. (Plenum Press, New York, 1998), p. 283.
[12] Naumis, G.G., *J. Phys. C: Condens. Matt.* **11** (1999) 7143.
[13] Kohmoto, M. and Sutherland, B., *Phys. Rev. Lett.* **56** (1986) 2740.

[14] Arai, M., Tokihiro, T., Fujiwara, T. and Kohmoto, M., *Phys. Rev. B* **38** (1988) 1621.
[15] Naumis, G.G., Barrio, R.A. and Wang, Ch., *Phys. Rev. B* **50** (1994) 9834.
[16] Cyrot-Lackmann, F., *J. Phys. Chem. Solids* **29** (1968) 1235.
[17] Sutton, A.P., *Electronic Structure of Materials* (Clarendon Press, Oxford, 1993), p. 66.
[18] Cohen, M., in *Topological Disorder in Condensed Matter*, eds. Yonezawa, F. and Ninomiya, T., Springer Series in Solid State Sciences, Vol. 46 (Springer, New York, 1983), p. 122.

CHAPTER 19

POWER TERMS IN THE CONSTRUCTION OF THERMODYNAMIC FUNCTIONS FOR MELTING DESCRIPTION

F.L. Castillo Alvarado and G. Ramirez Damaso

Escuela Superior de Fisica y Matemáticas,
Instituto Politecnico Nacional, Edif. 9 V.P. "ALM",
Zacatenco, DF 07738, Mexico

J.H. Rutkowski and L. Wojtczak

Department of Solid State Physics, University of Łódź,
Pomorska 149/153, 90-236 Łódź, Poland

A linear combination of power terms considered as an effective form of the Gibbs free energy for a homogeneous system is constructed in order to describe its behavior in the neighborhood of the melting temperature. The construction is based on a common solution of the equation of state, which is determined by the thermodynamic functions for solid and liquid phases, treated separately. The exact, *ab initio* Gibbs free energy given by the Los Alamos National Laboratory group is taken, as an example, into account. One of the conclusions seems to be that the relation between volume and the crystallinity order parameter is now generalized and, first of all, very well established due to the self-consistent character of the presented construction. In this case, the construction needs the effective form extended to a linear combination of power terms with respect to the crystallinity order parameter representation. Moreover, two examples of the considered construction are reported for the generic forms of the van der Waals model. The examples correspond then to the polynomial form of the constructed potential which belongs then to a narrower class of functions.

1. Introduction

As is well known, melting, or more precisely speaking, bulk melting refers to the phenomenon consisting of a transition from the solid phase to the liquid phase of a considered sample at the melting temperature $T_{\rm m}$, which is globally defined for a given material [1]. Bulk melting is an example of the phase transition reflecting the properties of the first order transition, e.g., characterized by a jump of the order parameter, which has a discontinuous character at the melting temperature.

Surface melting is considered as the appearance of a thin quasi-liquid layer on top of its own solid surface at the surface melting temperature $T_{\rm sm}$, which is below the bulk melting temperature $T_{\rm m}$ [1]. The thickness of the melt layer increases with gradual increasing temperature. It is an example of the second order phase transition. It is worthwhile stressing that the surface melting appears in the homogeneous thermodynamic bath without any gradient in temperature. In the last two decades, the surface melting has been a topical research area from the experimental and theoretical points of view [1–4] (and references therein). Its description needs, among others, the knowledge of the thermodynamic potential for thermodynamically homogeneous systems.

The aim of the present paper is to consider several examples of the Gibbs free energy (GFE) for bulk material in the context of its application to the surface melting discussion.

2. Gibbs Free Energy Construction for Homogeneous Systems

The GFE for the surface melting description was successfully introduced in the form of two intersecting parabolas of equal curvatures taken with respect to the crystallinity order parameter m [5]. In the present paper, we assume this potential as a reference for the forthcoming discussion.

This type of the GFE has been also used for the construction of the polynomial form found in [6] for a lead specimen. The construction consists of finding the polynomial shape of the GFE $f(m,T)$ with respect to the variable m in such a way as to lead to the equation of state $\delta f = 0$, equivalent to that obtained on the basis of the reference potential. It is worthwhile noticing that the polynomial form of the GFE $f(m,T)$ is common for both the phases while the reference potential describes each phase separately. The equation of state has minima at the same points, $m = 0$, $m = 1$, for both forms of the potential. Their maximum is also localized at the same value of $m^* \in (0,1)$, which corresponds to the intersecting point of two parabolas, $f_{\rm l}(m,T)$ and $f_{\rm c}(m,T)$, determining the reference GFE, where indices "l" and "c" stand for liquid and solid phases, respectively.

Now, in the present paper, the construction is discussed in its generalized form when a linear combination of the power terms of m (LCPT) instead of the polynomial shape with respect to m for the potential is taken into account. The extended version allows us to consider the reference functions $f_{\rm l}(m,T)$ and $f_{\rm c}(m,T)$ in more general forms than only parabolic ones.

Of course, the effective form of LCPT–GFE should have the minima at the same points, i.e., $m = 0$, $m = 1$. In this case, the simplest but sufficiently general form of the equation of state equivalent to the variational principle $\delta f(m) = 0$ can be given by

$$\eta(m) m^{\gamma} \left(m^{\gamma} - 1\right) \left(m^{\gamma} - m^{*\gamma}\right) = 0, \tag{1}$$

with an arbitrary function $\eta(m) > 0$ and a real, arbitrary power γ (the power terms m^{γ} and m are basic in order to constitute LCPT for GFE) while the value m^* satisfies the equation

$$f_{\rm l}(m^*,T) = f_{\rm c}\left(m^*,T\right), \tag{2}$$

where the normalization conditions are introduced:

$$f(m = 1, T\,) = 0 \tag{3}$$

and

$$f(m = 0, T) = \Lambda \tag{4}$$

with

$$\Lambda = f_{\rm l}\left(m = 0, T\right) - f_{\rm c}\left(m = 1, T\right). \tag{5}$$

From the relations of thermodynamics, the excess of the energy (Eq. (5)) in the melting area depends linearly on T, i.e.,

$$\Lambda = L_{\rm m}\left(1 - \frac{T}{T_{\rm m}}\right) \tag{6}$$

with $L_{\rm m}$ standing for the latent heat of melting. Confirming that Λ is a linear function of temperature, at least in the neighborhood of $T_{\rm m}$, we can determine the latent heat of melting $L_{\rm m}$ as well as the bulk melting temperature $T_{\rm m}$. In the present paper, the latent heat and the free energy is given in units of the energy per volume.

Equation (1) for integer γ reduces to the polynomial type of the potential, which will be considered in Sec. 3. In general, Eq. (1) is satisfied at

$m = 0$ and $m = 1$ independently of γ, so that the change of the trajectory m^γ in the interval $m \in (0, 1)$ can be chosen arbitrarily so far as γ is not determined. However, the parameter γ is connected with the behavior of m^*, whose structure is given by

$$(m^*)^\gamma = (m_0^*)^\gamma - \frac{2}{\gamma+1}\frac{\Lambda}{\alpha}, \tag{7}$$

where m_0^* does not depend on temperature and is determined at $T_{\rm m}$ while α is an adjustable parameter corresponding to the parabola curvature in the case of the reference potential. The temperature dependence of $(m^*)^\gamma$ is then linear. The verification of this property is a test for the self-consistence of the theory. We choose

$$\eta(m) = \frac{\alpha^*}{m^{(\gamma-1)}}, \tag{8}$$

where the constant α^* refers to the normalization of $f(m)$ given by conditions (3) and (4). In this case, the self-consistence of the theory requires one more condition, which is non-trivial. Namely, the term $f(m)$ independent of Λ or $m = 1$ should vanish. This requirement assures the non-vanishing constant α^* by the relation between γ and m_0^*, namely

$$(\gamma + 1)(m_0^*)^\gamma = 1, \tag{9}$$

which leads to γ given by the position m^* of the barrier separating both the phases at given temperature while the height of the barrier is determined by the maximum of $f(m^*, T)$ due to relation (2).

Taking into account condition (9), we can find the solution of (1) in the form

$$f(m,T) = \Lambda + \alpha^* \left[\frac{1}{2} m^{*\gamma} m^2 - \frac{1}{\gamma+2}\left(1 + m^{*\gamma}\right) m^{\gamma+2} + \frac{1}{2\gamma+2} m^{2\gamma+2}\right] \tag{10}$$

when relation (7) is satisfied, and

$$\alpha^* = \frac{(\gamma+1)(\gamma+2)}{\gamma}\alpha, \tag{11}$$

while m_0^* is given by (9).

The temperature dependence of $f(m,T)$ correctly represents the first order character of the bulk transition. For $T < T_{\rm m}$ the function $f(m,T)$ has its minimum at $m = 1$ and the behavior $f(m,T)$ describes the solid phase. For $T > T_{\rm m}$ the minimum of $f(m,T)$ appears at $m = 0$. This case corresponds to the liquid phase description. When T approaches $T_{\rm m}$ from below, the order parameter m jumps discontinuously from 1 to 0 at

$T = T_m$. The GFE (10) has its maximum inside the interval $m \in (0, 1)$ when the temperature $T \in (T_{\rm sm}, T_{\rm m})$ for $T_{\rm sm}$ calculated from the condition

$$\left(\frac{d^2 f(m,T)}{dm^2}\right)_{m=0} = 0. \tag{12}$$

This leads to

$$\frac{T_{\rm sm}}{T_{\rm m}} = 1 - \frac{\alpha}{2L_{\rm m}}. \tag{13}$$

Thus, we see that the minimal surface melting temperature $T_{\rm sm}$ given by (13) does not depend on γ, i.e., $T_{\rm sm}$ is independent of the power γ in GFE (10).

3. Examples of Polynomial Shapes for the Bulk Potential

Example 1. $\gamma = 1$.
In particular, for $\gamma = 1$, $\eta(m) = \alpha^*$, integrating Eq. (1) with respect to m, we obtain

$$f(m,T) = \Lambda + 6\alpha\left(\frac{1}{2}m^* m^2 - \frac{1}{3}(1+m^*)\, m^3 + \frac{1}{4}m^4\right), \tag{14}$$

which corresponds to the result reported in [6] for $\alpha^* = 6\alpha$. The self-consistency condition is satisfied when

$$m^* = \frac{1}{2} - \frac{\Lambda}{\alpha}, \tag{15}$$

i.e., $m_0^* = 1/2$, which takes place in the case of the reference GFE given in the form of two intersecting parabolas [5]:

$$f_{\rm l}(m,T) = \frac{1}{2}\alpha m^2 + \Lambda, \tag{16}$$

$$f_{\rm c}(m,T) = \frac{1}{2}\alpha(1-m)^2, \tag{17}$$

with the equal curvatures α.

We would like to stress that Eq. (1) for $\gamma = 1$ has no solution when $m_0^* \neq 1/2$ because condition (9) is not satisfied. This fact is of great importance in the context of the result for the potential of two intersecting parabolas with non-equal curvatures. We have then: $m_0^* = 1/(1+\varepsilon)$, where $\varepsilon = \alpha_{\rm l}/\alpha_{\rm c}$ for $\alpha_{\rm l}$ and $\alpha_{\rm c}$ replacing α and (16) and (17), respectively. Thus, we can see that $m_0^* = 1/2$ only for $\varepsilon = 1$, while $\varepsilon \neq 1$ leads to $m_0^* \neq 1/2$. This means that Eq. (14) is satisfied only for $\varepsilon = 1$, i.e., for the intersecting parabolas with equal curvatures. For $\varepsilon \neq 1$, we have to consider a more general approach in Eq. (10) for $\gamma \neq 1$.

Example 2. $\gamma = 2$.
In the case $\gamma = 2$, we have $\eta(m) = \alpha^*/m$. The solution of Eq. (1) then takes the form

$$f(m,T) = \Lambda + 6\alpha\left(\frac{1}{2}(m^*)^2 m^2 - \frac{1}{4}\left(1 + (m^*)^2\right) m^4 + \frac{1}{6}m^6\right) \tag{18}$$

with relation (7) given by

$$(m^*)^2 = (m_0^*)^2 - \frac{2}{3}\frac{\Lambda}{\alpha}. \tag{19}$$

This leads to $m_0^* = \sqrt{3}/3$ when the self-consistency condition is satisfied ($\alpha^* = 6\alpha$).

The GFE corresponds to the potential given in the form of two intersecting curves:

$$f_l(m,T) = \frac{1}{2}\alpha\left(m^4 + m^2\right) + \Lambda, \tag{20}$$

$$f_c(m,T) = \frac{1}{2}\alpha\left(1 - m^2\right)^2, \tag{21}$$

which represent a slightly modified potential discussed in [11].

4. The Linear Combination of Power Terms for the Gibbs Free Energy of Aluminium

The construction presented in Sec. 2 concerns the potential f dependent on m, i.e., $f = f(m,T)$. However, various functional forms of the thermodynamic Gibbs free energy depend on the volume V and temperature T, i.e., $f = f(V,T)$ (e.g. [1]). Thus, it is important for the compatibility of different kinds of variable dependences to find the relation between the volume V and the crystallinity order parameter m. For the sake of compatibility, we put the relation $V = V(m)$ in the form

$$V = V_l - (V_l - V_c)m^\gamma, \tag{22}$$

where $V_l = V_l(T)$ and $V_c = V_c(T)$ are the equilibrium volumes for liquid and solid phases, respectively. Relation (22) satisfies the conditions that $V = V_l$ for $m = 0$ and $V = V_c$ for $m = 1$.

For the construction of the polynomial potential we need to know several data characteristics for the considered system. In terms of the volume as an independent variable, they are: the equilibrium volume $V_l(T)$ and $V_c(T)$, the volume $V^*(T) \in (V_c(T), V_l(T))$ corresponding to the position of the barrier, as well as the bulk melting temperature T_m and the latent heat of melting L_m. The same characteristics are sufficient to construct the Gibbs free energy in the form of a linear combination of power terms (LCPT–GFE). The variational principle $\delta f(V) = 0$ now leads to the equation of state:

$$\eta(V)(V - V_l)(V - V_c)(V - V^*) = 0 \tag{23}$$

with an arbitrary function $\eta(V) > 0$. Equation (23) for relation (22) becomes (1) while relation (7) is now a simple consequence of the expected linear dependence of the volume on the temperature.

The best situation appears when the GFE $f_l(V,T)$ and $f_c(V,T)$ are taken in their exact forms. In this context, we discuss the melting characterization for aluminium on the basis of the results reported by the Los Alamos National Laboratory group (LANL) [7]. The potentials are obtained by means of *ab initio* calculations from the first principles, without recourse to any experimental data. The LANL potentials have been previously applied by us to the investigations of the surface melting for aluminium [8]. Now, we use them in order to construct the LCPT–GFE. Following [7], we keep for the units the energy and volume: mRy per atom and unit cell volume, respectively.

Figure 1 shows the behavior of the LANL GFE $f_l(V,T)$ and $f_c(V,T)$ with respect to V for several values of temperature. As we can expect, each potential has its own minimum which determines its equilibrium values $V_l(T)$ and $V_c(T)$. The positions of the minima increase with temperature.

The behavior of values $V_l(T)$ and $V_c(T)$ is presented in Fig. 2(a) in some interval of temperature. Next, we plot the functions $F_l(T) = F_l(V_l(T), T)$ and $F_c(T) = F_c(V_c(T), T)$ in Fig. 2(b) whose behavior corresponds to the phase diagram. The intersection point determines the bulk melting temperature T_m. In the considered case, $T_m = 962.19\,\mathrm{K}$. At the melting temperature, the volume behavior exhibits the jump $V_l(T_m)/V_c(T_m) = 1.0368$ while its experimental value is 1.065 [1].

In Fig. 3, we show the dependence $\Lambda(T)$ with respect to the temperature for LANL GFE's, i.e., we show

$$\Lambda(T) = F_l(V_l(T), T) - F_c(V_c(T), T). \tag{24}$$

We can see that relation (24) is linear, at least at the vicinity of T_m, which confirms the thermodynamic property (6). Fitting the results to the relation (6) in the interval of temperature $T \in (920\,\mathrm{K}, 1000\,\mathrm{K})$ we find

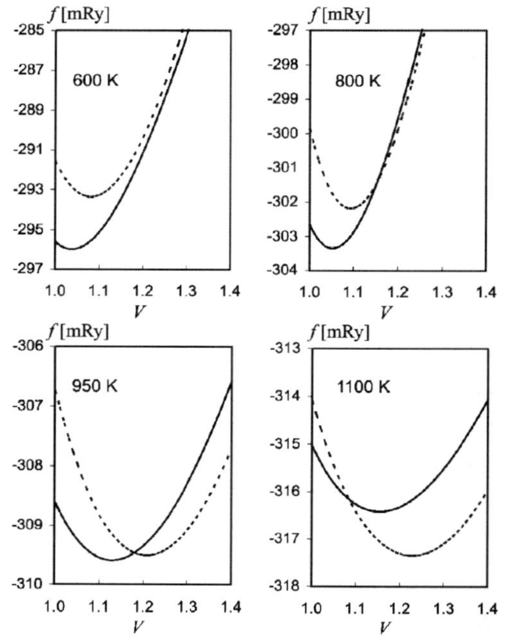

Fig. 1. The Gibbs free energies for the liquid phase $f_\mathrm{l}(V,T)$ (dashed curves) and the solid phase $f_\mathrm{c}(V,T)$ (solid curves) with respect to the volume V. The calculations are based on the exact, *ab initio* LANL procedure [7].

$L_\mathrm{m} = 0.57$ mRy and $T_\mathrm{m} = 962.22\,\mathrm{K}$. Our calculated melting temperature agrees well with the measured value of 933.45 K within possible numerical errors, estimated as less than $\pm 2\%$ [7].

Figure 4 presents the position V^* of the maximum of the barrier and its dependence on temperature. We can see that in the range of temperature $T \in (920\,\mathrm{K}, 1000\,\mathrm{K})$ the function $V^* = V^*(T)$ can be approximated by the linear function $V^* = aT + b$ with very high accuracy.

Taking into account the relation (22), we can put

$$(m^*)^\gamma = \frac{V_\mathrm{l}(T) - V^*(T)}{V_\mathrm{l}(T) - V_\mathrm{c}(T)}, \tag{25}$$

which is exhibited in Fig. 5, confirming the linear dependence of (7) with the condition (9), namely

$$(m^*)^\gamma = \frac{1}{\gamma + 1}\left(1 - 2\frac{\Lambda}{\alpha}\right). \tag{26}$$

In this way the graph in Fig. 5 allows us to determine α and γ.

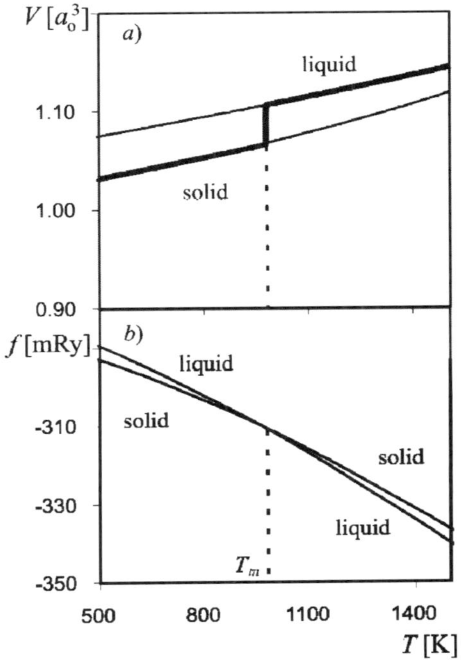

Fig. 2. (a) The equilibrium volumes $V_l(T)$ and $V_c(T)$ versus temperature T, and (b) the equilibrium Gibbs free energies $f_l(V_l(T),T)$ and $f_c(V_c(T),T)$ versus temperature T, calculated from the equation of state [7].

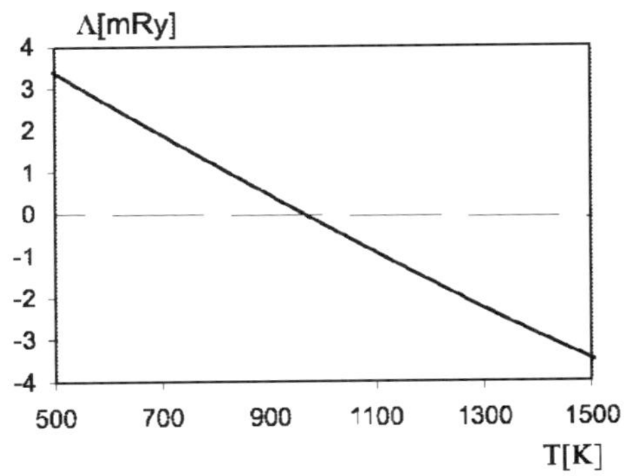

Fig. 3. The excess of the Gibbs free energies for the liquid phase with respect to the solid phase $\Lambda(T)$ in temperature T (Eq. (24)). The calculations are based on the paper [7].

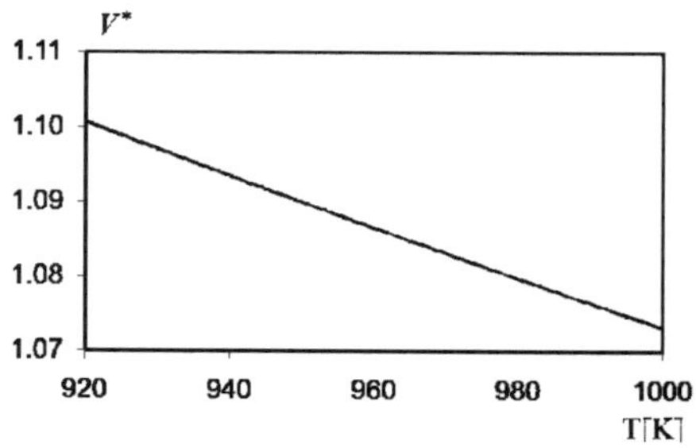

Fig. 4. The position V^* of the barrier maximum and its dependence on temperature: $V^* = V^*(T)$. The function V^* is an intersecting point for the Gibbs free energies derived in [7].

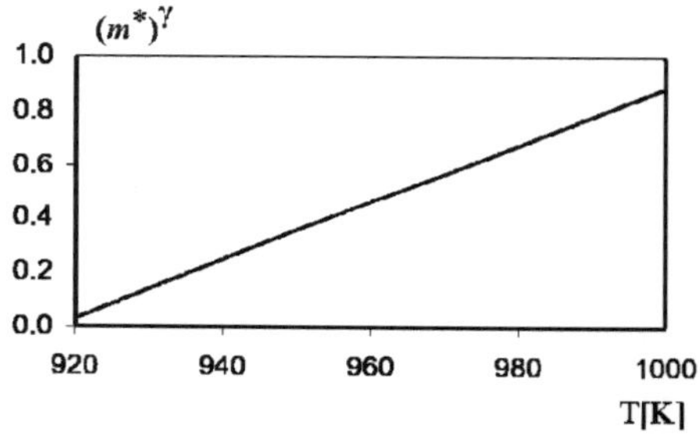

Fig. 5. The behavior of the barrier position $(m^*)^\gamma$ given by (25) and its dependence on temperature.

In the case of the LANL potential, we obtain: $\alpha = 0.6364$ mRy and $\gamma = 1.0592$. At $T = T_{\rm m}$, we find $m_0^* = 0.4856$.

Finally, the LCPT–GFE initiated by the LANL GFE and characterized by means of the above data in the case of aluminium can be written in the form

$$\frac{f(m,T)}{\alpha} = \frac{\Lambda}{\alpha} + \frac{(\gamma+1)(\gamma+2)}{\gamma}\left[\frac{1}{2}\frac{1}{\gamma+1}\left(1-2\frac{\Lambda}{\alpha}\right)m^2 - \frac{1}{\gamma+2}\left(1+\frac{1}{\gamma+1}\left(1-2\frac{\Lambda}{\alpha}\right)m^{\gamma+2}\right) + \frac{1}{2\gamma+2}m^{2\gamma+2}\right], \tag{27}$$

where Λ is given by (6).

The power function (27) for the parameter values corresponding to those of aluminium is shown in Fig. 6 for several temperatures.

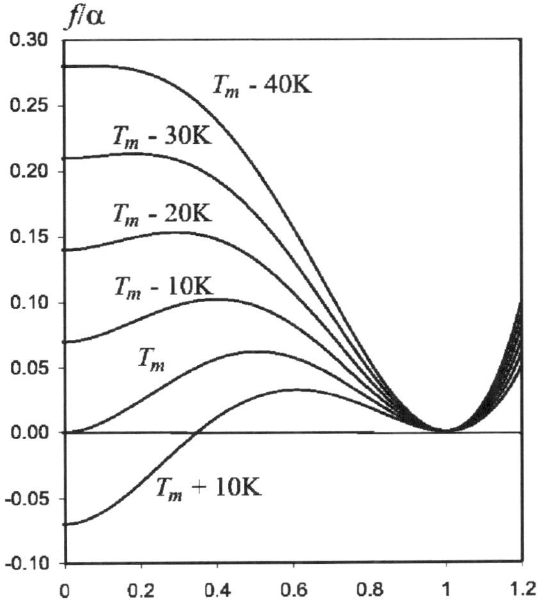

Fig. 6. The power form (27) of the Gibbs free energy $f(m)/\alpha$ with respect to m for several temperatures.

5. Applications

The polynomial form of the Gibbs free energy was mainly discussed in connection with the description of the surface melting phenomena. A detailed analysis was performed in the case of the reference potential in the form of two intersecting parabolas with equal curvatures, which was used to the description of lead specimen [9, 10].

For the same purpose the LCPT–GFE form was considered in the case of the generalization procedure proposed in the present paper with respect to the LANL GFE for aluminium. The LCPT which results from the LANL shape potentials can be equally applied to the surface melting description taking into account our previous experience [12]. The LANL GFE in its exact, *ab initio* form was introduced into the discussion of the surface melting properties in the case of aluminium [11, 12]. The results obtained can then be compared with the calculations based on the polynomial form of the van der Waals type potential also considered for the aluminium surface melting description [11] within the Landau model.

According to the results reported in our previous papers [8, 12], the discussion of the LCPT–GFE for the LANL GFE shows that the maximal temperature interval at which we can expect the surface melting appearance is equivalent to the interval containing the intersecting point of both the solid and liquid phase functions inside the interval $(V_l(T), V_c(T))$ for the equilibrium volumes. This interval corresponds to the crystallinity order parameter $m^* \in [0, 1]$.

The calculations show that the minimal value of temperature $T = 715\,\mathrm{K}$, at which the LANL GFE's for solid and liquid phases begin to have a common point cannot, however, be interpreted as the minimum of the surface melting temperature. The proper position of the intersection point V^* appears at $T \geq 917\,\mathrm{K}$. Therefore, this temperature can be interpreted as the lowest possible surface melting temperature T_{sm} (see also Fig. 1).

Thus, the potential constructed in the present paper leads to results comparable to those reported on the basis of the exact form of the LANL function [8], while these results differ from those given by means of the van der Waals potential in the form of the double parabola model [11].

Semi-infinite systems are also considered in the context of the phase diagrams discussed within the framework of Landau theory and can be applied to the phase of surface melting description. The analysis is particularly convenient when the potential in its polynomial form governs the order parameter profile [13, 14].

6. Conclusions

The present paper considers the construction of the linear combination of power terms based on m and m^γ for the Gibbs free energy of the homogeneous systems when the initiating functions are expressed by their exact forms given by LANL, separately for solid and liquid phases. The construction allows us to obtain the thermodynamic characteristics useful for the investigations of surface melting, namely: bulk melting temperature T_{m}, latent heat of melting L_{m}, and the energy normalization constant α. The lowest value for the surface melting temperature T_{sm} is also achieved.

The discussion presented above shows us that the polynomial form of the GFE can be obtained only when the intersection point m^*, taking its position at $T = T_{\mathrm{m}}$, is found in a self-consistent way with the parameter γ; this assures the non-vanishing form of the GFE. This self-consistency is not possible in every case; γ cannot always be an integer. Therefore, we have to extend the class of functions $f(m, T)$ to those also containing the rational power terms.

In the case of the LANL GFE, the parameter $\gamma = 1.0592$, i.e., it is almost 1. This result means that the exact potential is in fact close to that of the two intersecting parabolas with equal curvatures.

The self-consistent calculation of the parameter γ, which was introduced originally into the relation between the volume and crystallinity order parameter (22), allows us to justify this relation and to determine γ. This means that γ cannot be taken as an arbitrary value but it is univalent by its definition (9). The transformation of variable V to m is now evidently connected with the shape of the LCPT function, or more precisely, with the deviation of this function ($\gamma \neq 1$) from its reference shape, given as two intersecting parabolas ($\gamma = 1$).

The interesting conclusion of the present paper is that the relation between volume and crystallinity order parameter is now generalized and first of all very well established due to the self-consistent character of the presented construction.

Acknowledgments

The paper was prepared within the cooperation agreement between the University of Łódź and the Instituto Politecnico Nacional in Mexico City. The work is partially supported by the CONACyT, COFAA-IPN, EDD-IPN, Mexico and University Grant, UŁ 505/692/W. The authors are also grateful to Mrs T. Rychtelska for her assistance during the preparation of the paper.

References

[1] Castillo Alvarado, F.L., Ławrynowicz, J., Rutkowski, J.H. and Wojtczak, L., *Bull. Soc. Sci. Lett.* (Łódź), Rech. Deforms. **35** (2001) 7–25.

[2] Dash, J.G., *Contemp. Phys.* **30** (1989) 89.

[3] Van der Veen, J.F., in *Phase Transitions in Surface Films 2*, ed. Taub, H. (Plenum, New York, 1991).

[4] Romanowski, S., Rutkowski, J.H. and Wojtczak, L., *Bull. Soc. Sci. Lett.* (Łódź), Rech. Deforms. **27** (1999) 103–124.

[5] Pluis, B., Thesis, Rijksuniverseiteit te Leiden, 1990.

[6] Rutkowski, J.H., Wojtczak, L. and Castillo Alvarado, F.L., *Bull. Soc. Sci. Lett.* (Łódź), Rech. Deforms. **35** (2001) 33–41.

[7] Straub, G.K. *et al.*, *Phys. Rev. B* **50** (1994) 5055.

[8] Castillo Alvarado, F.L., Rutkowski, J.H., Urbaniak-Kucharczyk, A. and Wojtczak, L., *Thin Solid Films* **317** (1998) 43.

[9] Ramires Damaso, G., Diploma Thesis, Instituto Politecnico National, Mexico, D.F., 2003 (in Spanish).

[10] Zgid, B., Diploma Thesis, University of Łódź, Łódź, 2001 (in Polish).
[11] Wojtczak, L., Castillo Alvarado, F.L. and Rutkowski, J.H., *Tribologia*, XXVII, **4** (1996) 148 (in Polish).
[12] Rybka, K., Diploma Thesis, University of Łódź, Łódź, 1997 (in Polish).
[13] Lipowsky, R., *Phys. Rev Lett.* **49** (1982) 1575.
[14] Lipowsky, R. and Speth, W., *Phys. Rev. B* **28** (1983) 3983.

INDEX